Student Solutions Manual for Jones and Atkins's Chemistry

Molecules, Matter, and Change

FOURTH EDITION

Julie Henderleiter

Charles Trapp

W. H. FREEMAN AND COMPANY
New York

ISBN 0-7167-3437-0

Printed in the United States of America

First printing 1999

Contents

PREFACE

Welcome students. This solution manual contains worked out solutions and answers to the odd-numbered exercises in Loretta Jones and Peter Atkins's *Chemistry: Molecules, Matter, and Change*, Fourth Edition. This manual has been written to aid in your mastery of general chemistry topics. You should be aware of how to best use this manual. First and foremost, try to work all the exercises assigned from this text. If you find you have trouble with some types of exercises, work related exercises. Use this manual to check your answers, and, perhaps more importantly, to check the setup of your solution. If you are unsure of how to proceed with an exercise, try your best, then examine the solution. Though it is easy to look at the solution and think, "Of course, I can do that," it is to your advantage to work the exercise, then try another. By doing so, you will lessen the likelihood that, on an exam, you will recognize a question but have no recollection of how to arrive at an answer.

The rules of significant figures have been adhered to in reporting final numerical answers to the exercises. In some cases, extra digits are included in an answer; these are indicated by a line over the digit. For example, if an integral answer, such as 3×10 kg, is appropriate, the numerical value, 34 kg, will often be given. Lines over digits are also included in intermediate answers where rounding at that point would introduce significant error. This feature will also aid you in retracing steps in a solution.

Procedures and results that have full solutions in exercises found in earlier chapters may not be fully worked out in later chapters. For example, the conversion from °C to K and the determination of molar masses for compounds are not worked out in later chapters.

Some abbreviations have been included to save space. Common abbreviations include:

FU = formula unit
cmpd = compound
p = proton
n = neutron
e = electron

I wish to express my sincere thanks to Loretta Jones and to Peter Atkins for their help, suggestions, and support during the construction of this solutions manual. They, along with Carissa Bertin and Jessica Blunt from Grand Valley State University and Alice Allen, the proofreader, have made many valuable suggestions as they carefully reviewed the solutions. I would also like to thank the staff at W. H. Freeman and Company, especially Jodi Isman, for their patience and cooperation.

J. H.

CHAPTER 1
MATTER

EXERCISES

The Nuclear Atom and Isotopes

1.1 (a) charge = -1; mass = 9.109×10^{-28} g
(b) J. J. Thomson

1.3 A law is a statement about a basic relationship in nature for which there are no known exceptions. Theories are tested explanations or formal explanations of a law.

1.5 $\frac{2.00 \text{ g S}}{1.00 \text{ g S}} = 2{:}1; \frac{2.00 \text{ g S}}{0.667 \text{ g S}} = 3{:}1; \frac{1.00 \text{ g S}}{0.667 \text{ g S}} = 1.5 \times 2 = 3{:}2$
Each different mass of sulfur reacts with a given mass of oxygen to form compounds with small, whole number ratios of each element.

1.7 (a) As − 33 (b) S − 16 (c) Pd − 46 (d) Au − 79

1.9 (a) no error (b) molybdenum, Mo (c) yttrium, Y (d) strontium, Sr

1.11 (a) p = 6, n = 7, e = 6 (b) p = 17, n = 20, e = 17
(c) p = 17, n = 18, e = 17 (d) p = 92, n = 143, e = 92

1.13 (a) ^{111}Cd (b) ^{82}Kr (c) ^{11}B

1.15 (a) Each isotope has 6 protons and 6 electrons.
(b) Each isotope has a different number of neutrons.

The Periodic Table

1.17 (a) lithium, Group 1, metal (b) gallium, Group 13, metal
(c) xenon, Group 18, nonmetal (d) potassium, Group 1, metal

1.19 (a) Cl, nonmetal (b) Co, metal (c) As, metalloid

1.21 (a) I, nonmetal (b) Cr, metal (c) Hg, metal (d) Al, metal

1.23 lithium, Li, 3; sodium, Na, 11; potassium, K, 19; rubidium, Rb, 37; cesium, Cs, 55; francium, Fr, 87. The alkali metals react vigorously with water to form a basic solution and hydrogen gas.

Compounds

1.25 $C_{12}H_{20}O_2$

1.27 (a) anion, S^{2-} (b) cation, K^+ (c) cation, Sr^{2+} (d) anion, Cl^-

1.29 (a) group 16 (b) selenium, Se

1.31 (a) p = 1, n = 1, e = 0 (b) p = 4, n = 5, e = 2
(c) p = 35, n = 45, e = 36 (d) p = 16, n = 16, e = 18

1.33 (a) $^{19}F^-$ (b) $^{24}Mg^{2+}$ (c) $^{128}Te^{2-}$ (d) $^{86}Rb^+$

Substances and Mixtures

1.35 (a) element (b) element (diatomic) (c) compound

1.37 (a) mixture (b) single element

1.39 (a) physical (b) physical (c) chemical

1.41 (a) physical (b) physical (c) chemical

1.43 (a) physical (b) physical (c) chemical (d) physical

1.45 physical properties and changes: temperature, evaporation, and humidity.

1.47 (a) solubility differences
(b) abilities of substances to adsorb, or stick, to surfaces
(c) boiling point differences

1.49 (a) homogeneous, distillation
(b) heterogeneous, filtration (salt dissolves in water, chalk does not and can be filtered)
(c) homogeneous, evaporation or recrystallization

Chemical Nomenclature

1.51 (a) chloride ion (b) oxide ion (c) carbide ion (d) phosphide ion

1.53 (a) phosphate ion (b) sulfate ion (c) nitride ion
(d) sulfite ion (e) iodite ion (f) iodide ion

1.55 (a) ClO_3^- (b) NO_3^- (c) CO_3^{2-} (d) ClO^- (e) HSO_4^-

1.57 (a) plumbous ion; lead(II) ion (b) ferrous ion; iron(II) ion
(c) cobaltic ion; cobalt(III) ion (d) cuprous ion; copper(I) ion

1.59 (a) Cu^{2+} (b) ClO_2^- (c) P^{3-} (d) H^-

1.61 (a) MgO (b) $Ca_3(PO_4)_2$ (c) $Al_2(SO_4)_3$ (d) Ca_3N_2

1.63 (a) potassium phosphate (b) ferrous iodide; iron(II) iodide
(c) niobium(V) oxide (d) cupric sulfate; copper(II) sulfate

1.65 (a) copper(II) nitrate hexahydrate
(b) neodynium(III) chloride hexahydrate
(c) nickel(II) fluoride tetrahydrate

1.67 (a) ionic (b) $AlBr_3$

1.69 (a) $Na_2CO_3 \cdot H_2O$ (b) $In(NO_3)_3 \cdot 5H_2O$ (c) $Cu(ClO_4)_2 \cdot 6H_2O$

1.71 (a) SeO_3 (b) CCl_4 (c) CS_2 (d) SF_6 (e) As_2S_3 (f) PCl_5 (g) N_2O (h) ClF_3

1.73 (a) sulfur tetrafluoride (b) dinitrogen pentoxide
(c) nitrogen triiodide (d) xenon tetrafluoride
(e) arsenic tribromide (f) chlorine dioxide
(g) diphosphorus pentoxide

1.75 (a) hydrochloric acid (b) sulfuric acid
(c) nitric acid (d) acetic acid
(e) sulfurous acid (f) phosphoric acid

1.77 (a) Na_2O (b) K_2SO_4 (c) AgF (d) $Zn(NO_3)_2$ (e) Al_2S_3

SUPPLEMENTARY EXERCISES

1.79 (a) no (b) heterogeneous mixture

1.81

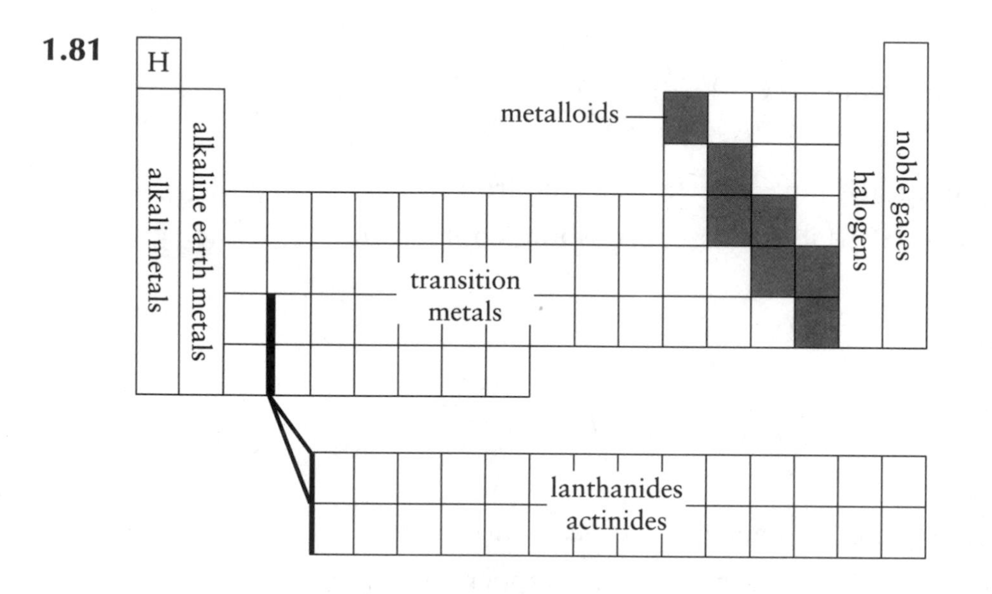

1.83 (a) Ionic compounds consist of an interlocking array of positive and negative ions held together by the attraction between their opposite charges. (b) Molecular compounds consist of molecules, which are discrete units of electrically neutral atoms bonded together. (c) Ionic compounds are hard, brittle solids at room temperature and have high melting and boiling points. Molecular compounds have lower melting and boiling points; many of them exist as gases or liquids at room temperature.

1.85 (a) silver sulfide (b) zinc chloride
(c) chlorine pentafluoride (d) magnesium hydroxide
(e) nickel(II) sulfate hexahydrate
(f) phosphorus pentachloride
(g) chromium(III) dihydrogen phosphate
(h) diarsenic trioxide (i) manganese(II) chloride

1.87 (a) $(NH_4)_2SO_3$ (b) Fe_2O_3 (c) $Cu(BrO_3)_2$
(d) PH_3 (e) $Ca(HCO_3)_2$ (f) HCN
(g) $LiHSO_4$ (h) SeF_4 (i) $FeSO_4 \cdot 7H_2O$

1.89 (a) H – 1p, 1e for each, O – 8p, 8n, 8e, so there are 10p, 8n, and 10e total.
(b) $10p(1.67262 \times 10^{-24} \text{ g/p}) + 8n(1.67492 \times 10^{-24} \text{ g/n})$
$+ 10e(9.10939 \times 10^{-28} \text{ g/e}) = 3.013 \times 10^{-23} \text{ g}$

(c) $\dfrac{\text{mass neutrons}}{\text{total mass}} = \dfrac{8n(1.67492 \times 10^{-24} \text{ g})}{3.013 \times 10^{-23} \text{ g}} = 0.4447$

so 0.4447 × (your mass) is due to neutrons

APPLIED EXERCISES

1.91 The ingot could be tested by pulsed fast-neutron analysis or spectrometry. X-ray fluorescence analysis may also be used.

1.93 (a) The substance must be HCl. The parent peak appears to be at 36, which is too heavy to be H_2S and not light enough to be HBr. Cl_2 has isotopes weighing 35 amu and 37 amu, which don't correspond to either major peak.
(b) The common isotopes for the atoms in HCl are as follows: ${}^{1}_{1}H$ – 99.985%, ${}^{2}_{1}H$ – 0.015%, ${}^{35}_{17}Cl$ – 75.77%, ${}^{37}_{17}Cl$ – 24.23%. The peak (parent) at 36 amu is thus ${}^{1}H^{35}Cl$. The peak at 38 amu is ${}^{1}H^{37}Cl$. The small peak at 37 amu is likely ${}^{37}Cl$, and the peak at 35 amu is ${}^{35}Cl$. The peak at 1 amu is ${}^{1}_{1}H$.

1.95 Distillation (or boiling the water), followed by condensation

1.97 Initial observational data might include the frequency and severity of headaches, and environmental conditions like food eaten, noise levels, or odors. Possible hypotheses include (1) food caused the headaches, (2) the room caused the headaches, (3) homework caused the headaches. Three experiments are (1) eliminate the food (ideally, one food at a time, so several experiments are required), (2) eliminate the room (stay with a friend or do homework elsewhere), (3) eliminate the homework. Other experimental variations are possible. The data to be collected are (1) food eaten, (2) room used, (3) homework begun (yes or no), (4) headache (yes or no).

CHAPTER 2
MEASUREMENTS AND MOLES

EXERCISES

International System (SI) of Units

2.1 (a) 5.556×10^{12} (b) $1.169\ 811 \times 10^{6}$ (c) 6×10^{-6} g (d) 1×10^{-7} m

2.3 (a) 4.3×10^{-1} (b) 1.492×10^{1} (c) 5.1×10^{-9} (d) 2.37×10^{14}

2.5 (a) $250.\ \text{g} \times \left(\dfrac{1\ \text{kg}}{10^{3}\ \text{g}}\right) = 0.250\ \text{kg}$

(b) $25.4\ \text{mm} \times \left(\dfrac{10^{-3}\ \text{m}}{1\ \text{mm}}\right) \times \left(\dfrac{1\ \text{cm}}{10^{-2}\ \text{m}}\right) = 2.54\ \text{cm}$

(c) $250.\ \mu\text{s} \times \left(\dfrac{10^{-6}\ \text{s}}{1\ \mu\text{s}}\right) \times \left(\dfrac{1\ \text{ms}}{10^{-3}\ \text{s}}\right) = 0.250\ \text{ms}$

(d) $1.49\ \text{cm} \times \left(\dfrac{10^{-2}\ \text{m}}{1\ \text{cm}}\right) \times \left(\dfrac{1\ \text{dm}}{10^{-1}\ \text{m}}\right) = 0.149\ \text{dm}$

(e) $2.48\ \text{cg} \times \left(\dfrac{10^{-2}\ \text{g}}{1\ \text{cg}}\right) = 2.48 \times 10^{-2}\ \text{g}$

(f) $28.35\ \text{g} \times \left(\dfrac{1\ \text{kg}}{10^{3}\ \text{g}}\right) = 2.835 \times 10^{-2}\ \text{kg}$

2.7 (a) $1\ \mu\text{m} \times \left(\dfrac{10^{-6}\ \text{m}}{1\ \mu\text{m}}\right) = 1 \times 10^{-6}\ \text{m}$

(b) $550.\ \text{nm} \times \left(\dfrac{10^{-9}\ \text{m}}{1\ \text{nm}}\right) \times \left(\dfrac{1\ \text{mm}}{10^{-3}\ \text{m}}\right) = 5.50 \times 10^{-4}\ \text{mm}$

(c) $0.10\ \text{g} \times \left(\dfrac{1\ \text{mg}}{10^{-3}\ \text{g}}\right) = 1.0 \times 10^{2}\ \text{mg}$

(d) $105\ \text{pm} \times \left(\dfrac{10^{-12}\ \text{m}}{1\ \text{pm}}\right) \times \left(\dfrac{1\ \mu\text{m}}{10^{-6}\ \text{m}}\right) = 1.05 \times 10^{-4}\ \mu\text{m}$

2.9 $\text{density} = \dfrac{\text{mass}}{\text{volume}}$

volume = volume displaced = 9.8 mL − 8.3 mL = 1.5 mL = 1.5 cm^3

$$\text{density} = \frac{3.60\text{ g}}{1.5\text{ cm}^3} = 2.4\text{ g/cm}^3$$

2.11 $\text{mass (g)} = 1.00\text{ ft}^3 \times \left(\frac{12\text{ in.}}{1\text{ ft}}\right)^3 \times \left(\frac{2.54\text{ cm}}{1\text{ in.}}\right)^3 \times 0.16\text{ g/cm}^3 = 4.5 \times 10^3\text{ g}$

2.13 Because density = mass/volume, volume = mass/density; hence volume = 0.300 carat × (200 mg/carat) × (10^{-3} g/1 mg) × (1 cm^3/3.51 g) = $1.71 \times 10^{-2}\text{ cm}^3$

Conversion Factors

2.15 (a) $25\text{ L} \times \left(\frac{1\text{ cm}^3}{10^{-3}\text{ L}}\right) \times \left(\frac{10^{-2}\text{ m}}{1\text{ cm}}\right)^3 = 2.5 \times 10^{-2}\text{ m}^3$

(b) $\frac{25\text{ g}}{1\text{ L}} \times \left(\frac{1\text{ mg}}{10^{-3}\text{ g}}\right) \times \left(\frac{10^{-1}\text{ L}}{1\text{ dL}}\right)^2 = 2.5 \times 10^2\text{ mg/dL}^2$

(c) $\frac{1.54\text{ mm}}{1\text{ s}} \times \left(\frac{10^{-3}\text{ m}}{1\text{ mm}}\right) \times \left(\frac{1\text{ pm}}{10^{-12}\text{ m}}\right) \times \left(\frac{10^{-6}\text{ s}}{1\text{ µs}}\right) = 1.54 \times 10^3\text{ pm/µs}$

(d) $\frac{2.66\text{ g}}{1\text{ cm}^3} \times \left(\frac{1\text{ µg}}{10^{-6}\text{ g}}\right) \times \left(\frac{1\text{ cm}}{10^{-2}\text{ m}}\right)^3 \times \left(\frac{10^{-6}\text{ m}}{1\text{ µm}}\right)^3 = 2.66 \times 10^{-6}\text{ µg/µm}^3$

(e) $\frac{4.2\text{ L}}{\text{h}^2} \times \left(\frac{1\text{ mL}}{10^{-3}\text{ L}}\right) \times \left(\frac{1\text{ h}}{3600\text{ s}}\right)^2 = 3.2 \times 10^{-4}\text{ mL/s}^2$

(f) $\frac{\$1.20}{\text{gallon}} \times \left(\frac{780\text{ peso}}{\$1.00}\right) \times \left(\frac{1\text{ gallon}}{3.785\text{ L}}\right) = 2.47 \times 10^2\text{ peso/L}$

2.17 These conversions make use of Eqs. 2 and 3 on pp. 57–58 or their inverse shown in Toolbox 2.2.

(a) $\text{Celsius temp.} = \left[\frac{(\text{Fahr. temp./°F}) - 32}{1.8}\right]\text{°C} = \left[\frac{98.6 - 32}{1.8}\right]\text{°C} = 37.0\text{°C}$

(b) $\text{Fahr. temp.} = \left(1.8 \times \frac{\text{Celsius temp.}}{\text{°C}} + 32\right)\text{°F}$

$= [1.8 \times (-40) + 32]\text{°F} = -40\text{°F}$

(c) $\text{Celsius temp.} = \left(\frac{\text{Kelvin temp.}}{\text{K}} - 273.15\right)\text{°C}$

$= (0 - 273.15)\text{°C} = -273.15\text{°C}$

Fahr. temp. = $[1.8 \times (-273.15) + 32]\text{°F} = -459.67\text{°F}$

(d) $\text{Kelvin temp.} = \left(\frac{\text{Celsius temp.}}{\text{°C}} + 273.15\right)\text{K}$

$= (-269 + 273.15)\text{K} = 4.15\text{ K} = 4\text{ K (1 sf)}$

2.19 (a) $1\ \text{cm}^3 \times \left(\frac{10^{-2}\ \text{m}}{1\ \text{cm}}\right)^3 = 1 \times 10^{-6}\ \text{m}^3$

(b) $30.\ \text{m/s} \times \left(\frac{1\ \text{cm}}{10^{-2}\ \text{m}}\right) \times \left(\frac{10^{-6}\ \text{s}}{1\ \mu\text{s}}\right) = 3.0 \times 10^{-3}\ \text{cm}/\mu\text{s}$

(c) $22\ \text{m}^2 \times \left(\frac{1\ \text{cm}}{10^{-2}\ \text{m}}\right)^2 = 2.2 \times 10^5\ \text{cm}^2$

(d) $25\ \text{cm}^3 \times \left(\frac{1\ \text{mL}}{1\ \text{cm}^3}\right) = 25\ \text{mL}$

2.21 Insert the following results at the brackets in the statement of the exercise.

$$1.0\ \text{cm}^2 \times \left(\frac{10^{-2}\ \text{m}}{1\ \text{cm}}\right)^2 \times \left(\frac{1\ \text{mm}}{10^{-3}\ \text{m}}\right)^2 = 1.0 \times 10^2\ \text{mm}^2$$

$$10.0\ \text{cm}^3 \times \left(\frac{10^{-2}\ \text{m}}{1\ \text{cm}}\right)^3 = 1.00 \times 10^{-5}\ \text{m}^3$$

$$100\ \text{mL} \times \left(\frac{10^{-3}\ \text{L}}{1\ \text{mL}}\right) = 1.00 \times 10^{-1}\ \text{L}$$

$$25.0\ \text{mL} \times \left(\frac{1\ \text{cm}^3}{1\ \text{mL}}\right) = 25.0\ \text{cm}^3$$

Uncertainty of Measurements and Calculations

2.23 (a) 3 (b) 3 (c) 3 (d) an integer, infinite (e) 2 (f) an exact number, infinite

2.25 (a) 6.60 mL (b) 26.0 mL

2.27 The sum 4.43 g rounds to 4.4 g (addition, so 1.4 g determines significant figures).

2.29 The sum 1.645 g rounds to 1.64 g (addition, so 0.21 g determines significant figures).

2.31 The factor 1.23 has the least number of significant figures, so the result should be reported to 3 significant figures. Note that (273.15 + 1.2) has 4 significant figures (not 5 and not 2).

2.33 The precision is very good; there is very little variation between the measurements. The accuracy is not as good, because there is a difference between the experimental results (all low) and the accepted value.

Fun with Atoms and Moles

2.35 moles of stars $= \dfrac{10^{22} \text{ stars}}{6 \times 10^{23} \text{ stars/mol}} = 10^{-2} \text{ mol}$

2.37 (a) moles of people $= 5.7 \times 10^{9} \text{ people} \times \dfrac{1 \text{ mol people}}{6.02 \times 10^{23} \text{ people}}$

$= 9.5 \times 10^{-15} \text{ mol people}$

(b) moles of peas/s $= 9.5 \times 10^{-15} \text{ mol people} \times \dfrac{1 \text{ mol peas/s}}{1 \text{ mol people}}$

$= 9.5 \times 10^{-15} \text{ mol peas/s}$

time $= 1 \text{ mol peas} \times \dfrac{1 \text{ s}}{9.5 \times 10^{-15} \text{ mol peas}} = 1.05 \times 10^{14} \text{ s}$

time (y) $= 1.05 \times 10^{14} \text{ s} \times \left(\dfrac{1 \text{ min}}{60 \text{ s}}\right) \times \left(\dfrac{1 \text{ h}}{60 \text{ min}}\right) \times \left(\dfrac{1 \text{ day}}{24 \text{ h}}\right) \times \left(\dfrac{1 \text{ y}}{365 \text{ day}}\right)$

$= 3.4 \times 10^{6} \text{ y}$

3.4 million years is greater than the life span of the human species, so more than 5.7 billion people would be required.

Moles and Molar Masses of Elements

2.39 $\dfrac{260. \text{ g}}{5.00 \text{ mol}} = 52.0 \text{ g/mol}$ The element is chromium, Cr.

2.41 average molar mass (g/mol)

$= \left(\dfrac{98.89}{100} \times 1.9926 \times 10^{-23} \text{ g} + \dfrac{1.11}{100} \times 2.1593 \times 10^{-23} \text{ g}\right) \times 6.022 \times 10^{23}/\text{mol}$

$= 12.01 \text{ g/mol}$

2.43 molar mass $= \left(\dfrac{50.54}{100}\right) \times 78.918 \text{ g/mol} + \left(\dfrac{49.46}{100}\right) \times 80.916 \text{ g/mol}$

$= 79.91 \text{ g/mol}$

2.45 (a) number of moles of ^{35}Cl atoms

$= 4.82 \times 10^{22} \text{ atoms } ^{35}\text{Cl} \times \dfrac{1 \text{ mol } ^{35}\text{Cl}}{6.022 \times 10^{23} \text{ atoms } ^{35}\text{Cl}} = 0.0800 \text{ mol } ^{35}\text{Cl}$

(b) number of moles of Cu atoms $= 2.22 \text{ g Cu} \times \dfrac{1 \text{ mol Cu}}{63.54 \text{ g Cu}} = 0.0349 \text{ mol Cu}$

(c) number of moles of He atoms

$$= 1.11 \times 10^{24} \text{ atoms He} \times \frac{1 \text{ mol He}}{6.022 \times 10^{23} \text{ atoms He}} = 1.84 \text{ mol He}$$

(d) number of moles of Fe atoms $= 8.96\ \mu\text{g Fe} \times \left(\frac{10^{-6} \text{ g Fe}}{1\ \mu\text{g Fe}}\right) \times \left(\frac{1 \text{ mol Fe}}{55.85 \text{ g Fe}}\right)$

$= 1.60 \times 10^{-7}$ mol Fe

2.47 (a) number of atoms $= 3.97 \text{ mol Xe} \times \left(\frac{6.022 \times 10^{23} \text{ atoms}}{1 \text{ mol Xe}}\right)$

$= 2.39 \times 10^{24}$ atoms

(b) number of atoms

$$= 18.3\ \mu\text{g Sc} \times \left(\frac{10^{-6} \text{ g}}{1\ \mu\text{g}}\right) \times \left(\frac{1 \text{ mol Sc}}{44.96 \text{ g}}\right) \times \left(\frac{6.022 \times 10^{23} \text{ atoms}}{1 \text{ mol Sc}}\right)$$

$= 2.45 \times 10^{17}$ atoms

(c) number of atoms

$$= 12.8 \text{ pg Li} \times \left(\frac{10^{-12} \text{ g}}{1 \text{ pg}}\right) \times \left(\frac{1 \text{ mol Li}}{6.94 \text{ g}}\right) \times \left(\frac{6.022 \times 10^{23} \text{ atoms}}{1 \text{ mol Li}}\right)$$

$= 1.11 \times 10^{12}$ atoms

(d) number of atoms $= 3.78 \times 10^{-4} \text{ mol Ar} \times \left(\frac{6.022 \times 10^{23} \text{ atoms}}{1 \text{ mol Ar}}\right)$

$= 2.28 \times 10^{20}$ atoms

2.49 (a) 12 g of C is one mole of C, so one mole of Ni contains the same number of atoms. One mole of Ni has a mass of 58.71 g, or 59 g (2 sf).

(b) One mole of any element contains the same number of atoms as one mole of any other element. That is, one mole of Cr atoms = one mol of Ni atoms, so the conversion factor is $\left(\frac{1 \text{ mol Ni atoms}}{1 \text{ mol Cr atoms}}\right)$ or, in general, $\left(\frac{1 \text{ mol A atoms}}{1 \text{ mol B atoms}}\right)$.

mass of Ni atoms

$$= 12 \text{ g Cr} \times \left(\frac{1 \text{ mol Cr atoms}}{52.00 \text{ g Cr}}\right) \times \left(\frac{1 \text{ mol Ni atoms}}{1 \text{ mol Cr atoms}}\right) \times \left(\frac{58.71 \text{ g Ni}}{1 \text{ mol Ni atoms}}\right)$$

$= 13.55$ g Ni $= 14$ g Ni (2 sf)

Mass Percentage Composition and Molar Masses of Compounds

2.51 (a) molar mass of $CaBr_2 = 1 \times 40.08 \text{ g/mol} + 2 \times 79.91 \text{ g/mol}$

$= 199.90$ g/mol

(b) molar mass of $C_8H_{18} = 8 \times 12.01 \text{ g/mol} + 18 \times 1.008 \text{ g/mol}$

$= 114.22$ g/mol

(c) molar mass of $NiSO_4 \cdot 6H_2O$ = 58.71 g/mol + 32.06 g/mol + 4 × 16.00 g/mol + 6(2 × 1.008 g/mol + 16.00 g/mol) = 262.86 g/mol

(d) molar mass of CO_2 = 12.01 g/mol + 2 × 16.00 g/mol = 44.01 g/mol

(e) molar mass of CH_4 = 12.01 g/mol + 4 × 1.008 g/mol = 16.04 g/mol

2.53 (a) Two-thirds of the atoms in N_2O_4 are oxygen atoms.

(b) 5.0 moles of oxygen gas, O_2, were produced.

(c) 1.000 mol Cl_2 contains 6.022×10^{23} chlorine molecules.

2.55 (a) molar mass CCl_4 = (12.01 + 4 × 35.45) g/mol = 153.81 g/mol

$$\text{number of moles of } CCl_4 = 10.0 \text{ g } CCl_4 \times \left(\frac{1 \text{ mol } CCl_4}{153.81 \text{ g } CCl_4}\right) = 0.0650 \text{ mol } CCl_4$$

$$0.0650 \text{ mol } CCl_4 \times \left(\frac{6.022 \times 10^{23} \text{ molecules } CCl_4}{1 \text{ mol } CCl_4}\right) = 3.92 \times 10^{22} \text{ molecules } CCl_4$$

(b) molar mass HI = (1.008 + 126.90) g/mol = 127.91 g/mol

$$\text{number of moles of HI} = 1.65 \text{ mg HI} \times \left(\frac{10^{-3} \text{ g HI}}{1 \text{ mg HI}}\right) \times \left(\frac{1 \text{ mol HI}}{127.91 \text{ g HI}}\right) = 1.29 \times 10^{-5} \text{ mol HI}$$

$$1.29 \times 10^{-5} \text{ mol HI} \times \left(\frac{6.022 \times 10^{23} \text{ molecules HI}}{1 \text{ mol HI}}\right) = 7.77 \times 10^{18} \text{ molecules HI}$$

(c) molar mass N_2H_4 = (2 × 14.01 + 4 × 1.008) g/mol = 32.05 g/mol

$$\text{number of moles of } N_2H_4 = 3.77 \text{ mg } N_2H_4 \times \left(\frac{10^{-3} \text{ g } N_2H_4}{1 \text{ mg } N_2H_4}\right) \times \left(\frac{1 \text{ mol } N_2H_4}{32.05 \text{ g } N_2H_4}\right) = 1.18 \times 10^{-4} \text{ mol } N_2H_4$$

$$1.18 \times 10^{-4} \text{ mol } N_2H_4 \times \left(\frac{6.022 \times 10^{23} \text{ molecules } N_2H_4}{1 \text{ mol } N_2H_4}\right) = 7.08 \times 10^{19} \text{ molecules } N_2H_4$$

(d) molar mass sucrose = (12 × 12.01 + 22 × 1.008 + 11 × 16.00) g/mol = 342.30 g/mol sucrose

$$\text{number of moles of sucrose} = 500. \text{ g sucrose} \times \left(\frac{1 \text{ mol sucrose}}{342.30 \text{ g sucrose}}\right) = 1.46 \text{ mol sucrose}$$

$$1.46 \text{ mol sucrose} \times \left(\frac{6.022 \times 10^{23} \text{ molecules}}{1 \text{ mol sucrose}}\right) = 8.79 \times 10^{23} \text{ molecules sucrose}$$

(e) $$\text{number of moles of O atoms} = 2.33 \text{ g O} \times \left(\frac{1 \text{ mol O}}{16.00 \text{ g O}}\right) = 0.146 \text{ mol O}$$

$$0.146 \text{ mol O} \times \left(\frac{6.022 \times 10^{23} \text{ atoms O}}{1 \text{ mol O}}\right) = 8.79 \times 10^{22} \text{ atoms O}$$

$$8.79 \times 10^{22} \text{ atoms O} \times \frac{1 \text{ molecule } O_2}{2 \text{ atoms O}} = 4.40 \times 10^{22} \text{ molecules } O_2$$

2.57 (a) molar mass AgCl = (107.87 + 35.45) g/mol = 143.32 g/mol

$$\text{number of moles of } Ag^+ = 2.00 \text{ g AgCl} \times \left(\frac{1 \text{ mol AgCl}}{143.32 \text{ g AgCl}}\right) \times \left(\frac{1 \text{ mol } Ag^+}{1 \text{ mol AgCl}}\right)$$

$$= 0.0140 \text{ mol } Ag^+$$

(b) molar mass UO_3 = (238.03 + 3 × 16.00) g/mol = 286.03 g/mol

$$\text{number of moles of } UO_3 = 600. \text{ g } UO_3 \times \left(\frac{1 \text{ mol } UO_3}{286.03 \text{ g } UO_3}\right) = 2.10 \text{ mol } UO_3$$

(c) molar mass $FeCl_3$ = (55.85 + 3 × 35.45) g/mol = 162.20 g/mol

number of moles of Cl^-

$$= 4.19 \times 10^{-3} \text{ g } FeCl_3 \times \left(\frac{1 \text{ mol } FeCl_3}{162.20 \text{ g } FeCl_3}\right) \times \left(\frac{3 \text{ mol } Cl^-}{1 \text{ mol } FeCl_3}\right)$$

$$= 7.75 \times 10^{-5} \text{ mol } Cl^-$$

(d) molar mass $AuCl_3 \cdot 2H_2O$

$$= [196.97 + 3 \times 35.45 + 2(2 \times 1.008 + 16.00)] \text{ g/mol} = 339.35 \text{ g/mol}$$

$$\text{amount (moles) of } H_2O = 1.00 \text{ g } AuCl_3 \cdot 2H_2O \times \left(\frac{1 \text{ mol } AuCl_3 \cdot 2H_2O}{339.35 \text{ g } AuCl_3 \cdot 2H_2O}\right)$$

$$\times \left(\frac{2 \text{ mol } H_2O}{1 \text{ mol } AuCl_3 \cdot 2H_2O}\right) = 5.89 \times 10^{-3} \text{ mol } H_2O$$

2.59 FU = formula unit, # = number sign

(a)
$$\text{\# of FU of } AgNO_3 = 0.670 \text{ mol } AgNO_3 \times \left(\frac{6.022 \times 10^{23} \text{ FU of } AgNO_3}{1 \text{ mol } AgNO_3}\right)$$

$$= 4.03 \times 10^{23} \text{ FU of } AgNO_3$$

(b)
$$\text{mass (mg) of } Rb_2SO_4 = 2.39 \times 10^{20} \text{ FU } Rb_2SO_4 \times \left(\frac{1 \text{ mol } Rb_2SO_4}{6.022 \times 10^{23} \text{ FU } Rb_2SO_4}\right)$$

$$\times \left(\frac{267.00 \text{ g } Rb_2SO_4}{1 \text{ mol } Rb_2SO_4}\right) \times \left(\frac{1 \text{ mg } Rb_2SO_4}{10^{-3} \text{ g } Rb_2SO_4}\right) = 1.06 \times 10^{2} \text{ mg } Rb_2SO_4$$

(c)
$$\text{\# of FU of } NaHCO_2 = 6.66 \text{ kg } NaHCO_2 \times \left(\frac{10^3 \text{ g } NaHCO_2}{1 \text{ kg } NaHCO_2}\right)$$

$$\times \left(\frac{1 \text{ mol } NaHCO_2}{68.01 \text{ g } NaHCO_2}\right) \times \left(\frac{6.022 \times 10^{23} \text{ FU } NaHCO_2}{1 \text{ mol } NaHCO_2}\right)$$

$$= 5.90 \times 10^{25} \text{ FU of } NaHCO_2$$

2.61 (a) molar mass testosterone $= (19 \times 12.01 + 28 \times 1.008 + 2 \times 16.00)$ g/mol
$= 288.41$ g/mol

$$\text{amount (moles) of testosterone} = 1 \text{ mg} \times \left(\frac{10^{-3} \text{ g testosterone}}{1 \text{ mg testosterone}}\right)$$

$$\times \left(\frac{1 \text{ mol testosterone}}{288.41 \text{ g testosterone}}\right) = 3.5 \times 10^{-6} \text{ mol testosterone}$$

(b) $$\text{mass \% C} = \frac{19 \times 12.01 \text{ g/mol}}{288.41 \text{ g/mol}} \times 100\% = 79.1\% \text{ C}$$

$$\text{mass \% H} = \frac{28 \times 1.008 \text{ g/mol}}{288.41 \text{ g/mol}} \times 100\% = 9.8\% \text{ H}$$

$$\text{mass \% O} = \frac{2 \times 16.00 \text{ g/mol}}{288.41 \text{ g/mol}} \times 100\% = 11.1\% \text{ O}$$

2.63 (a) molar mass water = 18.02 g/mol
mass of 1 molecule H_2O

$$= 1 \text{ molecule } H_2O \times \left(\frac{1 \text{ mol } H_2O}{6.022 \times 10^{23} \text{ molecules } H_2O}\right) \times \left(\frac{18.02 \text{ g } H_2O}{1 \text{ mol } H_2O}\right)$$

$= 2.99 \times 10^{-23}$ g H_2O

(b) number of H_2O molecules

$$= 1.00 \text{ g } H_2O \times \left(\frac{1 \text{ mol } H_2O}{18.02 \text{ g } H_2O}\right) \times \left(\frac{6.022 \times 10^{23} \text{ molecules } H_2O}{1 \text{ mol } H_2O}\right)$$

$= 3.34 \times 10^{22}$ molecules H_2O

2.65 (a) molar mass $CuBr_2 \cdot 4H_2O$

$= [63.54 + 2 \times 79.91 + 4(2 \times 1.008 + 16.00]$ g/mol $= 295.42$ g/mol

number of moles of $CuBr_2 \cdot 4H_2O$

$$= 5.50 \text{ g } CuBr_2 \cdot 4H_2O \times \left(\frac{1 \text{ mol } CuBr_2 \cdot 4H_2O}{295.42 \text{ g } CuBr_2 \cdot 4H_2O}\right)$$

$= 0.0186$ mol $CuBr_2 \cdot 4H_2O$

(b) $$\text{number of moles of } Br^- = 0.0186 \text{ mol } CuBr_2 \cdot 4H_2O \times \left(\frac{2 \text{ mol } Br^-}{1 \text{ mol } CuBr_2 \cdot 4H_2O}\right)$$

$= 0.0372$ mol Br^-

(c) $$0.0186 \text{ mol } CuBr_2 \cdot 4H_2O \times \left(\frac{4 \text{ mol } H_2O}{1 \text{ mol } CuBr_2 \cdot 4H_2O}\right)$$

$$\times \left(\frac{6.022 \times 10^{23} \text{ molecules } H_2O}{1 \text{ mol } H_2O}\right) = 4.48 \times 10^{22} \text{ molecules } H_2O$$

(d) mass fraction of Cu

$$= \left(\frac{1 \text{ mol Cu}}{1 \text{ mol } CuBr_2 \cdot 4H_2O}\right) \times \left(\frac{63.54 \text{ g Cu}}{1 \text{ mol Cu}}\right) \times \left(\frac{1 \text{ mol } CuBr_2 \cdot 4H_2O}{295.42 \text{ g } CuBr_2 \cdot 4H_2O}\right)$$

$= 0.215$

Determining Chemical Formulas

2.67 (a) molecular: $C_4H_6Cl_2$; empirical: C_2H_3Cl
(b) molecular: $C_3H_8O_3$; empirical: $C_3H_8O_3$

2.69 Assume 100. g of sample in each case.

(a) number of moles of Na $= 32.79 \text{ g Na} \times \left(\dfrac{1 \text{ mol Na}}{22.99 \text{ g Na}}\right) = 1.426 \text{ mol Na}$

number of moles of Al $= 13.02 \text{ g Al} \times \left(\dfrac{1 \text{ mol Al}}{26.98 \text{ g Al}}\right) = 0.4826 \text{ mol Al}$

number of moles of F $= 54.19 \text{ g F} \times \left(\dfrac{1 \text{ mol F}}{19.00 \text{ g F}}\right) = 2.852 \text{ mol F}$

The ratio is Na : Al : F = 1.426 : 0.4826 : 2.852. Dividing through by 0.4826 gives 2.955 : 1 : 5.910, which is close to the ratio 3 : 1 : 6. Therefore, the empirical formula is Na_3AlF_6.

(b) number of moles of K $= 31.91 \text{ g K} \times \left(\dfrac{1 \text{ mol K}}{39.10 \text{ g K}}\right) = 0.8161 \text{ mol K}$

number of moles of Cl $= 28.93 \text{ g Cl} \times \left(\dfrac{1 \text{ mol Cl}}{35.45 \text{ g Cl}}\right) = 0.8161 \text{ mol Cl}$

mass O $= (100 - 31.91 - 28.93) \text{ g} = 39.16 \text{ g O}$

number of moles of O $= 39.16 \times \left(\dfrac{1 \text{ mol O}}{16.00 \text{ g O}}\right) = 2.448 \text{ mol O}$

The ratio is K : Cl : O = 0.8161 : 0.8161 : 2.448. Dividing through by 0.8161 gives 1 : 1 : 3.00. Therefore, the empirical formula is $KClO_3$.

(c) number of moles of N $= 12.2 \text{ g N} \times \left(\dfrac{1 \text{ mol N}}{14.01 \text{ g N}}\right) = 0.871 \text{ mol N}$

number of moles of H $= 5.26 \text{ g H} \times \left(\dfrac{1 \text{ mol H}}{1.008 \text{ g H}}\right) = 5.21\overline{8} \text{ mol H}$

number of moles of P $= 26.9 \text{ g P} \times \left(\dfrac{1 \text{ mol P}}{30.97 \text{ g P}}\right) = 0.868 \text{ mol P}$

number of moles of O $= 55.6 \text{ g O} \times \left(\dfrac{1 \text{ mol O}}{16.00 \text{ g O}}\right) = 3.47\overline{5} \text{ mol O}$

The ratio is N : H : P : O = $0.871 : 5.21\overline{8} : 0.869 : 3.47\overline{5}$. Dividing through by 0.869 gives 1.00 : 6.00 : 1.00 : 4.00. Therefore, the empirical formula is NH_6PO_4 or $NH_4H_2PO_4$.

2.71 number of moles of P $= \dfrac{4.14 \text{ g}}{30.97 \text{ g/mol}} = 0.133\overline{7} \text{ mol}$

mass of Cl in compound $= 27.8 \text{ g} - 4.14 \text{ g} = 23.6\overline{6} \text{ g}$

number of moles of Cl $= \dfrac{23.6\overline{6} \text{ g}}{35.45 \text{ g/mol}} = 0.667\overline{4} \text{ mol}$

The ratio is P : Cl = $0.133\bar{7}:0.667\bar{4}$. Dividing by $0.133\bar{6}$ gives 1.00 : 5.00. Therefore, the empirical formula is PCl_5.

2.73 Assume 100. g of material.

$$\text{number of moles of C} = 24.78 \text{ g C} \times \left(\frac{1 \text{ mol C}}{12.01 \text{ g C}}\right) = 2.063 \text{ mol C}$$

$$\text{number of moles of H} = 2.08 \text{ g H} \times \left(\frac{1 \text{ mol H}}{1.008 \text{ g H}}\right) = 2.06\bar{3} \text{ mol H}$$

$$\text{number of moles of Cl} = 73.14 \text{ g Cl} \times \left(\frac{1 \text{ mol Cl}}{35.45 \text{ g Cl}}\right) = 2.063 \text{ mol Cl}$$

The ratio is C : H : Cl = $2.063:2.06\bar{3}:2.063$, or 1 : 1 : 1. Therefore, the empirical formula is CHCl. The molar mass of CHCl = (12.01 + 1.008 + 35.45) g/mol = 48.47 g/mol.

$$\frac{\text{molar mass of lindane}}{\text{molar mass of CHCl}} = \frac{290.85 \text{ g/mol}}{48.47 \text{ g/mol}} = 6.001 \approx 6$$

Therefore, the molecular formula of lindane is $C_6H_6Cl_6$.

2.75 Assume 100. g of material.

$$\text{number of moles of C} = 49.48 \text{ g C} \times \frac{1 \text{ mol C}}{12.01 \text{ g C}} = 4.120 \text{ mol C}$$

$$\text{number of moles of H} = 5.19 \text{ g H} \times \frac{1 \text{ mol H}}{1.008 \text{ g H}} = 5.14\bar{9} \text{ mol H}$$

$$\text{number of moles of N} = 28.85 \text{ g N} \times \frac{1 \text{ mol N}}{14.01 \text{ g N}} = 2.059 \text{ mol N}$$

$$\text{number of moles of O} = 16.48 \text{ g O} \times \frac{1 \text{ mol O}}{16.00 \text{ g O}} = 1.030 \text{ mol O}$$

The ratio is C : H : N : O = 4.120 : 5.149 : 2.059 : 1.030. Dividing by 1.030 gives 4.000 : 4.999 : 1.999 : 1, which is very close to 4 : 5 : 2 : 1. Therefore, the empirical formula = $C_4H_5N_2O$. The molar mass of $C_4H_5N_2O$ = (4 × 12.01 + 5 × 1.008 + 2 × 14.01 + 16.00) g/mol = 97.10 g/mol.

$$\frac{\text{molar mass caffeine}}{\text{molar mass } C_4H_5N_2O} = \frac{194.19 \text{ g/mol}}{97.10 \text{ g/mol}} = 2$$

Therefore, the molecular formula = 2 × ($C_4H_5N_2O$) = $C_8H_{10}N_4O_2$.

SUPPLEMENTARY EXERCISES

2.77 (a) kg (b) pm (c) g (d) μm

2.79 (a) $12\ \text{u}\left(\frac{1.6605 \times 10^{-24}\ \text{g}}{1\ \text{u}}\right)\left(\frac{1\ \text{mol}\ ^{12}\text{C}}{12\ \text{g}}\right)\left(\frac{6.022 \times 10^{23}\ \text{atoms}}{1\ \text{mol}\ ^{12}\text{C}}\right) = 1\ \text{atom}$

(b) $\frac{1}{1.6605 \times 10^{-24}\ \text{g}} = 6.022 \times 10^{23}$, which is Avogadro's number.

2.81 $26\ \text{miles} \times \frac{1760\ \text{yd}}{1\ \text{mile}} = 4.576 \times 10^4\ \text{yd}$

$$\text{marathon distance (yd)} = 4.576 \times 10^4\ \text{yd} + 385\ \text{yd}$$
$$= 4.614\bar{5} \times 10^4\ \text{yd}$$

$$4.614\bar{5} \times 10^4\ \text{yd} \times \left(\frac{36\ \text{in}}{1\ \text{yd}}\right) \times \left(\frac{2.54\ \text{cm}}{1\ \text{in}}\right) \times \left(\frac{10^{-2}\ \text{m}}{1\ \text{cm}}\right) \times \left(\frac{1\ \text{km}}{10^3\ \text{m}}\right) = 42.4\ \text{km}$$

2.83 $P_{sn} = \left(\frac{1\ \text{in.}}{2\ \text{min}}\right) \times \left(\frac{2.54\ \text{cm}}{1\ \text{in.}}\right) \times \left(\frac{1\ \text{min}}{60\ \text{s}}\right) = 2 \times 10^{-2}\ \text{cm/s}$

2.85 Day side:

Kelvin temp. = $(127 + 273.15)\ \text{K} = 400.\ \text{K}$

Fahrenheit temp. = $(1.8 \times 127 + 32)°\text{F} = 261°\text{F}$

Night side:

Kelvin temp. = $(-183 + 273.15)\ \text{K} = 90.\ \text{K}$

Fahrenheit temp. = $[1.8 \times (-183) + 32]°\text{F} = -297°\text{F}$

2.87 Given that moles $CuSO_4 \cdot 5H_2O$ = moles Cu,

$$\text{moles } CuSO_4 \cdot 5H_2O = 10.0\ \text{g}\ CuSO_4 \cdot 5H_2O \times \left(\frac{1\ \text{mol}\ CuSO_4 \cdot 5H_2O}{249.68\ \text{g}\ CuSO_4 \cdot 5H_2O}\right)$$
$$= .0400\bar{5} = 4.00\bar{5} \times 10^{-2}\ \text{mol}\ CuSO_4 \cdot 5H_2O$$

$$\text{mass Cu} = 4.00\bar{5} \times 10^{-2}\ \text{mol Cu} \times \left(\frac{63.54\ \text{g Cu}}{1\ \text{mol Cu}}\right) = 2.54\ \text{g Cu}$$

2.89 Assume a 100. g sample.

$$\text{number of moles of C} = 54.82\ \text{g C} \times \frac{1\ \text{mol}}{12.01\ \text{g C}} = 4.565\ \text{mol C}$$

$$\text{number of moles of H} = 5.62\ \text{g H} \times \frac{1\ \text{mol H}}{1.008\ \text{g H}} = 5.57\bar{5}\ \text{mol H}$$

$$\text{number of moles of N} = 7.10\ \text{g N} \times \frac{1\ \text{mol N}}{14.01\ \text{g N}} = 0.506\bar{8}\ \text{mol N}$$

$$\text{number of moles of O} = 32.64\ \text{g O} \times \frac{1\ \text{mol O}}{16.00\ \text{g O}} = 2.040\ \text{mol O}$$

The ratio is C:H:N:O = $4.56\overline{5}:5.57\overline{5}:0.506\overline{8}:2.040$. Dividing by $0.506\overline{8}$ gives $9.005:11.00:1:4.025 \approx 9:11:1:4$. Hence, the empirical formula is $C_9H_{11}NO_4$.

2.91 (a) 100°C − 0°C = 250°X − 50°X = 200°X

That is, 100°C = 200°X or 1°C = 2°X, hence

$$\text{new temperature} = \frac{2°\text{X}}{1°\text{C}} \times \text{Celsius temperature} + 50°\text{X}$$

or
$$\frac{\text{new temperature}}{°\text{X}} = 2 \times \frac{\text{Celsius temperature}}{°\text{C}} + 50$$

(b)
$$\text{new temperature} = \frac{2°\text{X}}{1°\text{C}} \times 22°\text{C} + 50°\text{X} = 94°\text{X}$$

2.93 Let x = fraction abundance of ^{10}B, then $(1 - x)$ = fraction abundance of ^{11}B.

Then $x \times 10.013$ g/mol + $(1 - x)11.093$ g/mol = 10.81 g/mol

Solving for x

$$x = \frac{10.81 - 11.093}{10.013 - 11.093} = 0.26$$

Therefore ^{10}B = 26% in abundance and ^{11}B = (100 − 26)% = 74% in abundance

APPLIED EXERCISES

2.95
$$\text{number of moles of C} = \frac{1.55 \times 10^{-3}\ \text{g C}}{12.01\ \text{g C/mol}} = 1.29 \times 10^{-4}\ \text{mol}$$

$$\text{number of moles of H} = \frac{0.204 \times 10^{-3}\ \text{g H}}{1.008\ \text{g H/mol}} = 2.02 \times 10^{-4}\ \text{mol}$$

$$\text{number of moles of N} = \frac{0.209 \times 10^{-3}\ \text{g N}}{14.01\ \text{g N/mol}} = 0.149 \times 10^{-4}\ \text{mol}$$

$$\text{number of moles of O} = \frac{0.557 \times 10^{-3}\ \text{g O}}{16.00\ \text{g O/mol}} = 0.348 \times 10^{-4}\ \text{mol}$$

The ratio is C:H:N:O = 1.29:2.02:0.149:0.348. Dividing by 0.149 gives 8.66:13.56:1:2.34. We need to convert to whole number ratios. A factor of 6 will accomplish this, except for H, yielding C:H:N:O = 52:81.4:6:14. Chemical analysis for H is notoriously imprecise, so it is not unreasonable to round 81.4 to 81. The formula is then $C_{52}H_{81}N_6O_{14}$.

which has a molar mass of 1014.2 g/mol. Note that any factor other than 6 would have yielded a molar mass much different from the known value.

2.97 (a) mass of carbon $= 2.492\ \text{g CO}_2 \left(\dfrac{12.01\ \text{g C}}{44.01\ \text{g CO}_2}\right) = 0.6800\ \text{g C}$

(b) mass of hydrogen $= 0.6495\ \text{g H}_2\text{O} \left(\dfrac{2.016\ \text{g H}}{18.02\ \text{g H}_2\text{O}}\right) = 0.072\ 66\ \text{g H}$

(c) mass of oxygen $= 1.000 - (0.6800 + 0.072\ 66)\text{g} = 0.2473\ \text{g O}$

The compound is: $\dfrac{0.6800\ \text{g C}}{1.000\ \text{g sample}} \times 100 = 68.00\%\ \text{C}$

$$\frac{0.072\ 66\ \text{g H}}{1.000\ \text{g sample}} \times 100 = 7.266\%\ \text{H}$$

$$\frac{0.2473\ \text{g O}}{1.000\ \text{g sample}} \times 100 = 24.73\%\ \text{O}$$

(d) number of moles of C $= \dfrac{68.00\ \text{g C}}{12.01\ \text{g C/mol}} = 5.662\ \text{mol}$

number of moles of H $= \dfrac{7.266\ \text{g H}}{1.008\ \text{g H/mol}} = 7.208\ \text{mol}$

number of moles of O $= \dfrac{24.73\ \text{g O}}{16.00\ \text{g O/mol}} = 1.546\ \text{mol}$

The ratio is C:H:O = 5.662:7.208:1.546. Dividing by 1.546 gives 3.662:4.662:1. We need to convert to whole number ratios. A factor of 3 accomplishes this, yielding C:H:O = 10.98:13.99:3 or 11:14:3. The empirical formula is $C_{11}H_{14}O_3$.

(e) The empirical mass is thus $(12.01 \times 11 + 1.008 \times 14 + 16.00 \times 3)\text{g/mol} = 194.22\ \text{g/mol}$, and $\dfrac{388.46\ \text{g/mol}}{194.22\ \text{g/mol}} = 2$, so the molecular formula is $C_{22}H_{28}O_6$.

INTEGRATED EXERCISES

2.99 (a) $100 - 88.8 = 11.2\%\ \text{O}$

$11.2\ \text{g O} \left(\dfrac{1\ \text{mol O}}{16.00\ \text{g O}}\right) = 0.700\ \text{mol O}$

The formula suggests that there should be 2(0.700 mol) = 1.40 mol M.

$88.8\ \text{g M} \left(\dfrac{1\ \text{mol M}}{x\ \text{g M}}\right) = 1.40\ \text{mol M}$. Find x.

$x = \dfrac{88.8\ \text{g M}}{1.40\ \text{mol M}} = 63.4\ \text{g/mol}$

(b) copper(I) oxide

2.101 (a) If 1 atom of silicon-28 weighs $4.645\,67 \times 10^{-23}$ g, then Avogadro's number becomes $\left(\dfrac{1 \text{ atom Si}}{4.64567 \times 10 \text{ g}}\right) \times 28 \text{ g Si} = 6.027\,12 \times 10^{23}$ atoms (as in problem 2.102).

The molar mass of carbon-12 becomes

$$\left(\frac{1.9926 \times 10^{-23} \text{ g C}}{1 \text{ atom C}}\right) \times \left(\frac{6.027\,12 \times 10^{23} \text{ atoms C}}{1 \text{ mol C}}\right) = 12.010 \text{ g/mol}$$

(b) $\left(\dfrac{35.45 \text{ g Cl}}{6.022 \times 10^{23} \text{ atoms}}\right) \times 6.027\,12 \times 10^{23} \text{ atoms Cl} = 35.48 \text{ g/mol}$

2.103 (a) $Cu(HSO_4)_2$ (b) Cu^{2+} and HSO_4^-

(c) one Cu^{2+} ion and two HSO_4^- ions

(d) one Cu atom, two H atoms, two S atoms, and eight O atoms

(e) four mole of Cu, eight moles each of H and S, 32 moles of O

(f) $[63.54 + 2(1.008 + 32.06 + 4 \times 16.00)]\text{g/mol} = 257.68 \text{ g/mol}$

(g) $\text{FU} = 3.45 \text{ g} \left(\dfrac{1 \text{ mol } Cu(HSO_4)_2}{257.68 \text{ g}}\right) \times \left(\dfrac{6.022 \times 10^{23} \text{ FU}}{1 \text{ mol } Cu(HSO_4)_2}\right) = 8.06 \times 10^{21} \text{ FU}$

(h) $8 \times 16.00 \text{ g O} = 128.0 \text{ g O}$

$$\frac{128.0 \text{ g O}}{257.68 \text{ g } Cu(HSO_4)_2} \times 100 = 49.67\%$$

(i) $\text{percent of atoms} = \dfrac{8 \text{ O atoms}}{13 \text{ atoms}} \times 100 = 61.54\%$

The answers in (h) and (i) differ because (h) asks about the mass of oxygen; the mass of the other elements differ from oxygen. Part (i) asks for the percent of atoms in the compound; all atoms count equally.

CHAPTER 3
CHEMICAL REACTIONS

EXERCISES

Balancing Equations

3.1

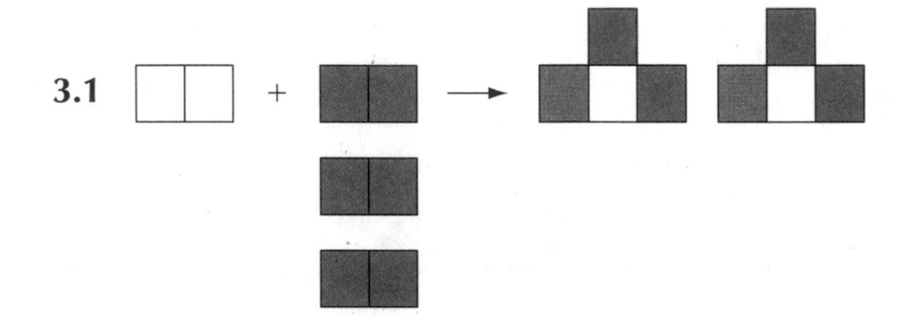

3.3 (a) $P_4O_{10}(s) + 6\ H_2O(l) \longrightarrow 4\ H_3PO_4(l)$
(b) $Cd(NO_3)_2(aq) + Na_2S(aq) \longrightarrow CdS(s) + 2\ NaNO_3(aq)$
(c) $4\ KClO_3(s) \xrightarrow{\Delta} 3\ KClO_4(s) + KCl(s)$
(d) $2\ HCl(aq) + Ca(OH)_2(aq) \longrightarrow CaCl_2(aq) + 2\ H_2O(l)$

3.5 (a) $2\ Na(s) + 2\ H_2O(l) \longrightarrow H_2(g) + 2\ NaOH(aq)$
(b) $Na_2O(s) + H_2O(l) \longrightarrow 2\ NaOH(aq)$
(c) $6\ Li(s) + N_2(g) \longrightarrow 2\ Li_3N(s)$
(d) $Ca(s) + 2\ H_2O(l) \longrightarrow Ca(OH)_2(aq) + H_2(g)$

3.7 1st stage: $3\ Fe_2O_3(l) + CO(g) \longrightarrow 2\ Fe_3O_4(l) + CO_2(g)$
2nd stage: $Fe_3O_4(l) + 4\ CO(g) \longrightarrow 3\ Fe(l) + 4\ CO_2(g)$

3.9 engine: $N_2(g) + O_2(g) \longrightarrow 2\ NO(g)$
atmosphere: $2\ NO(g) + O_2(g) \longrightarrow 2\ NO_2(g)$

3.11 $4\ HF(aq) + SiO_2(s) \longrightarrow SiF_4(g) + 2\ H_2O(l)$

Precipitation Reactions

3.13 First identify all soluble ionic compounds. Then write the complete ionic equation by writing all the soluble ionic compounds in ionic form. Then cancel all ions common to both sides of the reaction. What remains is the net ionic equation.

3.15 (a) soluble, all nitrates are soluble
(b) slightly soluble
(c) soluble, all nitrates are soluble
(d) soluble, Group 1 compounds are soluble

3.17 (a) $Na^+(aq)$, $I^-(aq)$
(b) insoluble, but very slight amounts of $Ag^+(aq)$ and $CO_3^{2-}(aq)$ form
(c) $NH_4^+(aq)$, $PO_4^{3-}(aq)$
(d) $Fe^{2+}(aq)$, $SO_4^{2-}(aq)$

3.19 (a) $FeCl_3(aq) + 3\ NaOH(aq) \longrightarrow Fe(OH)_3(s) + 3\ NaCl(aq)$

$Fe^{3+}(aq) + 3\ Cl^-(aq) + 3\ Na^+(aq) + 3\ OH^-(ag) \longrightarrow$
$Fe(OH)_3(s) + 3\ Na^+(aq) + 3\ Cl^-(aq)$

$Fe^{3+}(aq) + 3\ OH^-(aq) \longrightarrow Fe(OH)_3(s)$

Na^+ and Cl^- are spectator ions.

(b) $AgNO_3(aq) + KI(aq) \longrightarrow AgI(s) + KNO_3(aq)$

$Ag^+(aq) + NO_3^-(aq) + K^+(aq) + I^-(aq) \longrightarrow AgI(s) + K^+(aq) + NO_3^-(aq)$

$Ag^+(aq) + I^-(aq) \longrightarrow AgI(s)$

K^+ and NO_3^- are spectator ions.

(c) $Pb(NO_3)_2(aq) + K_2SO_4(aq) \longrightarrow PbSO_4(s) + 2\ KNO_3(aq)$

$Pb^{2+}(aq) + 2\ NO_3^-(aq) + 2\ K^+(aq) + SO_4^{2-}(aq) \longrightarrow$
$PbSO_4(s) + 2\ K^+(aq) + 2\ NO_3^-(aq)$

$Pb^{2+}(aq) + SO_4^{2-}(aq) \longrightarrow PbSO_4(s)$

K^+ and NO_3^- are spectator ions.

(d) $Na_2CrO_4(aq) + Pb(NO_3)_2(aq) \longrightarrow PbCrO_4(s) + 2\ NaNO_3(aq)$

$2\ Na^+(aq) + CrO_4^{2-}(aq) + Pb^{2+}(aq) + 2\ NO_3^-(aq) \longrightarrow$
$PbCrO_4(s) + 2\ Na^+(aq) + 2\ NO_3^-(aq)$

$Pb^{2+}(aq) + CrO_4^{2-}(aq) \longrightarrow PbCrO_4(s)$

Na^+ and NO_3^- are spectator ions.

(e) $Hg_2(NO_3)_2(aq) + K_2SO_4(aq) \longrightarrow Hg_2SO_4(s) + 2\ KNO_3(aq)$

$Hg_2^{2+}(aq) + 2\ NO_3^-(aq) + 2\ K^+\ (aq) + SO_4^{2-}(aq) \longrightarrow$
$Hg_2SO_4(s) + 2\ K^+(aq) + 2\ NO_3^-(aq)$

$Hg_2^{2+}(aq) + SO_4^{2-}(aq) \longrightarrow Hg_2SO_4(s)$

K^+ and NO_3^- are spectator ions.

3.21 (a) $(NH_4)_2CrO_4(aq) + BaCl_2(aq) \longrightarrow BaCrO_4(s) + 2\ NH_4Cl(aq)$

$2\ NH_4^+(aq) + CrO_4^{2-}(aq) + Ba^{2+}(aq) + 2\ Cl^-(aq) \longrightarrow$
$BaCrO_4(s) + 2\ NH_4^+(aq) + 2\ Cl^-(aq)$

$Ba^{2+}(aq) + CrO_4^{2-}(aq) \longrightarrow BaCrO_4(s)$

NH_4^+ and Cl^- are spectator ions.

(b) $CuSO_4(aq) + Na_2S(aq) \longrightarrow CuS(s) + Na_2SO_4(aq)$

$Cu^{2+}(aq) + SO_4^{2-}(aq) + 2\ Na^+(aq) + S^{2-}(aq) \longrightarrow$

$CuS(s) + 2\ Na^+(aq) + SO_4^{2-}(aq)$

$Cu^{2+}(aq) + S^{2-}(aq) \longrightarrow CuS(s)$

Na^+ and SO_4^{2-} are spectator ions.

(c) $3\ FeCl_2(aq) + 2\ (NH_4)_3PO_4(aq) \longrightarrow Fe_3(PO_4)_2(s) + 6\ NH_4Cl(aq)$

$3\ Fe^{2+}(aq) + 6\ Cl^-(aq) + 6\ NH_4^+(aq) + 2\ PO_4^{3-}(aq) \longrightarrow$

$Fe_3(PO_4)_2(s) + 6\ NH_4^+(aq) + 6\ Cl^-(aq)$

$3\ Fe^{2+}(aq) + 2\ PO_4^{3-}(aq) \longrightarrow Fe_3(PO_4)_2(s)$

NH_4^+ and Cl^- are spectator ions.

(d) $K_2C_2O_4(aq) + Ca(NO_3)_2(aq) \longrightarrow 2\ KNO_3(aq) + CaC_2O_4(s)$

$2\ K^+(aq) + C_2O_4^{2-}(aq) + Ca^{2+}(aq) + 2\ NO_3^-(aq) \longrightarrow$

$2\ K^+(aq) + 2\ NO_3^-(aq) + CaC_2O_4(s)$

$Ca^{2+}(aq) + C_2O_4^{2-}(aq) \longrightarrow CaC_2O_4(s)$

K^+ and NO_3^- are spectator ions.

(e) $NiSO_4(aq) + Ba(NO_3)_2(aq) \longrightarrow Ni(NO_3)_2(aq) + BaSO_4(s)$

$Ni^{2+}(aq) + SO_4^{2-}(aq) + Ba^{2+}(aq) + 2\ NO_3^-(aq) \longrightarrow$

$Ni^{2+}(aq) + 2\ NO_3^-(aq) + BaSO_4(s)$

$Ba^{2+}(aq) + SO_4^{2-}(aq) \longrightarrow BaSO_4(s)$

Ni^{2+} and NO_3^- are spectator ions.

3.23 (a) $Pb^{2+}(aq) + 2\ ClO_4^-(aq) + 2\ Na^+(aq) + 2\ Br^-(aq) \longrightarrow$

$PbBr_2(s) + 2\ Na^+(aq) + 2\ ClO_4^-(aq)$

$Pb^{2+}(aq) + 2\ Br^-(aq) \longrightarrow PbBr_2(s)$

(b) $Ag^+(aq) + NO_3^-(aq) + NH_4^+(aq) + Cl^-(aq) \longrightarrow$

$AgCl(s) + NH_4^+(aq) + NO_3^-(aq)$

$Ag^+(aq) + Cl^-(aq) \longrightarrow AgCl(s)$

(c) $2\ Na^+(aq) + 2\ OH^-(aq) + Cu^{2+}(aq) + 2\ NO_3^-(aq) \longrightarrow$

$Cu(OH)_2(s) + 2\ Na^+(aq) + 2\ NO_3^-(aq)$

$Cu^{2+}(aq) + 2\ OH^-(aq) \longrightarrow Cu(OH)_2(s)$

3.25 (a) $Pb^{2+}(aq) + SO_4^{2-}(aq) \longrightarrow PbSO_4(s)$

(b) $Cu^{2+}(aq) + S^{2-}(aq) \longrightarrow CuS(s)$

(c) $Co^{2+}(aq) + CO_3^{2-}(aq) \longrightarrow CoCO_3(s)$

(d) In each case, use Table 3.1 to match the cation with an anion from the table that results in an insoluble compound, and the anion with a cation from the table

that results in an insoluble compound.
Thus, for (a), $Pb(NO_3)_2$, Na_2SO_4 [spectators Na^+, NO_3^-];
for (b), $Cu(NO_3)_2$, Na_2S [spectators Na^+, NO_3^-];
for (c), $Co(NO_3)_2$, Na_2CO_3 [spectators Na^+, NO_3^-]

3.27 Ag^+ is present; the white precipitate formed when hydrochloric acid is added is AgCl(s). Zn^{2+} is also present; the black precipitate formed by adding hydrogen sulfide is ZnS(s).

Acids and Bases

3.29 See Section 3.7. Acids are molecules or ions that contain hydrogen and produce hydronium ions, H_3O^+, in water. Bases are molecules or ions that produce hydroxide ions, OH^-, in water. A base does not need to contain the hydroxide ion.

3.31 In each case, consider how the substance exists in aqueous solution; and, if it reacts with water, determine the products.
(a) $NH_3(aq) + H_2O(l) \longrightarrow NH_4^+(aq) + OH^-(aq)$
Because OH^- is produced, NH_3 is a base.
(b) $HCl(aq) + H_2O(l) \longrightarrow H_3O^+(aq) + Cl^-(aq)$
Because H_3O^+ is produced, HCl is an acid.
(c) $NaOH(aq) \longrightarrow Na^+(aq) + OH^-(aq)$; a base
(d) $H_2SO_4(aq) + H_2O(l) \longrightarrow H_3O^+(aq) + HSO_4^-(aq)$
$HSO_4^-(aq) + H_2O(l) \longrightarrow H_3O^+(aq) + SO_4^{2-}(aq)$
Because H_3O^+ is produced, H_2SO_4 is an acid.
(e) $Ba(OH)_2(aq) \longrightarrow Ba^{2+}(aq) + 2\ OH^-(aq)$; a base

3.33 (a) $HCl(aq) + NaOH(aq) \longrightarrow H_2O(l) + NaCl(aq)$
$H_3O^+(aq) + Cl^-(aq) + Na^+(aq) + OH^-(aq) \longrightarrow 2\ H_2O(l) + Na^+(aq) + Cl^-(aq)$
$H_3O^+(aq) + OH^-(aq) \longrightarrow 2\ H_2O(l)$
(b) $NH_3(aq) + HNO_3(aq) \longrightarrow NH_4NO_3(aq)$
$NH_3(aq) + H_3O^+(aq) + NO_3^-(aq) \longrightarrow NH_4^+(aq) + NO_3^-(aq) + H_2O(l)$
$NH_3(aq) + H_3O^+(aq) \longrightarrow NH_4^+(aq) + H_2O(l)$
(c) $CH_3NH_2(aq) + HI(aq) \longrightarrow CH_3NH_3I(aq)$
$CH_3NH_2(aq) + H_3O^+(aq) + I^-(aq) \longrightarrow CH_3NH_3^+(aq) + I^-(aq) + H_2O(l)$
$CH_3NH_2(aq) + H_3O^+(aq) \longrightarrow CH_3NH_3^+(aq) + H_2O(l)$

3.35 In each case, we need to identify a base that provides the cation of the salt and an acid that provides its anion.

(a) HBr and KOH

$HBr(aq) + KOH(aq) \longrightarrow KBr(aq) + 2\ H_2O(l)$

$H_3O^+(aq) + Br^-(aq) + K^+(aq) + OH^-(aq) \longrightarrow K^+(aq) + Br^-(aq) + 2\ H_2O(l)$

$H_3O^+(aq) + OH^-(aq) \longrightarrow 2\ H_2O(l)$

(b) HNO_2 and $Ba(OH)_2$

$2\ HNO_2(aq) + Ba(OH)_2(aq) \longrightarrow Ba(NO_2)_2(aq) + 2\ H_2O(l)$

$2\ H_3O^+(aq) + 2\ NO_2^-(aq) + Ba(OH)_2(aq) \longrightarrow Ba^{2+}(aq) + 2\ NO_2^-(aq) + 4\ H_2O(l)$

$H_3O^+(aq) + OH^-(aq) \longrightarrow 2\ H_2O(l)$

(c) HCN and $Ca(OH)_2$

$2\ HCN(aq) + Ca(OH)_2(aq) \longrightarrow Ca(CN)_2(aq) + 2\ H_2O(l)$

$2\ HCN(aq) + Ca^{2+}(aq) + 2\ OH^-(aq) \longrightarrow Ca^{2+}(aq) + 2\ CN^-(aq) + 2\ H_2O(l)$

$HCN(aq) + OH^-(aq) \longrightarrow CN^-(aq) + H_2O(l)$

(d) H_3PO_4 and KOH

$H_3PO_4(aq) + 3\ KOH(aq) \longrightarrow K_3PO_4(aq) + 3\ H_2O(l)$

$3\ H_3O^+(aq) + PO_4^{3-}(aq) + 3\ K^+(aq) + 3\ OH^-(aq) \longrightarrow$

$3\ K^+(aq) + PO_4^{3-}(aq) + 6\ H_2O(l)$

$3\ H_3O^+(aq) + 3\ OH^-(aq) \longrightarrow 6\ H_2O(l)$ or $H_3O^+(aq) + OH^-(aq) \longrightarrow 2\ H_2O(l)$

3.37 In each case, the acid is the substance that has donated a proton, H^+, and the base is the substance that has accepted it.

(a) $CH_3NH_2(aq)$ [base], $H_3O^+(aq)$ [acid]

(b) $C_2H_5NH_2(aq)$ [base], HCl(aq) [acid]

(c) HI(aq) [acid], CaO(s) [base]

3.39 In each case, we use the periodic table to determine whether the element combined with oxygen is a metal or a nonmetal. Metallic oxides form bases in aqueous solution and nonmetallic oxides form acids.

(a) basic, $Ca(OH)_2$ (b) acidic, H_2SO_4 (c) acidic, HNO_2

(d) basic, TlOH

Redox Reactions

3.41 Oxidation is electron loss.

Reduction is electron gain.

3.43 In each case, first determine the oxidation numbers of all elements on both sides of the equation. Then balance, with appropriate multiplicative factors, if required, the number of electrons lost by one element against the number gained by another.

(a) $2\ P(s) + 3\ Br_2(l) \longrightarrow 2\ PBr_3(s)$

P: $2 \times 3 = 6$ lost; Br: $2 \times 3 \times 1 = 6$ gained

(b) $2\ Fe^{2+}(aq) + Sn^{4+}(aq) \longrightarrow 2\ Fe^{3+}(aq) + Sn^{2+}(aq)$

Fe: $2 \times 1 = 2$ lost; Sn: $1 \times 2 = 2$ gained

(c) $8\ H_2(g) + S_8(s) \longrightarrow 8\ H_2S(g)$

H: $8 \times 2 \times 1 = 16$ lost; S: $8 \times 2 = 16$ gained

(d) $2\ NO(g) + O_2(g) \longrightarrow 2\ NO_2(g)$

N: $2 \times 2 = 4$ lost; O: $2 \times 2 = 4$ gained

3.45 (a) $Mg(s) + Cu^{2+}(aq) \longrightarrow Mg^{2+}(aq) + Cu(s)$

(b) $2\ Fe^{2+}(aq) + Pb^{4+}(aq) \longrightarrow 2\ Fe^{3+}(aq) + Pb^{2+}(aq)$

(c) $H_2(g) + Cl_2(g) \longrightarrow 2\ HCl(g)$

(d) $4\ Fe(s) + 3\ O_2(g) \longrightarrow 2\ Fe_2O_3(s)$

Fe: $4 \times 3 = 12$ lost; O: $3 \times 2 \times 2 = 12$ gained

Oxidation Numbers

3.47 The oxidation number of an element is a number assigned on the basis of a set of rules and is used to monitor whether an element has been oxidized or reduced.

3.49 Let x represent the oxidation number of the italicized element.

(a) $x + 4 \times (-1) = 0$; therefore, $x = +4$ for *Xe*

(b) $1 + x - 2 = 0$; therefore, $x = +1$ for *Cl*

(c) $x - 2 = 0$; therefore, $x = +2$ for *N*

(d) $1 + x + 3 \times (-2) = 0$; therefore, $x = +5$ for *N*

(e) $x + 2 \times (-2) = 0$; therefore, $x = +4$ for *S*

(f) $2 \times 1 + x = 0$; therefore, $x = -2$ for *S*

3.51 Let x represent the oxidation number of the italicized element.

(a) $x + 4 \times (-2) = -1$; therefore, $x = +7$ for *Mn*

(b) $2 \times x + 3 \times (-2) = -2$; therefore, $x = +2$ for *S*

(c) $x + 4 \times (-2) = -2$; therefore, $x = +6$ for *S*

(d) $x + 4 \times (-2) = -2$; therefore, $x = +6$ for *Mn*

(e) $2 \times x + 7 \times (-2) = -2$; therefore, $x = +6$ for *Cr*

3.53 (a) This is a substitution reaction; no oxidation or reduction occurs.

$$\overset{-2}{}\quad\overset{-1}{}\quad\overset{+1}{}\quad\overset{0}{}\quad\overset{+1-2}{}$$

(b) $\underset{+5}{BrO_3^-}(aq) + 5\,Br^-(aq) + 6\,H^+(aq) \longrightarrow 3\,Br_2(l) + 3\,H_2O(l)$

Oxidation numbers: BrO_3^-: Br +5, O −2; Br^-: −1; H^+: +1; Br_2: 0; H_2O: +1 −2

BrO_3^- is reduced and Br^- is oxidized.

(c) $2\,F_2(g) + 2\,H_2O(l) \longrightarrow 4\,HF(aq) + O_2(g)$

Oxidation numbers: F_2: 0; H_2O: +1 −2; HF: +1 −1; O_2: 0

F_2 is reduced and water is oxidized.

Oxidizing and Reducing Agents

3.55 Oxidizing agents contain an element that gains electrons and decreases its oxidation number. Reducing agents lose electrons to the oxidizing agent. A loss of electrons means an increase in oxidation number for an element in the reducing agent.

3.57 The higher (in the algebraic sense) the oxidation number of an element, the stronger it is as an oxidizing agent. It has the greater ability to gain electrons.
(a) Cl_2 (0) is stronger than Cl^- (−1), because $0 > -1$
(b) N_2O_5 (N = +5) is stronger than N_2O (N = +1), because $+5 > +1$

3.59 In each case, identify the substances that contain an element that undergoes a change in oxidation number. If the change is positive, in the algebraic sense, that substance is the reducing agent. If the change is negative, in the algebraic sense, that substance is the oxidizing agent.
(a) Oxidation number changes are
Zn ($0 \longrightarrow +2$); so Zn is the reducing agent.
H ($+1 \longrightarrow 0$); so HCl is the oxidizing agent.
(b) Oxidation number changes are
S (in H_2S, $-2 \longrightarrow 0$); so H_2S is the reducing agent.
S (in SO_2, $+4 \longrightarrow 0$); so SO_2 is the oxidizing agent.
(c) Oxidation number changes are
B ($+3 \longrightarrow 0$); so B_2O_3 is the oxidizing agent.
Mg ($0 \longrightarrow +2$); so Mg is the reducing agent.

3.61 In each case, decide whether the change is oxidation or reduction. If oxidation, an oxidizing agent is required. If reduction, a reducing agent is required. Let x = oxidation number.

(a) $x(\text{Br}) = -1 \longrightarrow x(\text{Br}) = +5$; so oxidation has occurred and an oxidizing agent is required.
(b) $x(\text{S}) = +5 \longrightarrow x(\text{S}) = +6$; so oxidation has occurred and an oxidizing agent is required.
(c) $x(\text{N}) = +5 \longrightarrow x(\text{N}) = +2$; so reduction has occurred and a reducing agent is required.
(d) $x(\text{C}) = 0 \longrightarrow x(\text{C}) = -2$; so reduction has occurred and a reducing agent is required.
Rationale:

$$\left.\begin{aligned} 2 \times x(\text{H}) + x(\text{C}) + x(\text{O}) &= 0 \\ 2 + x(\text{C}) + (-2) &= 0 \\ x(\text{C}) &= 0 \end{aligned}\right\} \text{for HCHO}$$

$$\left.\begin{aligned} 4 \times x(\text{H}) + x(\text{C}) + x(\text{O}) &= 0 \\ 4 + x(\text{C}) + (-2) &= 0 \\ x(\text{C}) &= -2 \end{aligned}\right\} \text{for } CH_3OH$$

3.63 $2\ NaCl(l) \xrightarrow{\text{electrolysis at 600°C}} 2\ Na(l) + Cl_2(g)$
Oxidation number changes are
Na ($+1 \longrightarrow 0$); so reduction has occurred.
Cl ($-1 \longrightarrow 0$); so oxidation has occurred.
Therefore, chlorine is produced by oxidation and sodium by reduction.

SUPPLEMENTARY EXERCISES

3.65 Use Tables 3.1 and 3.2. If an acid or a base is not in Table 3.2, it is assumed to be weak.
(a) strong acid (b) base (c) base (d) soluble ionic
(e) weak acid (f) insoluble ionic (g) insoluble ionic (h) strong acid

3.67 (a) HNO_3, $Ba(OH)_2$ (b) H_2SO_4, NaOH (c) $HClO_4$, KOH (d) HCl, CsOH

3.69 (a) nonelectrolyte (b) strong electrolyte (c) strong electrolyte

3.71 (a) acid-base neutralization; HCl, acid; $Mg(OH)_2$, base
(b) acid-base neutralization; H_2SO_4, acid; $Ba(OH)_2$, base, or precipitation;
$Ba^{2+}(aq) + SO_4^{2-}(aq) \longrightarrow BaSO_4(s)$

(c) redox; O_2 is the oxidizing agent, because oxidation number of O is decreased ($0 \longrightarrow -2$); SO_2 is the reducing agent, because oxidation number of S in SO_2 is increased ($+4 \longrightarrow +6$).

3.73 (a) redox; I_2O_5 is the oxidizing agent, because oxidation number of I decreases ($+5 \longrightarrow 0$); CO is the reducing agent, because oxidation number of C increases ($+2 \longrightarrow +4$).

(b) redox; I_2 is the oxidizing agent, because oxidation number of I decreases ($0 \longrightarrow -1$); $S_2O_3^{2-}(aq)$ is the reducing agent, because oxidation number of S increases ($+2 \longrightarrow 2.5$).

(c) precipitation; $Ag^+(aq) + Br^-(aq) \longrightarrow AgBr(s)$

(d) redox; UF_4 is the oxidizing agent, because oxidation number of U decreases ($+4 \longrightarrow 0$); Mg is the reducing agent, because oxidation number of Mg increases ($0 \longrightarrow +2$).

3.75 $2\ C_8H_{18}(l) + 25\ O_2(g) \longrightarrow 16\ CO_2(g) + 18\ H_2O(g)$

3.77 $4\ C_{10}H_{15}N(s) + 55\ O_2(g) \longrightarrow 40\ CO_2(g) + 30\ H_2O(l) + 2\ N_2(g)$

3.79 Let x = oxidation number of H or O.

(a) $K = +1$, then $+1 + 2 \times x = 0$, $x(O) = -\frac{1}{2}$

(b) $Li = +1$, $Al = +3$, then $+1 + 3 + 4 \times x = 0$, $x(H) = -1$

(c) $Na = +1$, then $2 \times (+1) + 2 \times x = 0$, $x(O) = -1$

(d) $Na = +1$, then $+1 + x = 0$, $x(H) = -1$

(e) $K = +1$, then $+1 + 3 \times x = 0$, $x(O) = -\frac{1}{3}$

APPLIED EXERCISES

3.81 $2\ CO_2(g) + CaSiO_3(s) + H_2O(l) \longrightarrow SiO_2(s) + Ca(HCO_3)_2(aq)$

3.83 The strategy is to write the balanced reaction between each metal hydroxide and CO_2 to form a metal carbonate and water. Then determine the mass of metal hydroxide needed per mole of CO_2.

$CO_2(g) + 2\ NaOH(s) \longrightarrow Na_2CO_3(s) + H_2O(l)$

$$1 \text{ mol } CO_2 \times \left(\frac{2 \text{ mol NaOH}}{1 \text{ mol } CO_2}\right) \times \left(\frac{40.0 \text{ g NaOH}}{1 \text{ mol NaOH}}\right) = 80.0 \text{ g NaOH/mol } CO_2$$

$CO_2(g) + 2\ KOH(s) \longrightarrow K_2CO_3(s) + H_2O(l)$

$$1\ \text{mol } CO_2 \times \left(\frac{2\ \text{mol KOH}}{1\ \text{mol } CO_2}\right) \times \left(\frac{56.11\ \text{g KOH}}{1\ \text{mol KOH}}\right) = 112.2\ \text{g KOH/mol } CO_2$$

$CO_2(g) + 2\ CsOH(s) \longrightarrow Cs_2CO_3(s) + H_2O(l)$

$$1\ \text{mol } CO_2 \times \left(\frac{2\ \text{mol CsOH}}{1\ \text{mol } CO_2}\right) \times \left(\frac{149.9\ \text{g CsOH}}{1\ \text{mol CsOH}}\right) = 299.8\ \text{g CsOH/mol } CO_2$$

NaOH would be the best option if lithium is in short supply.

3.85 $H_2S(g) + 2\ NaOH(aq) \longrightarrow Na_2S(aq) + 2\ H_2O(l)$

$4\ H_2S(g) + Na_2S(alc) \longrightarrow Na_2S_5(alc) + 4\ H_2(g)$

$10\ H_2O(l) + 9\ O_2(g) + 2\ Na_2S_5(alc) \longrightarrow 2\ Na_2S_2O_3 \cdot 5H_2O + 6\ SO_2(g)$

INTEGRATED EXERCISES

3.87 (a) $100. - (26.68 + 2.239) = 71.08\%$ O

$$\text{mol C: } 26.68\ \text{g C} \times \left(\frac{1\ \text{mol C}}{12.01\ \text{g C}}\right) = 2.221\ \text{mol C}$$

$$2.293\ \text{g H} \times \left(\frac{1\ \text{mol H}}{1.008\ \text{g H}}\right) = 2.275\ \text{mol H}$$

$$71.08\ \text{g O} \times \left(\frac{1\ \text{mol O}}{16.00\ \text{g O}}\right) = 4.443\ \text{mol O}$$

Divide the number of moles of C:H:O by 2.275 to get small, whole number ratios of atoms. C:H:O is thus 1:1:2. The empirical formula is CHO_2.

(b) The empirical formula weight is $(12.01 \times 1 + 1.008 \times 1 + 16.00 \times 2)$g/mol $= 45.02$ g/mol. Divide the molecular weight by the empirical weight to determine the actual molecular formula:

$$\frac{90.0\ \text{g/mol}}{45.02\ \text{g/mol}} \approx 2 \quad \text{so the molecular formula is } H_2C_2O_4.$$

(c) $H_2C_2O_4(aq) + 2\ NaOH(aq) \longrightarrow Na_2C_2O_4(aq) + 2\ H_2O(l)$

$H_2C_2O_4(aq) + 2\ OH^-(aq) \longrightarrow C_2O_4^{2-}(aq) + 2\ H_2O(l)$

(d) X is oxalic acid. The products of the reaction are water and sodium oxalate.

(e) This is an acid-base reaction.

3.89 (a) potassium ion, chloride ion

(b) copper(II) ion, chloride ion

(c) silver ion, nitrate ion

3.91 (a)

Group	Maximum oxidation number	Minimum oxidation number
1	+1	0
2	+2	0
13	+3	0
14	+4	−4
15	+5	−3
16	+6	−2
17	+7	−1

(b) The maximum oxidation number increases as group number increases. In most cases, the minimum oxidation number decreases according to this same pattern.

CHAPTER 4
CHEMISTRY'S ACCOUNTING: REACTION STOICHIOMETRY

EXERCISES

Mole Calculations

4.1 (a) Stoichiometric relation = 3 mol $NO_2 \simeq$ 2 mol HNO_3

$$\text{number of moles of } NO_2 = 7.33 \text{ mol } HNO_3 \times \left(\frac{3 \text{ mol } NO_2}{2 \text{ mol } HNO_3}\right) = 11.0 \text{ mol } NO_2$$

(b) Stoichiometric relation = 2 mol $MnO_4^- \simeq$ 10 mol I^-

$$\text{number of moles of } MnO_4^- = 0.042 \text{ mol } I^- \times \left(\frac{2 \text{ mol } MnO_4^-}{10 \text{ mol } I^-}\right)$$

$$= 8.4 \times 10^{-3} \text{ mol } MnO_4^-$$

4.3 The stoichiometric relation, as determined from the chemical equation, is 12 mol $CO_2 \simeq$ 2 mol C_6H_{14}. Hexane is $C_6H_{14}(l)$.

$$\text{number of moles of } CO_2 = 1.5 \text{ mol hexane} \times \left(\frac{12 \text{ mol } CO_2}{2 \text{ mol hexane}}\right) = 9.0 \text{ mol } CO_2$$

4.5 $C_{11}H_{22}O_{11}(s) \xrightarrow{H_2SO_4} 11\ C(s) + 11\ H_2O(g)$

$$0.500 \text{ mol } C_{11}H_{22}O_{11} \left(\frac{11 \text{ mol } H_2O}{1 \text{ mol } C_{11}H_{22}O_{11}}\right) = 5.50 \text{ mol } H_2O$$

Mass-Mole Relationships

4.7 (a) 6 mol $H_2O \simeq$ 4 mol NH_3

$$\text{number of moles of } H_2O = 1.0 \text{ g } NH_3 \times \left(\frac{1 \text{ mol } NH_3}{17.03 \text{ g } NH_3}\right) \times \left(\frac{6 \text{ mol } H_2O}{4 \text{ mol } NH_3}\right)$$

$$= 0.088 \text{ mol } H_2O$$

(b) 3 mol $O_2 \simeq$ 4 mol NH_3

$$\text{mass of } O_2 = 13.7 \text{ mol } NH_3 \times \left(\frac{3 \text{ mol } O_2}{4 \text{ mol } NH_3}\right) \times \left(\frac{32.00 \text{ g } O_2}{1 \text{ mol } O_2}\right) = 329 \text{ g } O_2$$

4.9 (a) 3 mol $CO_2 \simeq$ 1 mol C_3H_8

$$\text{mass of } CO_2 = 1.55 \text{ mol } C_3H_8 \times \left(\frac{3 \text{ mol } CO_2}{1 \text{ mol } C_3H_8}\right) \times \left(\frac{44.01 \text{ g } CO_2}{1 \text{ mol } CO_2}\right)$$
$$= 205 \text{ g } CO_2$$

(b) $4 \text{ mol } H_2O \simeq 3 \text{ mol } CO_2$

$$\text{number of moles of } H_2O = 4.40 \text{ g } CO_2 \times \left(\frac{1 \text{ mol } CO_2}{44.01 \text{ g } CO_2}\right) \times \left(\frac{4 \text{ mol } H_2O}{3 \text{ mol } CO_2}\right)$$
$$= 0.133 \text{ mol } H_2O$$

Mass-Mass Relationships

4.11 (a) $$\text{mass of } Al_2O_3 = 10.0 \text{ g Al} \times \left(\frac{1 \text{ mol Al}}{26.98 \text{ g Al}}\right) \times \left(\frac{2 \text{ mol } Al_2O_3}{4 \text{ mol Al}}\right) \times \left(\frac{101.96 \text{ g } Al_2O_3}{1 \text{ mol } Al_2O_3}\right) = 18.9 \text{ g } Al_2O_3$$

(b) $$\text{mass of } O_2 = 10.0 \text{ g Al} \times \left(\frac{1 \text{ mol Al}}{26.98 \text{ g Al}}\right) \times \left(\frac{3 \text{ mol } O_2}{4 \text{ mol Al}}\right) \times \left(\frac{32.00 \text{ g } O_2}{1 \text{ mol } O_2}\right)$$
$$= 8.90 \text{ g } O_2$$

4.13 (a) $$\text{mass of Al} = 1.5 \times 10^4 \text{ kg } NH_4ClO_4 \times \left(\frac{10^3 \text{ g } NH_4ClO_4}{1 \text{ kg } NH_4ClO_4}\right)$$
$$\times \left(\frac{1 \text{ mol } NH_4ClO_4}{117.49 \text{ g } NH_4ClO_4}\right) \times \left(\frac{10 \text{ mol Al}}{6 \text{ mol } NH_4ClO_4}\right) \times \left(\frac{26.98 \text{ g Al}}{1 \text{ mol Al}}\right) \times \left(\frac{1 \text{ kg}}{1000 \text{ g}}\right)$$
$$= 5.7 \times 10^3 \text{ kg Al}$$

(b) $$\text{mass of } Al_2O_3 = 5000. \text{ kg Al} \times \left(\frac{10^3 \text{ g Al}}{1 \text{ kg Al}}\right) \times \left(\frac{1 \text{ mol Al}}{26.98 \text{ g Al}}\right)$$
$$\times \left(\frac{5 \text{ mol } Al_2O_3}{10 \text{ mol Al}}\right) \times \left(\frac{101.96 \text{ g } Al_2O_3}{1 \text{ mol } Al_2O_3}\right) = 9.448 \times 10^6 \text{ g } Al_2O_3$$
$$= 9.448 \times 10^3 \text{ kg } Al_2O_3$$

4.15 (a) $$\text{mass of } H_2O = 2.5 \times 10^3 \text{ g tristearin} \times \left(\frac{1 \text{ mol tristearin}}{891.45 \text{ g tristearin}}\right)$$
$$\times \left(\frac{110 \text{ mol } H_2O}{2 \text{ mol tristearin}}\right) \times \left(\frac{18.02 \text{ g } H_2O}{1 \text{ mol } H_2O}\right) = 2.8 \times 10^3 \text{ g } H_2O$$

(b) $$\text{mass of } O_2 = 2.5 \text{ g tristearin} \times \left(\frac{1 \text{ mol tristearin}}{891.45 \text{ g tristearin}}\right) \times \left(\frac{163 \text{ mol } O_2}{2 \text{ mol tristearin}}\right)$$
$$\times \left(\frac{32.00 \text{ g } O_2}{1 \text{ mol } O_2}\right) = 7.3\text{g } O_2$$

4.17 There are two approaches. We can first calculate the mass of gasoline in 1.0 L of gasoline and then use that result to calculate the mass of water produced. Alternatively, we can calculate the mass of water directly from the volume of gasoline. Both approaches are illustrated below.

$$(1)\ \text{mass of gasoline} = 1.0\ \text{L gasoline} \times \left(\frac{1000\ \text{mL gasoline}}{1\ \text{L gasoline}}\right) \times \left(\frac{0.79\ \text{g gasoline}}{1\ \text{mL gasoline}}\right)$$

$$= 79\bar{0}\ \text{g gasoline}$$

$$\text{mass of } H_2O = 79\bar{0}\ \text{g gasoline} \times \left(\frac{1\ \text{mol gasoline}}{114.22\ \text{g gasoline}}\right) \times \left(\frac{18\ \text{mol } H_2O}{2\ \text{mol gasoline}}\right)$$

$$\times \left(\frac{18.02\ \text{g } H_2O}{1\ \text{mol } H_2O}\right) = 1.1 \times 10^3\ \text{g } H_2O$$

$$(2)\ \text{mass of } H_2O = 1.0\ \text{L gasoline} \times \left(\frac{1000\ \text{mL gasoline}}{1\ \text{L gasoline}}\right) \times \left(\frac{0.79\ \text{g gasoline}}{1\ \text{mL gasoline}}\right)$$

$$\times \left(\frac{1\ \text{mol gasoline}}{114.22\ \text{g gasoline}}\right) \times \left(\frac{18\ \text{mol } H_2O}{2\ \text{mol gasoline}}\right) \times \left(\frac{18.02\ \text{g } H_2O}{1\ \text{mol } H_2O}\right)$$

$$= 1.1 \times 10^3\ \text{g } H_2O$$

Reaction Yield

4.19 $$\text{percentage yield} = \frac{\text{actual yield}}{\text{theoretical yield}} \times 100\%$$

$$\text{theoretical yield of } CO_2 = 30.7\ \text{g } CaCO_3 \times \left(\frac{1\ \text{mol } CaCO_3}{100.09\ \text{g } CaCO_3}\right)$$

$$\times \left(\frac{1\ \text{mol } CO_2}{1\ \text{mol } CaCO_3}\right) \times \left(\frac{44.01\ \text{g } CO_2}{1\ \text{mol } CO_2}\right) = 13.5\ \text{g } CO_2$$

$$\text{percentage yield } CO_2 = \frac{11.7\ \text{g } CO_2}{13.5\ \text{g } CO_2} \times 100\% = 86.7\%$$

Limiting Reactants

4.21

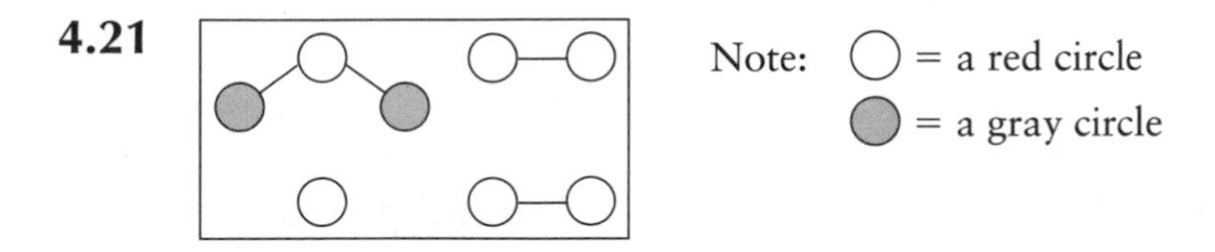

4.23 number of moles of C_2H_5OH supplied =

$$2.00\ \text{g } C_2H_5OH \times \left(\frac{1\ \text{mol } C_2H_5OH}{46.07\ \text{g } C_2H_5OH}\right) = 0.0434\ \text{mol } C_2H_5OH$$

$$\text{number of moles of } O_2 \text{ supplied} = 1.00\ \text{g } O_2 \times \left(\frac{1\ \text{mol } O_2}{32.00\ \text{g } O_2}\right) = 0.0313\ \text{mol } O_2$$

$$\text{number of moles of } O_2 \text{ required} = 0.0434 \text{ mol } C_2H_5OH \times \left(\frac{1 \text{ mol } O_2}{1 \text{ mol } C_2H_5OH}\right)$$

$$= 0.0434 \text{ mol } O_2$$

Because the number of moles of O_2 supplied is less than the number required, O_2 is the limiting reactant.

4.25 (a) mass of SO_2 supplied = 2.86 g/L SO_2 × 4.0 L = $11.\bar{4}$ g SO_2

mass of H_2S supplied = 1.52 g/L H_2S × 4.0 L = $6.0\bar{8}$ g H_2S

$$\text{number of moles of } SO_2 \text{ supplied} = 11.\bar{4} \text{ g } SO_2 \times \left(\frac{1 \text{ mol } SO_2}{64.06 \text{ g } SO_2}\right)$$

$$= 0.17\bar{8} \text{ mol } SO_2$$

$$\text{number of moles of } H_2S \text{ supplied} = 6.0\bar{8} \; H_2S \times \left(\frac{1 \text{ mol } H_2S}{34.02 \text{ g } H_2S}\right)$$

$$= 0.17\bar{8} \text{ mol } H_2S$$

$$\text{number of moles of } H_2S \text{ required} = 0.17\bar{8} \text{ mol } SO_2 \times \left(\frac{2 \text{ mol } H_2S}{1 \text{ mol } SO_2}\right)$$

$$= 0.35\bar{6} \text{ mol } H_2S$$

Because 0.356 mol of H_2S is required and there is only $0.17\bar{8}$ mol of H_2S available, H_2S is the limiting reactant.

(b) number of moles of excess SO_2 =

$$\left(0.17\bar{8} \text{ mol } SO_2 - 0.17\bar{8} \text{ mol } H_2S \times \frac{1 \text{ mol } SO_2}{2 \text{ mol } H_2S}\right) = 0.089 \text{ mol } SO_2$$

$$\text{mass of excess } SO_2 = 0.089 \text{ mol } SO_2 \times \left(\frac{64.06 \text{ g } SO_2}{1 \text{ mol } SO_2}\right) = 5.7 \text{ g } SO_2$$

$$\text{(c) mass of S} = 0.17\bar{8} \text{ mol } H_2S \times \left(\frac{3 \text{ mol S}}{2 \text{ mol } H_2S}\right) \times \left(\frac{32.06 \text{ g S}}{1 \text{ mol S}}\right) = 8.6 \text{ g S}$$

$$\text{mass of } H_2O = 0.17\bar{8} \text{ mol } H_2S \left(\frac{2 \text{ mol } H_2O}{2 \text{ mol } H_2S}\right) \times \left(\frac{18.02 \text{ g } H_2O}{1 \text{ mol } H_2O}\right)$$

$$= 3.2 \text{ g } H_2O$$

(d) $11.\bar{4}$ g SO_2 + $6.0\bar{8}$ g H_2S = $17.\bar{5}$ g of reactants and 5.7 g SO_2 + 8.6 g S + 3.2 g H_2O = 17.5 g of excess reactant and products; so everything checks with the law of the conservation of mass.

4.27 (a) Here, there are three reactants. The principles involved in determining the limiting reactant are the same as those employed for two reactants. The difference is only in the number of comparisons of moles supplied to moles required that must be considered.

$$\text{moles of } Al_2O_3 \text{ supplied} = 100.\ g\ Al_2O_3 \times \left(\frac{1\ mol\ Al_2O_3}{101.95\ g\ Al_2O_3}\right) = 0.981\ mol\ Al_2O_3$$

$$\text{moles of C supplied} = 40.\ g\ C \times \left(\frac{1\ mol\ C}{12.01\ g\ C}\right) = 3.33\ mol\ C$$

$$\text{moles of } Cl_2 \text{ supplied} = 160.\ g\ Cl_2 \times \left(\frac{1\ mol\ Cl_2}{70.90\ g\ Cl_2}\right) = 2.25\bar{6}\ mol\ Cl_2$$

Inspection of the balanced equation shows that three times as many moles of C and Cl_2 are required compared to Al_2O_3. We see that

$$\frac{mol\ C}{mol\ Al_2O_3} = \frac{3.33}{0.981} = 3.39 > 3.00$$

$$\frac{mol\ Cl_2}{mol\ Al_2O_3} = \frac{2.25\bar{6}}{0.981} = 2.30 < 3.00$$

Therefore, C is present in excess relative to Al_2O_3 and Al_2O_3 is present in excess relative to Cl_2. So, Cl_2 is the limiting reactant.

(b) $$\text{mass of } AlCl_3 = 2.25\bar{6}\ mol\ Cl_2 \times \left(\frac{2\ mol\ AlCl_3}{3\ mol\ Cl_2}\right) \times \left(\frac{133.33\ g\ AlCl_3}{1\ mol\ AlCl_3}\right) = 200.\ g\ AlCl_3$$

Combustion Analysis

4.29 $$\text{number of moles of C} = 3.03\ g\ C \times \left(\frac{1\ mol\ C}{12.01\ g\ C}\right) = 0.252\ mol\ C$$

$$\text{number of moles of O} = 1.14\ g\ O \times \left(\frac{1\ mol\ O}{16.00\ g\ O}\right) = 0.0712\ mol\ O$$

$$\text{number of moles of H} = (4.39 - 3.03 - 1.14)\ g\ H \times \left(\frac{1\ mol\ H}{1.008\ mol\ H}\right) = 0.218\ mol\ H$$

The ratio is C:O:H = 0.252:0.0712:0.218. Dividing by 0.0712 gives C:O:H = 3.54:1.00:3.06. Multiplying by 2 gives C:O:H = 7.08:2.00:6.12, which is close to C:O:H = 7:2:6. Therefore, the empirical formula of benzoic acid is $C_7H_6O_2$.

4.31 $$\text{number of moles of C} = 0.714\ g\ C \times \left(\frac{1\ mol\ C}{12.01\ g\ C}\right) = 0.0595\ mol\ C$$

$$\text{number of moles of N} = 0.138\ g\ N \times \left(\frac{1\ mol\ N}{14.01\ g\ N}\right) = 9.85 \times 10^{-3}\ mol\ N$$

$$\text{mass of H} = (0.922 - 0.714 - 0.138)\ g\ H = 0.070\ g\ H$$

$$\text{number of moles of H} = 0.070\ g\ H \times \left(\frac{1\ mol\ H}{1.008\ g\ H}\right) = 0.069\ mol\ H$$

The ratio of amounts (in moles) is C:N:H = 0.0595:0.009 85:0.069. Dividing by 0.009 85 gives C:N:H = 6.04:1:7.00. Therefore, the empirical formula is C_6H_7N.

4.33 number of moles of C $= 0.318 \text{ g } CO_2 \times \left(\frac{1 \text{ mol } CO_2}{44.01 \text{ g } CO_2}\right) \times \left(\frac{1 \text{ mol C}}{1 \text{ mol } CO_2}\right)$

$= 0.007\ 23 \text{ mol C}$

mass of C $= 0.007\ 23 \text{ mol C} \times \left(\frac{12.01 \text{ g C}}{1 \text{ mol C}}\right) = 0.0868 \text{ g C}$

number of moles of H $= 0.084 \text{ g } H_2O \times \left(\frac{1 \text{ mol } H_2O}{18.02 \text{ g } H_2O}\right) \times \left(\frac{2 \text{ mol H}}{1 \text{ mol } H_2O}\right)$

$= 0.0093 \text{ mol H}$

mass of H $= 0.0093 \text{ mol H} \times \left(\frac{1.008 \text{ g H}}{1 \text{ mol H}}\right) = 9.4 \times 10^{-3} \text{ g H}$

number of moles of N $= 0.0145 \text{ g N} \times \left(\frac{1 \text{ mol N}}{14.01 \text{ g N}}\right) = 1.03 \times 10^{-3} \text{ mol N}$

sum of masses of C, H, and N = (0.0868 + 0.0094 + 0.0145) g = 0.1107 g

mass of O = (0.152 − 0.111) g O = 0.041 g O

number of moles of O $= 0.041 \text{ g O} \times \left(\frac{1 \text{ mol O}}{16.00 \text{ g O}}\right) = 2.5\overline{6} \times 10^{-3} \text{ mol O}$

The ratio of amounts (in moles) is C:H:N:O = 0.007 23:0.0093:0.001 03:0.002 $5\overline{6}$.

Dividing by 0.001 03 gives C:H:N:O = 7.02:9.0:1.00:2.5. Multiplying by 2 gives C:H:N:O = 14:18:2:5. The empirical formula is $C_{14}H_{18}N_2O_5$, which has a molar mass of 294 g/mol; consequently, this is also the molecular formula.

4.35 number of moles of C $= 0.682 \text{ g } CO_2 \times \left(\frac{1 \text{ mol } CO_2}{44.01 \text{ g } CO_2}\right) \times \left(\frac{1 \text{ mol C}}{1 \text{ mol } CO_2}\right)$

$= 0.0155 \text{ mol C}$

mass of C $= 0.0155 \text{ mol C} \times \left(\frac{12.01 \text{ g C}}{1 \text{ mol C}}\right) = 0.186 \text{ g C}$

number of moles of H $= 0.174 \text{ g } H_2O \times \left(\frac{1 \text{ mol } H_2O}{18.02 \text{ g } H_2O}\right) \times \left(\frac{2 \text{ mol H}}{1 \text{ mol } H_2O}\right)$

$= 0.0193 \text{ mol H}$

mass of H $= 0.0193 \text{ mol H} \times \left(\frac{1.008 \text{ g H}}{1 \text{ mol H}}\right) = 0.0195 \text{ g H}$

number of moles of N $= 0.110 \text{ g N} \times \left(\frac{1 \text{ mol N}}{14.01 \text{ g N}}\right) = 0.007\ 85 \text{ mol N}$

mass of O = (0.376 − 0.186 − 0.0195 − 0.110) g O = 0.060 g O

$$\text{number of moles of O} = 0.060 \text{ g O} \times \left(\frac{1 \text{ mol O}}{16.00 \text{ g O}}\right) = 0.0038 \text{ mol O}$$

The ratio of amounts (in moles) is C:H:N:O = 0.0155:0.0193:0.007 85:0.0038.

Dividing by 0.0038 gives C:H:N:O = 4.1:5.1:2.1:1, which is close to C:H:N:O = 4:5:2:1. Therefore, the empirical formula is $C_4H_5N_2O$; its molar mass = (4 × 12.01 + 5 × 1.008 + 2 × 14.01 + 16.00) g/mol = 97.1 g/mol.

We note that $\frac{194 \text{ g/mol}}{97.1 \text{ g/mol}} \approx 2$. Therefore, the molecular formula is $C_8H_{10}N_4O_2$.

The equation for the combustion is

$$2\, C_8H_{10}N_4O_2(s) + 19\, O_2(g) \longrightarrow 16\, CO_2(g) + 10\, H_2O(g) + 4\, N_2(g)$$

Solutions

4.37 (a) $\text{molarity of } AgNO_3 = \frac{1.567 \text{ mol}}{0.2500 \text{ L}} = 6.268 \text{ mol/L}$

(b) $\text{molarity of NaCl} = \left(\frac{2.11 \text{ g NaCl}}{1.500 \text{ L}}\right) \times \left(\frac{1 \text{ mol}}{58.44 \text{ g NaCl}}\right) = 0.0241 \text{ mol/L}$

4.39 $\text{mass of } AgNO_3 = 0.025\,00 \text{ L} \times \left(\frac{0.155 \text{ mol } AgNO_3}{1 \text{ L}}\right) \times \left(\frac{169.88 \text{ g } AgNO_3}{1 \text{ mol } AgNO_3}\right)$

$= 0.658 \text{ g } AgNO_3$

4.41 $\text{molarity of } Ba(OH)_2 \text{ solution} = \left(\frac{2.577 \text{ g } Ba(OH)_2}{0.2500 \text{ L}}\right) \times \left(\frac{1 \text{ mol } Ba(OH)_2}{171.36 \text{ g } Ba(OH)_2}\right)$

$= 0.060\,15 \text{ mol/L}$

(a) $\text{volume} = \frac{n}{M}$

$$\text{volume of solution} = \frac{1.0 \times 10^{-3} \text{ mol}}{0.0602 \text{ mol/L}} = 0.017 \text{ L} = 17 \text{ mL}$$

(b) $\text{moles of } Ba(OH)_2 = 3.5 \times 10^{-3} \text{ mol } OH^- \times \left(\frac{1 \text{ mol } Ba(OH)_2}{2 \text{ mol } OH^-}\right)$

$= 1.7\bar{5} \times 10^{-3} \text{ mol } Ba(OH)_2$

$$\text{volume of solution} = \frac{1.7\bar{5} \times 10^{-3} \text{ mol}}{0.0602 \text{ mol/L}} = 0.029 \text{ L} = 29 \text{ mL}$$

(c) $\text{volume of solution} = 50.0 \text{ mg } Ba(OH)_2 \times \left(\frac{1 \text{ g } Ba(OH)_2}{1.0 \times 10^3 \text{ mg } Ba(OH)_2}\right)$

$\times \left(\frac{1 \text{ mol } Ba(OH)_2}{171.36 \text{ g } Ba(OH)_2}\right) = 2.92 \times 10^{-4} \text{ mol } Ba(OH)_2$

$$\text{volume of solution} = \frac{2.92 \times 10^{-4}\ \text{mol}}{0.0602\ \text{mol/L}} = 0.0048\ \text{L} = 4.8\ \text{mL}$$

4.43 (a) We would determine the mass of 0.010 mole of $KMnO_4$

$\left(= 0.010\ \text{mol KMnO}_4 \times \dfrac{158.04\ \text{g KMnO}_4}{1\ \text{mol KMnO}_4} = 1.58\ \text{g KMnO}_4\right)$ and then dissolve that mass of $KMnO_4$ in enough water to make 1.00 liter of solution.

(b) We dilute the 0.050 M $KMnO_4$ solution by a factor of 5 $\left(= \dfrac{0.050\ \text{M}}{0.010\ \text{M}}\right)$; for example, we could take 10 mL of 0.050 M $KMnO_4$, put it into a 50-mL volumetric flask, and add water up to the calibration mark.

4.45 In both parts, we use the procedure described in Toolbox 4.5.

$$V_{\text{initial}} = \frac{M_{\text{final}}\, V_{\text{final}}}{M_{\text{initial}}}$$

(a) $V_{\text{initial}} = \dfrac{0.0234\ \text{mol/L} \times 0.1500\ \text{L}}{0.778\ \text{mol/L}} = 4.51 \times 10^{-3}\ \text{L} = 4.51\ \text{mL}$

(b) $V_{\text{initial}} = \dfrac{0.50\ \text{mol/L} \times 0.0600\ \text{L}}{2.5\ \text{mol/L}} = 0.012\ \text{L} = 12\ \text{mL NaOH}$

A 60-mL volumetric flask is not normally available, but a 100-mL flask can be found. So, $12\ \text{mL} \times \left(\dfrac{100\ \text{mL}}{60\ \text{mL}}\right) = 20$ mL of the initial solution can be added to a 100-mL volumetric flask and diluted with water to the mark. 60 mL of this solution can then be used.

Titrations

4.47 The stoichiometric point is the point at which the exact amount of one reactant, the titrant, has been added to complete the reaction with the other reactant, the analyte, according to the balanced chemical equation. An example is an acid-base reaction:

$CH_3COOH(aq) + NaOH(aq) \longrightarrow NaCH_3CO_2(aq) + H_2O(l)$

The stoichiometric point of this reaction occurs when the number of moles of NaOH added equals the number of moles of CH_3COOH originally present.

4.49 $NaOH(aq) + HCl(aq) \longrightarrow NaCl(aq) + H_2O(l)$

(Hereafter, "soln" means solution.)

(a) number of moles of NaOH $= 0.017\ 40\ \text{L HCl} \times \left(\frac{0.234\ \text{mol HCl}}{1\ \text{L HCl}}\right)$

$$\times \left(\frac{1\ \text{mol NaOH}}{1\ \text{mol HCl}}\right) = 4.07 \times 10^{-3}\ \text{mol NaOH}$$

molarity of NaOH soln $= \left(\frac{4.07 \times 10^{-3}\ \text{mol NaOH}}{0.015\ 00\ \text{L NaOH}}\right) = 0.271\ \text{mol/L}$

(b) mass of NaOH $= 0.015\ \text{L NaOH} \times \left(\frac{0.271\ \text{mol NaOH}}{1\ \text{L NaOH}}\right)$

$$\times \left(\frac{40.0\ \text{g NaOH}}{1\ \text{mol NaOH}}\right) = 0.163\ \text{g NaOH}$$

4.51 $2\ HNO_3(aq) + Ba(OH)_2(aq) \longrightarrow Ba(NO_3)_2(aq) + 2\ H_2O(l)$

molar concentration of $Ba(OH)_2$ soln $= \left(\frac{9.670\ \text{g}\ Ba(OH)_2}{0.2500\ \text{L}}\right) \times \left(\frac{1\ \text{mol}\ Ba(OH)_2}{171.36\ \text{g}\ Ba(OH)_2}\right)$

$$= 0.2257\ \text{mol/L}$$

(a) number of moles of HNO_3 soln

$$= 0.011\ 56\ \text{L}\ Ba(OH)_2 \times \left(\frac{0.2257\ \text{mol}\ Ba(OH)_2}{1\ \text{L}\ Ba(OH)_2}\right) \times \left(\frac{2\ \text{mol}\ HNO_3}{1\ \text{mol}\ Ba(OH)_2}\right)$$

$$= 5.219 \times 10^{-3}\ \text{mol}\ HNO_3$$

molarity of HNO_3 soln $= \left(\frac{5.219 \times 10^{-3}\ \text{mol}\ HNO_3}{0.025\ 00\ \text{L}\ HNO_3}\right) = 0.2087\ \text{mol/L}$

(b) mass of $HNO_3 = 0.025\ 00\ \text{L}\ HNO_3 \times \left(\frac{0.2087\ \text{mol}\ HNO_3}{1\ \text{L}\ HNO_3}\right)$

$$\times \left(\frac{63.02\ \text{g}\ HNO_3}{1\ \text{mol}\ HNO_3}\right) = 0.3289\ \text{g}\ HNO_3$$

4.53 $HX(aq) + NaOH(aq) \longrightarrow NaX(aq) + H_2O(l)$

number of moles of acid $= 0.0688\ \text{L NaOH} \times \left(\frac{0.750\ \text{mol NaOH}}{1\ \text{L NaOH}}\right)$

$$\times \left(\frac{1\ \text{mol acid}}{1\ \text{mol NaOH}}\right) = 0.0516\ \text{mol acid}$$

molar mass of acid $= \left(\frac{3.25\ \text{g acid}}{0.0516\ \text{mol acid}}\right) = 63.0\ \text{g/mol}$

SUPPLEMENTARY EXERCISES

4.55 The theoretical yield is the maximum quantity of product(s) that can be obtained, according to the reaction stoichiometry, from a given quantity of a specified reactant (the limiting reactant).

The percentage yield of a product is the percentage of its theoretical yield that is actually achieved. It is calculated as

$$\text{percentage yield} = \frac{\text{actual yield}}{\text{theoretical yield}} \times 100\%$$

The percentage yield may be less than 100% for a specified product, because alternate reactions may take place in addition to the desired reaction, forming products other than the desired product(s).

4.57 In (b) and (c), the limiting reactant is H_2O_2, because N_2H_4 is present in excess.

(a) $\text{number of moles of } H_2O_2 = 0.477 \text{ mol } N_2H_4 \times \left(\frac{7 \text{ mol } H_2O_2}{1 \text{ mol } N_2H_4}\right)$

$= 3.34 \text{ mol } H_2O_2$

(b) $\text{number of moles of } HNO_3 = 6.77 \text{ g } H_2O_2 \times \left(\frac{1 \text{ mol } H_2O_2}{34.02 \text{ g } H_2O_2}\right)$

$\times \left(\frac{2 \text{ mol } HNO_3}{7 \text{ mol } H_2O_2}\right) = 5.69 \times 10^{-2} \text{ mol } HNO_3$

(c) $\text{mass of } H_2O = 0.0496 \text{ g } H_2O_2 \times \left(\frac{1 \text{ mol } H_2O_2}{34.02 \text{ g } H_2O_2}\right)$

$\times \left(\frac{8 \text{ mol } H_2O}{7 \text{ mol } H_2O_2}\right) \times \left(\frac{18.02 \text{ g } H_2O}{1 \text{ mol } H_2O}\right) = 3.00 \times 10^{-2} \text{ g } H_2O$

4.59 (a) $\text{mass of Mn} = 2.935 \times 10^{-3} \text{ g Al} \times \left(\frac{1 \text{ mol Al}}{26.98 \text{ g Al}}\right) \times \left(\frac{3 \text{ mol Mn}}{4 \text{ mol Al}}\right)$

$\times \left(\frac{54.94 \text{ g Mn}}{1 \text{ mol Mn}}\right) = 4.482 \times 10^{-3} \text{ g Mn} = 4.482 \text{ mg Mn}$

(b) $\text{percentage yield} = \frac{2.386 \text{ mg}}{4.482 \text{ mg}} \times 100\% = 53.24\%$

4.61 (a) $\text{volume of } O_2(g) = 2.27 \times 10^{-3} \text{ g } C_8H_{18} \times \left(\frac{1 \text{ mol } C_8H_{18}}{114.22 \text{ g } C_8H_{18}}\right)$

$\times \left(\frac{25 \text{ mol } O_2}{2 \text{ mol } C_8H_{18}}\right) \times \left(\frac{32.00 \text{ g } O_2}{1 \text{ mol } O_2}\right) \times \left(\frac{1 \text{ L } O_2}{1.43 \text{ g } O_2}\right) = 5.56 \times 10^{-3} \text{ L } O_2$

(b) $\text{volume of air} = 5.56 \times 10^{-3} \text{ L } O_2 \times \left(\frac{100 \text{ L air}}{21 \text{ L } O_2}\right) = 2.6 \times 10^{-2} \text{ L air}$

4.63 (a) See figure. Two molecules of ammonia can form; two molecules of N_2 remain unreacted.

(b) 3 molecules H_2 + 1 molecule $N_2 \longrightarrow$ 2 molecules NH_3

As all molecules of H_2 are used up and two molecules of N_2 remain, H_2 is the limiting reagent.

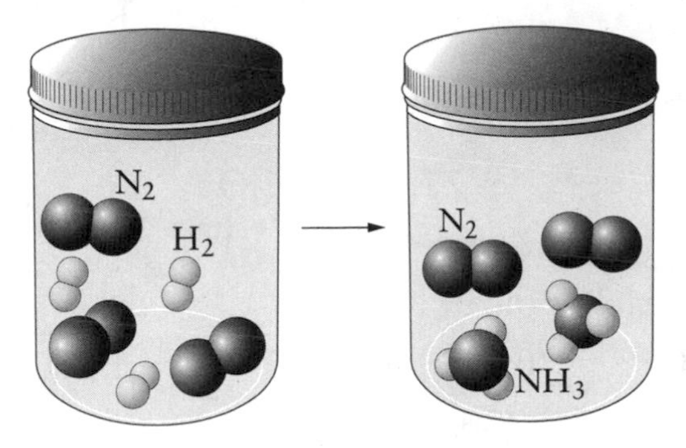

4.65 mass of gold (in kg) $= 3.23 \times 10^{11}\ \text{km}^3 \times \left(\frac{10^3\ \text{m}}{1\ \text{km}}\right)^3 \times \left(\frac{1\ \text{cm}}{10^{-2}\ \text{m}}\right)^3$

$$\times \left(\frac{1\ \text{L}}{10^3\ \text{cm}^3}\right) \times \left(\frac{0.011\ \text{mg}}{1\ \text{L}}\right) \times \left(\frac{10^{-3}\ \text{g}}{1\ \text{mg}}\right) \times \left(\frac{1\ \text{kg}}{10^3\ \text{g}}\right) = 3.6 \times 10^{15}\ \text{kg}$$

4.67 (a) mass of $CuSO_4 = 0.250\ \text{L} \times \left(\frac{0.20\ \text{mol}\ CuSO_4}{1\ \text{L}}\right) \times \left(\frac{159.6\ \text{g}\ CuSO_4}{1\ \text{mol}\ CuSO_4}\right)$

$$= 8.0\ \text{g}\ CuSO_4$$

(b) mass of $CuSO_4 \cdot 5H_2O = 0.250\ \text{L} \times \left(\frac{0.20\ \text{mol}\ CuSO_4}{1\ \text{L}}\right)$

$$\times \left(\frac{1\ \text{mol}\ CuSO_4 \cdot 5H_2O}{1\ \text{mol}\ CuSO_4}\right) \times \left(\frac{249.68\ \text{g}\ CuSO_4 \cdot 5H_2O}{1\ \text{mol}\ CuSO_4 \cdot 5H_2O}\right) = 12\ \text{g}\ CuSO_4 \cdot 5H_2O$$

4.69 Each sandwich is 2 bread + 3 cheese + 1 pickle + 1 tomato + peanut butter. Sixteen bread makes 8 sandwiches, 21 cheese makes 7 sandwiches, 10 pickle makes 10 sandwiches, 12 tomato makes 12 sandwiches, so the cheese limits; 7 sandwiches can be made.

4.71 (a) molar mass $(NH_4)_3PO_4 \cdot 3H_2O = [3 \times (14.01 + 4 \times 1.008) + 30.97$

$$+ 4 \times 16.00 + 3 \times 18.02]\ \text{g/mol} = 203.16\ \text{g/mol}$$

molar mass $K_3PO_4 = (3 \times 39.10 + 30.97 + 4 \times 16.00)\ \text{g/mol}$

$$= 212.27\ \text{g/mol}$$

molar mass water = 18.02 g/mol

number of moles of $(NH_4)_3PO_4 \cdot 3H_2O = 2.50\ \text{g}\ (NH_4)_3PO_4 \cdot 3H_2O$

$$\times \frac{1\ \text{mol}\ (NH_4)_3PO_4 \cdot 3H_2O}{203.16\ \text{g}\ (NH_4)_3PO_4 \cdot 3H_2O} = 0.0123\ \text{mol}\ (NH_4)_3PO_4 \cdot 3H_2O$$

$$\text{number of moles of } K_3PO_4 = 1.50 \text{ g } K_3PO_4 \times \frac{1 \text{ mol } K_3PO_4}{212.27 \text{ g } K_3PO_4}$$
$$= 7.07 \times 10^{-3} \text{ mol } K_3PO_4$$
$$\text{number of moles of } H_2O = 500. \text{ g } H_2O \times \frac{1 \text{ mol } H_2O}{18.02 \text{ g } H_2O}$$
$$= 27.7 \text{ mol } H_2O$$

(b) 1 mol PO_4^{3-} for each mole of $(NH_4)_3PO_4 \cdot 3H_2O$

1 mol PO_4^{3-} for each mole of K_3PO_4

$\therefore$ total moles $PO_4^{3-} = (0.0123 + 0.007\ 06)$ mol $PO_4^{3-} = 0.0194$ mol PO_4^{3-}

(c) molar mass $PO_4^{3-} = (30.97 + 4 \times 16.00)$ g/mol $= 94.97$ g/mol

mass of $PO_4^{3-} = 0.0194 \text{ mol} \times 94.97 \text{ g/mol} = 1.84$ g

(d) water has two sources, $(NH_4)_3PO_4 \cdot 3H_2O$ and from the 500. g of water added

mass of water in $(NH_4)_3PO_4 \cdot 3H_2O = 0.0123$ mol $(NH_4)_3PO_4 \cdot 3H_2O$

$$\times \frac{3 \text{ mol } H_2O}{1 \text{ mol } (NH_4)_3PO_4 \cdot 3H_2O} \times \frac{18.02 \text{ g } H_2O}{1 \text{ mol } H_2O} = 0.665 \text{ g } H_2O$$

total mass water $= (0.665 + 500.)$ g $H_2O = 500.665$ g $H_2O = 501$ g H_2O

4.73 $H_2SO_4(aq) + 2\ NaOH(aq) \longrightarrow Na_2SO_4(aq) + 2\ H_2O(l)$

(a) This error would have no effect assuming that the volume of the sample was accurately measured before it was added to the flask. The number of moles of sulfuric acid in the flask is not affected.

(b) This error would give an incorrectly high volume of NaOH(aq) required for neutralization. Therefore, the calculated concentration of acid in the solution would also be incorrectly high.

(c) This error would also give an incorrectly high volume of NaOH(aq) required. The calculated concentration of acid in the solution would be too high. The justification behind the conclusions stated in (b) and (c) is as follows:

$$\text{molarity } (H_2SO_4) = \frac{n\ (H_2SO_4)}{V\ (H_2SO_4)}$$

$V(H_2SO_4)$ is assumed to be measured accurately.

$n(H_2SO_4) = V(NaOH) \times \text{molarity (NaOH)}$

Molarity (NaOH) is assumed to be measured accurately.

$$\text{Then, molarity } (H_2SO_4) = \frac{V(NaOH) \times \text{molarity (NaOH)}}{V(H_2SO_4)}$$

Anything that causes V(NaOH) to be too high, as do the conditions of (b) and (c), will result in the molarity (H_2SO_4) of the sample to be too high.

APPLIED EXERCISES

4.75 The density of 98% H_2SO_4 is 1.841 g/mL.

Thus $\left(\frac{1.841 \text{ g } H_2SO_4}{\text{mL}}\right) \times \left(\frac{1 \text{ mol } H_2SO_4}{98.08 \text{ g } H_2SO_4}\right) \times \left(\frac{10^3 \text{ mL}}{\text{L}}\right) \times 0.98 = 18.4 \text{ mol/L}$

4.77 $1.00 \times 10^3 \text{ kg } CaCO_3 \times \left(\frac{10^3 \text{ g}}{1 \text{ kg}}\right) \times \left(\frac{1 \text{ mol } CaCO_3}{100.09 \text{ g } CaCO_3}\right) \times \left(\frac{1 \text{ mol } CO_2}{1 \text{ mol } CaCO_3}\right)$

$$\times \left(\frac{44.01 \text{ g } CO_2}{1 \text{ mol } CO_2}\right) = 4.40 \times 10^5 \text{ g } CO_2 = 440. \text{ kg } CO_2$$

4.79 (a) $\frac{(2 \times 12.01) \text{ g/mol}}{(2 \times 12.01 + 4 \times 1.008) \text{ g/mol}} \times 100 = 85.63\% \text{ C}$

(b) $\frac{(3 \times 12.01) \text{ g/mol}}{(3 \times 12.01 + 8 \times 1.008 + 16.00) \text{ g/mol}} \times 100 = 59.96\% \text{ C}$

(c) $\frac{(7 \times 12.01) \text{ g/mol}}{(7 \times 12.01 + 16 \times 1.008) \text{ g/mol}} \times 100 = 83.90\% \text{ C}$

(d) $C_2H_4(g) + 3\ O_2(g) \longrightarrow 2\ CO_2(g) + 2\ H_2O(g)$

$2\ C_3H_7OH(l) + 9\ O_2(g) \longrightarrow 6\ CO_2(g) + 8\ H_2O(g)$

$C_7H_{16}(l) + 11\ O_2(g) \longrightarrow 7\ CO_2(g) + 8\ H_2O(g)$

(e) $1.0 \text{ g } C_2H_4 \times \left(\frac{1 \text{ mol } C_2H_4}{28.05 \text{ g } C_2H_5}\right) \times \left(\frac{2 \text{ mol } CO_2}{1 \text{ mol } C_2H_4}\right) = 7.1 \times 10^{-2} \text{ mol } CO_2$

$1.0 \text{ g } C_3H_7OH \times \left(\frac{1 \text{ mol } C_3H_7OH}{60.09 \text{ g } C_3H_7OH}\right) \times \left(\frac{6 \text{ mol } CO_2}{2 \text{ mol } C_3H_7OH}\right)$

$$= 5.0 \times 10^{-2} \text{ mol } CO_2$$

$1.0 \text{ g } C_7H_{16} \times \left(\frac{1 \text{ mol } C_7H_{16}}{100.20 \text{ g } C_7H_{16}}\right) \times \left(\frac{7 \text{ mol } CO_2}{1 \text{ mol } C_7H_{16}}\right) = 7.0 \times 10^{-2} \text{ mol } CO_2$

The C_2H_4 produces the greatest amount of CO_2 per gram of fuel.

4.81 $H_2SO_4(aq) + 2\ NaOH(aq) \longrightarrow Na_2SO_4(aq) + 2\ H_2O(l)$

(a) number of moles $H_2SO_4 = 0.0174 \text{ L NaOH} \times \left(\frac{0.100 \text{ mol NaOH}}{1 \text{ L NaOH}}\right)$

$$\times \left(\frac{1 \text{ mol } H_2SO_4}{2 \text{ mol NaOH}}\right) = 8.70 \times 10^{-4} \text{ mol } H_2SO_4$$

(b) mass percentage $S = 8.70 \times 10^{-4} \text{ mol } H_2SO_4 \times \left(\frac{1 \text{ mol S}}{1 \text{ mol } H_2SO_4}\right)$

$$\times \left(\frac{32.06 \text{ g S}}{1 \text{ mol S}}\right) \times \left(\frac{1}{6.43 \text{ g coal}}\right) \times 100\% = 0.434\%$$

4.83 (a) molarity of I_3^- soln $= 0.010\ 00\ \text{L}\ H_3AsO_3 \times \left(\frac{0.1235\ \text{mol}\ H_3AsO_3}{\text{L}}\right)$

$$\times \left(\frac{1\ \text{mol}\ I_3^-}{1\ \text{mol}\ H_3AsO_4}\right) \times \left(\frac{1}{0.010\ 42\ \text{L}}\right) = 0.1185\ \text{M}\ I_3^-$$

(b) molarity of HCN $= 5.21 \times 10^{-3}\ \text{L}\ I_3^- \times \left(\frac{0.1185\ \text{mol}\ I_3^-}{1\ \text{L}\ I_3^-}\right)$

$$\times \left(\frac{1\ \text{mol HCN}}{1\ \text{mol}\ I_3^-}\right) \times \left(\frac{1}{0.015\ 00\ \text{L blood}}\right) = 0.0412\ \text{M HCN}$$

INTEGRATED EXERCISES

4.85 (a) $12.0 \times 10^3\ \text{g}\ SO_2 \times \left(\frac{1\ \text{mol}\ SO_2}{64.06\ \text{g}\ SO_2}\right) = 1.87 \times 10^2\ \text{mol}\ SO_2$

$8.0 \times 10^3\ \text{g}\ H_2S \times \left(\frac{1\ \text{mol}\ H_2S}{34.08\ \text{g}\ H_2S}\right) \times \left(\frac{8\ \text{mol}\ SO_2}{16\ \text{mol}\ H_2S}\right) = 1.17 \times 10^2\ \text{mol}\ SO_2$

H_2S is the limiting reagent.

(b) $8\ SO_2(g) + 16\ H_2S(g) \longrightarrow 3\ S_8(s) + 16\ H_2O(l)$

(oxidation numbers: S in SO_2: +4, −4; S in H_2S: +2, −2; S in S_8: 0)

The oxidizing agent is SO_2; the reducing agent is H_2S.

4.87 (a) This is a redox reaction.

(b) $5.0\ \text{mol Sb} \times \left(\frac{3\ \text{mol}\ O_2}{4\ \text{mol Sb}}\right) = 3.75\ \text{mol}\ O_2$, so Sb limits.

(c) $5.0\ \text{mol}\ O_2 - 3.75\ \text{mol}\ O_2 = 1.2\ \text{mol}\ O_2$ remains.

(d) $5.0\ \text{mol Sb} \times \left(\frac{2\ \text{mol}\ Sb_2O_3}{4\ \text{mol Sb}}\right) = 2.5\ \text{mol}\ Sb_2O_3$

(e) $\left(\frac{2.0\ \text{mol}\ Sb_2O_3}{2.5\ \text{mol}\ Sb_2O_3}\right) \times 100 = 80.\%$

4.89 (a) reactant: $62.6\ \text{g Sn} \times \left(\frac{1\ \text{mol Sn}}{118.69\ \text{g Sn}}\right) = 0.527\ \text{mol Sn}$

$$37.4\ \text{g Cl} \times \left(\frac{1\ \text{mol Cl}}{35.45\ \text{g Cl}}\right) = 1.06\ \text{mol Cl}$$

Divide by 0.527 to get a ratio of Sn : Cl = 1 : 2, so the reactant compound is $SnCl_2$. The oxidation state of Sn in $SnCl_2$ is +2. Sn also has a common oxidation state of +4, so the other compound is likely to be $SnCl_4$.

(b) $I_3^-(aq) + Sn^{2+}(aq) \longrightarrow 3\ I^-(aq) + Sn^{4+}(aq)$

(c) Sn in $SnCl_2$ = +2; Sn in $SnCl_4$ = +4

CHAPTER 5
THE PROPERTIES OF GASES

EXERCISES

Pressure

5.1 Pressure is a force exerted divided by the area of surface. Air pressure exerted on the surface of the mercury in the dish (see Figure 5.5) is transmitted through the liquid mercury and supports the mercury column. Equilibrium requires that the pressure at the base of the column of mercury is balanced by the pressure on the surface of the mercury in the dish. Thus, the height of the column is proportional to the pressure. The space above the mercury is a vacuum, so it adds no pressure. A column of mercury 760 mm high represents, by definition, one standard atmosphere. As in a thermometer, calibration makes it a true measuring device.

5.3 (a) $\left.\begin{array}{r} 1.00\ \text{bar} = 100\ \text{kPa} \\ 1\ \text{atm} = 101.3\ \text{kPa} \end{array}\right\}$ Table 5.2

$$(1.00\ \text{bar}) \times \left(\frac{100\ \text{kPa}}{1\ \text{bar}}\right) \times \left(\frac{1\ \text{atm}}{101.3\ \text{kPa}}\right) = 0.987\ \text{atm}$$

(b) $$(1.00\ \text{mg Hg}) \times \left(\frac{1\ \text{Torr}}{1\ \text{mm Hg}}\right) \times \left(\frac{1\ \text{atm}}{760\ \text{Torr}}\right) \times \left(\frac{101.3\ \text{kPa}}{1\ \text{atm}}\right) \times \left(\frac{10^3\ \text{Pa}}{1\ \text{kPa}}\right) = 1.33 \times 10^2\ \text{Pa}$$

(c) $$(1.00\ \text{kPa}) \times \left(\frac{1\ \text{atm}}{101.3\ \text{kPa}}\right) = 9.87 \times 10^{-3}\ \text{atm}$$

5.5 (a) $$(8 \times 10^4\ \text{atm}) \times \left(\frac{101.3\ \text{kPa}}{1\ \text{atm}}\right) \times \left(\frac{1\ \text{bar}}{100\ \text{kPa}}\right) \times \left(\frac{1\ \text{kbar}}{10^3\ \text{bar}}\right) = 8 \times 10^1\ \text{kbar}$$

(b) $$(8 \times 10^4\ \text{atm}) \times \left(\frac{101.3\ \text{kPa}}{1\ \text{atm}}\right) \times \left(\frac{10^3\ \text{Pa}}{1\ \text{kPa}}\right) = 8 \times 10^9\ \text{Pa}$$

5.7 $P = dgh$ (See section 5.2 for equations)

Because the pressure, P, is the same no matter what the measuring device, h is inversely proportional to d. That is,

$$h = \frac{P}{dg}, \text{ or } hd = \frac{P}{g} = \text{constant}$$

Then

$$h(\text{water})\, d(\text{water}) = h(\text{Hg})\, d(\text{Hg})$$

or

$$h(\text{water}) = \frac{h(\text{Hg})\, d(\text{Hg})}{d(\text{water})}$$

$$= \frac{77.5 \text{ cm} \times 13.5 \text{ g/cm}^3}{1.10 \text{ g/cm}^3}$$

$$= 9.51 \times 10^2 \text{ cm}$$

The Gas Laws

5.9 In each case, $V_2 = V_1 \times \frac{P_1}{P_2}$

(a) $V_2 = (1.00 \text{ L}) \times \left(\frac{2.20 \text{ kPa}}{3.00 \text{ atm}}\right) \times \left(\frac{1 \text{ atm}}{101.3 \text{ kPa}}\right) = 7.24 \times 10^{-3} \text{ L}$

(b) $V_2 = (25.0 \text{ mL}) \times \left(\frac{200. \text{ Torr}}{0.500 \text{ atm}}\right) \times \left(\frac{1 \text{ atm}}{760 \text{ Torr}}\right) = 13.2 \text{ mL}$

5.11 In each case, $P_2 = P_1 \times \frac{V_1}{V_2}$

(a) $P_2 = 105 \text{ kPa} \times \left(\frac{1.0 \times 10^{-3} \text{ L}}{1.0 \text{ L}}\right) = 0.10\overline{5} \text{ kPa} = 1.0 \times 10^2 \text{ Pa}$

(b) $P_2 = 600. \text{ Torr} \times \left(\frac{30.0 \text{ cm}^3}{5.0 \text{ cm}^3}\right) = 3.6 \times 10^3 \text{ Torr}$

5.13 (a) $P_{gas} = P_{Hg} + P_{atm}$, so in (1), (29.75 + 12.0) inHg = 41.8 inHg and in (2), (30.0 + 29.75) inHg = 59.8 inHg

(b) If we assume pressure is given by the height of the mercury and volume by the height of air, we get Boyle's Law: $P_1V_1 = P_2V_2$.

(12 inHg) × (32 in. air) = (30.0 inHg) × V_2, V_2 = 12.8 in. air

5.15 $V_2 = V_1 \times \frac{T_2}{T_1}$

$T_1 = (273 + 85) \text{ K} = 358 \text{ K}$

$T_2 = 273 \text{ K}$

$V_2 = 255 \text{ mL} \times \left(\frac{273 \text{ K}}{358 \text{ K}}\right) = 194 \text{ mL}$

5.17 (a) $V_1 = 1.00\text{ L}; T_1 = 20. + 273 = 293\text{ K}$

$V_2 = 1.03\text{ L}; T_2 = ?$ $\qquad T_2 = T_1 \times \frac{V_2}{V_1}$

$$T_2 = 293\text{ K} \times \left(\frac{1.03\text{ L}}{1.00\text{ L}}\right) = 302\text{ K or } 29°\text{C}$$

(b) $T_2 = T_1 \times \frac{V_2}{V_1}$ $\qquad T_2 = 293\text{ K} \times \left(\frac{0.95\text{ L}}{1.00\text{ L}}\right) = 280\text{ K (2 sf) or } 7°\text{C}$

5.19 (a) $P_2 = P_1 \times \frac{T_2}{T_1}$

$T_1 = (273 + 10)\text{ K} = 283\text{ K}$

$T_2 = (273 + 30)\text{ K} = 303\text{ K}$

$$P_2 = 1.5\text{ atm} \times \left(\frac{303\text{ K}}{283\text{ K}}\right) = 1.6\text{ atm}$$

5.21 $T_2 = T_1 \times \frac{P_2}{P_1}$, $\quad T_1 = (273 + 20)\text{ K} = 293\text{ K}$

$P_1 = (30. + 14.7)\text{ lb/inch}^2 = 44.\bar{7}\text{ lb/inch}^2$, $\quad P_2 = (34 + 14.7\text{ lb/inch}^2)$

$= 48.\bar{7}\text{ lb/inch}^2$

$$T_2 = 293\text{ K} \times \frac{48.\bar{7}\text{ lb/inch}^2}{44.\bar{7}\text{ lb/inch}^2} = 319\text{ K} = (319 - 273)°\text{C} = 46°\text{C}$$

5.23 Because half the gas molecules have been removed, $n_2 = \frac{1}{2}n_1$. To obtain the new temperature, we rearrange the ideal gas law.

$PV = nRT$

$$nT = \left(\frac{PV}{R}\right) = \text{constant}$$

Thus, $n_1T_1 = n_2T_2$, or

$$T_2 = T_1 \times \frac{n_1}{n_2}$$

$$T_2 = T_1 \times \frac{n_1}{\frac{1}{2}n_1} = 2T_1$$

Therefore, the temperature must be doubled.

The Ideal Gas Law

5.25 Boyle, Charles, and Avogadro demonstrated the following proportionalities:

Boyle: $V \propto \frac{1}{P}$ or $P \propto \frac{1}{V}$ at constant n and T

Charles: $V \propto T$ at constant n and P

Avogadro: $V \propto n$ at constant P and T

These proportions can be combined into one equation by introducing a constant of proportionality, R.

$$V = R \times \left(\frac{1}{P}\right) \times T \times n$$

$$= \frac{nRT}{P}$$

or $\quad PV = nRT$

which is the ideal gas law.

5.27 (a) $PV = nRT$; therefore, amount (moles) $= n = \dfrac{PV}{RT}$

$$n = \frac{1.3 \text{ atm} \times 0.100 \text{ L}}{8.206 \times 10^{-2} \text{ L·atm/K·mol} \times 350 \text{ K}} = 4.5 \times 10^{-3} \text{ mol}$$

(b) $n = \dfrac{2.7 \times 10^{-6} \text{ g}}{32.00 \text{ g/mol}} = 8.4\overline{4} \times 10^{-8} \text{ mol}$

$$P = \frac{nRT}{V} = \frac{8.4\overline{4} \times 10^{-8} \text{ mol} \times 62.36 \text{ L·Torr/K·mol} \times 290 \text{ K}}{0.120 \text{ L}}$$

$$= 1.3 \times 10^{-2} \text{ Torr}$$

(c) $n = \dfrac{16.7 \text{ g}}{83.80 \text{ g/mol}} = 0.199\overline{3} \text{ mol}$

$$V = \frac{nRT}{P} = \frac{0.199\overline{3} \text{ mol} \times 62.36 \text{ L·Torr/K·mol} \times (44 + 273) \text{ K}}{0.100 \text{ Torr}}$$

$$= 3.94 \times 10^{4} \text{ L}$$

5.29 (a) $n = \dfrac{PV}{RT} = \dfrac{20. \text{ Torr} \times 20. \text{ L}}{62.36 \text{ L·Torr/K·mol} \times 200. \text{ K}} = 0.032\overline{1} \text{ mol}$

mass $= 0.032\overline{1} \text{ mol} \times 28.02 \text{ g/mol} = 0.90 \text{ g}$

(b) $n = \dfrac{PV}{RT} = \dfrac{2.00 \text{ Torr} \times 2.6 \times 10^{-6} \text{ L}}{62.36 \text{ L·Torr/K·mol} \times 288 \text{ K}} = 2.9 \times 10^{-10} \text{ mol}$

number of Xe atoms $= 2.9 \times 10^{-10} \text{ mol} \times 6.022 \times 10^{23}/\text{mol} = 1.7 \times 10^{14}$

5.31 $n_{H_2S} = (12 \text{ mg } H_2S) \times \left(\dfrac{10^{-3} \text{ g}}{1 \text{ mg}}\right) \times \left(\dfrac{1 \text{ mol } H_2S}{34.0 \text{ g } H_2S}\right) = 3.5 \times 10^{-4} \text{ mol } H_2S$

According to Avogadro's law, n_{H_2S} must equal n_{NH_3}.

Therefore, $(3.5 \times 10^{-4} \text{ mol } NH_3) \times \left(\dfrac{17.0 \text{ g}}{1 \text{ mol } NH_3}\right) = 6.0 \times 10^{-3} \text{ g } NH_3$ or $6.0 \text{ mg } NH_3$

5.33 (a) $n = \dfrac{PV}{RT} = \dfrac{1.22 \text{ atm} \times 0.250 \text{ L}}{8.206 \times 10^{-2} \text{ L·atm/K·mol} \times 298 \text{ K}} = 1.24\overline{7} \times 10^{-2} \text{ mol}$

Let M = molar mass; then mass = $n \times M = 1.24\overline{7} \times 10^{-2} \text{ mol} \times 28.97 \text{ g/mol}$

$= 0.361 \text{ g}$

(b) mass of nitrogen = $n \times M = 1.24\overline{7} \times 10^{-2} \text{ mol} \times 28.02 \text{ g/mol} = 0.349 \text{ g}$

5.35 $\dfrac{P_1V_1}{T_1} = \dfrac{P_2V_2}{T_2} \qquad P_2 = \dfrac{P_1V_1T_2}{V_2T_1}$

$P_2 = \dfrac{(750. \text{ Torr} \times 0.150 \text{ L} \times 795 \text{ K})}{(0.500 \text{ L} \times 283 \text{ K})} = 632 \text{ Torr}$

5.37 $\dfrac{P_1V_1}{T_1} = \dfrac{P_2V_2}{T_2} \qquad V_1 = \dfrac{P_2V_2T_1}{P_1T_2}$

$V_2 = \dfrac{(1.08 \text{ atm} \times 0.350 \text{ L} \times 296 \text{ K})}{(0.958 \text{ atm} \times 310. \text{ K})} = 0.377 \text{ L}$

Molar Volume

5.39 (a) $V = \dfrac{nRT}{P} = \dfrac{1.00 \text{ mol} \times 8.206 \times 10^{-2} \text{ L·atm/K·mol} \times 298 \text{ K}}{1.00 \text{ atm}}$

$= 24.4\overline{5} \text{ L} = 24.4 \text{ L}$

(b) $V = \dfrac{27.0 \text{ g} \times 8.206 \times 10^{-2} \text{ L·atm/K·mol} \times 298 \text{ K}}{70.91 \text{ g/mol} \times 1.00 \text{ atm}} = 9.31 \text{ L}$

(c) $V = 3 \text{ mol} \times 24.4\overline{5} \text{ L/mol} = 73.4 \text{ L}$

(d) $V = \dfrac{0.0148 \text{ g} \times 8.206 \times 10^{-2} \text{ L·atm/K·mol} \times 298 \text{ K}}{64.06 \text{ g/mol} \times 1.00 \text{ atm}}$

$= 5.65 \times 10^{-3} \text{ L} = 5.65 \text{ mL}$

5.41 $n = \dfrac{PV}{RT} = V\left(\dfrac{P}{RT}\right)$

$\dfrac{P}{RT}$ is a common factor in (a) $\longrightarrow$ (d)

$\dfrac{P}{RT} = \dfrac{1.00 \text{ atm}}{8.206 \times 10^{-2} \text{ L·atm/K·mol} \times 298 \text{ K}} = 0.040\ 8\overline{9} \text{ mol/L}$

mass $= n \times M = V \times \left(\dfrac{P}{RT}\right) \times M = V \times 0.040\ 8\overline{9} \text{ mol/L} \times M$

(a) mass $= 2.45 \text{ L} \times 0.040\ 8\overline{9} \text{ mol/L} \times 32.00 \text{ g/mol} = 3.21 \text{ g}$

(b) mass $= 1.94 \times 10^{-3} \text{ L} \times 0.040\ 8\overline{9} \text{ mol/L} \times 80.06 \text{ g/mol}$

$= 6.35 \times 10^{-3} \text{ g} = 6.35 \text{ mg}$

(c) mass $= 6000 \text{ L} \times 0.040\ 8\bar{9} \text{ mol/L} \times 16.04 \text{ g/mol}$
$= 3.94 \times 10^3 \text{ g} = 3.94 \text{ kg}$

(d) mass $= 1.44 \times 10^{-3} \text{ L} \times 0.040\ 8\bar{9} \text{ mol/L} \times 44.01 \text{ g/mol}$
$= 2.59 \times 10^{-3} \text{ g} = 2.59 \text{ mg}$

Stoichiometry of Reacting Gases

5.43 $n_{O_2} = 1.00 \text{ g KClO}_3 \times \left(\frac{1 \text{ mol KClO}_3}{122.6 \text{ g KClO}_3}\right) \times \left(\frac{3 \text{ mol O}_2}{2 \text{ mol KClO}_3}\right)$
$= 1.22 \times 10^{-2} \text{ mol O}_2$

$$V_{O_2} = \frac{n_{O_2}RT}{P} = \frac{1.22 \times 10^{-2} \text{ mol O}_2 \times 8.206 \times 10^{-2} \text{ L·atm/K·mol} \times 298 \text{ K}}{1.00 \text{ atm}}$$
$= 0.299 \text{ L}$

5.45 (a) $V_{H_2} = 127 \text{ mL} \times \left(\frac{(737.7 - 14.53) \text{ Torr}}{760 \text{ Torr}}\right) \times \left(\frac{298 \text{ K}}{290 \text{ K}}\right) = 124 \text{ mL}$

(b) $\text{Zn(s)} + 2 \text{ HCl(aq)} \longrightarrow \text{ZnCl}_2\text{(aq)} + \text{H}_2\text{(g)}$

amount (moles) of Zn reacted = amount (moles) of H_2 formed

$$n_{H_2} = \frac{PV}{RT} = \frac{1.00 \text{ atm} \times 0.124 \text{ L}}{0.082\ 06 \text{ L·atm/K·mol} \times 298 \text{ K}} = 5.07 \times 10^{-3} \text{ mol}$$

(c) mass of Zn reacted $= 5.07 \times 10^{-3} \text{ mol} \times \left(\frac{65.37 \text{ g Zn}}{\text{mol}}\right) = 0.331 \text{ g Zn}$

$$\% \text{ purity of Zn} = \frac{\text{mass of Zn reacted}}{\text{mass of impure Zn}} \times 100\% = \frac{0.331 \text{ g}}{0.40 \text{ g}} \times 100\% = 83\%$$

5.47 Haber process: $\text{N}_2\text{(g)} + 3 \text{ H}_2\text{(g)} \longrightarrow 2 \text{ NH}_3\text{(g)}$

(a) $n_{H_2} = 1.0 \times 10^3 \text{ kg NH}_3 \times \left(\frac{10^3 \text{ g NH}_3}{1 \text{ kg NH}_3}\right) \times \left(\frac{1 \text{ mol NH}_3}{17.03 \text{ g NH}_3}\right)$
$\times \left(\frac{3 \text{ mol H}_2}{2 \text{ mol NH}_3}\right) = 8.8\bar{1} \times 10^4 \text{ mol H}_2$

$$V_{H_2} = \frac{n_{H_2}RT}{P} = \frac{8.8\bar{1} \times 10^4 \text{ mol H}_2 \times 0.082\ 06 \text{ L·atm/K·mol} \times 298 \text{ K}}{1.00 \text{ atm}}$$
$= 2.15 \times 10^6 \text{ L}$

(b) $V_{H_2} = 2.15 \times 10^6 \text{ L} \times \left(\frac{1.00 \text{ atm}}{200 \text{ atm}}\right) \times \left(\frac{(273 + 400) \text{ K}}{298 \text{ K}}\right) = 2.43 \times 10^6 \text{ L}$

5.49 $n_{CO_2} = (2.50 \text{ kg urea}) \times \left(\frac{10^3 \text{ g}}{1 \text{ kg}}\right) \times \left(\frac{1 \text{ mol urea}}{60.1 \text{ g urea}}\right) \times \left(\frac{1 \text{ mol CO}_2}{1 \text{ mol urea}}\right)$
$= 41.6 \text{ mol CO}_2$

$$n_{NH_3} = (2.50 \text{ kg urea}) \times \left(\frac{10^3 \text{ g}}{1 \text{ kg}}\right) \times \left(\frac{1 \text{ mol urea}}{60.1 \text{ g urea}}\right) \times \left(\frac{2 \text{ mol } NH_3}{1 \text{ mol urea}}\right) = 83.2 \text{ mol } NH_3$$

$$V_{CO_2} = \frac{nRT}{P} = (41.6 \text{ mol } CO_2) \times \left(\frac{0.082\ 06 \text{ L·atm}}{\text{K·mol}}\right) \times (450. + 273) \text{ K} \times \left(\frac{1}{200. \text{ atm}}\right) = 12.4 \text{ L}$$

Because $V \propto n$ at constant T and P, $V_{NH_3} = 24.7$ L

5.51 (b) $n_{CH_4} = 2.00 \text{ g } CH_4 \times \left(\frac{1 \text{ mol } CH_4}{16.04 \text{ g } CH_4}\right) = 0.125 \text{ mol } CH_4$

$$V_{CH_4} = \frac{n_{CH_4}RT}{P} = \frac{0.125 \text{ mol} \times 0.082\ 06 \text{ L·atm/K·mol} \times 348 \text{ K}}{1.00 \text{ atm}} = 3.57 \text{ L}$$

Because 3.57 L > 2.00 L, starting condition (b) would produce the larger volume of CO_2 by combustion.

Density

5.53 Let M = molar mass, then $n = \frac{\text{mass}}{M}$.

Assume gases are ideal, then $PV = nRT = \frac{\text{mass}}{V}RT$.

Rearranging, density $= d = \frac{\text{mass}}{V} = \frac{PM}{RT}$, or $d = \left(\frac{P}{RT}\right)M$ = constant $\times M$ (at constant T and P). Therefore, at constant T and P, the most dense gas is the one with the largest molar mass.

(a) $M_{N_2} = 28.02$ g/mol

(b) $M_{NH_3} = 17.03$ g/mol

(c) $M_{NO_2} = 46.01$ g/mol

NO_2 is the most dense.

5.55 density $= \frac{\text{molar mass}}{\text{molar volume}} = \frac{M}{\left(\frac{RT}{P}\right)} = \left(\frac{P}{RT}\right)M$

See Exercise 5.53 for a derivation of this relation from the ideal gas law.

(a) density $= d = \left(\frac{1.00 \text{ atm}}{0.082\ 06 \text{ L·atm/K·mol} \times 298 \text{ K}}\right) \times 119.37$ g/mol

$= 4.88$ g/L

(b) $d = 4.88 \text{ g/L} \times \left(\frac{298 \text{ K}}{373 \text{ K}}\right) = 3.90$ g/L

5.57 It is convenient to work part (b) first. Rearrange $d = \left(\frac{P}{RT}\right) M$ to give

$$M = \left(\frac{RT}{P}\right) \times d = \left(\frac{RT}{P}\right)\left(\frac{\text{mass}}{V}\right)$$

(b) $\text{molar mass} = M = \left(\frac{0.082\ 06\ \text{L·atm/K·mol} \times 303\ \text{K}}{0.880\ \text{atm}}\right) \times \left(\frac{21.3\ \text{g}}{7.73\ \text{L}}\right)$

$= 77.8\ \text{g/mol}$

(a) $d = \left(\frac{1.00\ \text{atm}}{0.082\ 06\ \text{L·atm/K·mol} \times 298\ \text{K}}\right) \times 77.8\ \text{g/mol} = 3.18\ \text{g/L}$

5.59 (a) $M\text{-molar mass} = \left(\frac{RT}{P}\right) \times d$ [see Exercise 5.57]

$= \left(\frac{62.36\ \text{L·Torr/K·mol} \times 420\ \text{K}}{727\ \text{Torr}}\right) \times 3.60\ \text{g/L} = 130.\ \text{g/mol}$

(b) $d = 3.60\ \text{g/L} \times \left(\frac{760\ \text{Torr}}{727\ \text{Torr}}\right) \times \left(\frac{420\ \text{K}}{298\ \text{K}}\right) = 5.30\ \text{g/L}$

Mixtures of Gases

5.61 (a) There are nine ● HCl molecules and one ○ benzene molecule shown. Assume there are nine moles of HCl and one mole of benzene (instead of using Avogadro's number to convert).

$\chi_{HCl} = \frac{9}{(9+1)} = 0.9$ and $\chi_{benzene} = \frac{1}{(9+1)} = 0.1$

(b) $P_{HCl} = \chi_{HCl}P$ $\quad P_{HCl} = (0.9) \times (0.80\ \text{atm}) = 0.7\ \text{atm}$ (1 sf)

$P_{benzene} = \chi_{benzene}P = (0.1) \times (0.80\ \text{atm}) = 0.08\ \text{atm}$

5.63 (a) $n_{N_2} = 0.020\ \text{mol}\ N_2$; $n_{O_2}(2.33\ \text{g}\ O_2) \times \left(\frac{1\ \text{mol}\ O_2}{32.0\ \text{g}\ O_2}\right) = 0.0728\ \text{mol}\ O_2$

Then, $P = \frac{nRT}{V}$

$P_{N_2} = (0.020\ \text{mol}\ N_2) \times \left(\frac{0.082\ 06\ \text{L·atm}}{\text{K·mol}}\right) \times (273\ \text{K}) \times \left(\frac{1}{500.\ \text{mL}} \times \frac{1\ \text{mL}}{10^{-3}\ \text{L}}\right)$

$= 0.90\ \text{atm}$

$P_{O_2} = (0.0728\ \text{mol}\ O_2) \times \left(\frac{0.082\ 06\ \text{L·atm}}{\text{K·mol}}\right) \times (273\ \text{K}) \times \left(\frac{1}{500.\ \text{mL}} \times \frac{1\ \text{mL}}{10^{-3}\ \text{L}}\right)$

$= 3.26\ \text{atm}$

$P_{total} = P_{N_2} + P_{O_2} = 0.90\ \text{atm} + 3.26\ \text{atm} = 4.16\ \text{atm}$ (3 sf)

(b) $n_{H_2} = 0.015 \text{ mol } H_2$; $n_{NH_3} = 0.030 \text{ mol } NH_3$

$$n_{He} = (4.22 \text{ mg}) \times \left(\frac{10^{-3} \text{ g}}{1 \text{ mg}}\right) \times \left(\frac{1 \text{ mol He}}{4.00 \text{ g He}}\right) = 1.06 \times 10^{-3} \text{ mol He}$$

Then, $P = \dfrac{nRT}{V}$

$$P_{H_2} = (0.015 \text{ mol } H_2) \times \left(\frac{0.082\ 06 \text{ L}\cdot\text{atm}}{\text{K}\cdot\text{mol}}\right) \times (273 \text{ K}) \times \left(\frac{1}{500. \text{ mL}} \times \frac{1 \text{ mL}}{10^{-3} \text{ L}}\right)$$

$$= 0.67 \text{ atm}$$

$$P_{NH_3} = (0.030 \text{ mol } NH_3) \times \left(\frac{0.082\ 06 \text{ L}\cdot\text{atm}}{\text{K}\cdot\text{mol}}\right) \times (273 \text{ K}) \times \left(\frac{1}{500 \text{ mL}} \times \frac{1 \text{ mL}}{10^{-3} \text{ L}}\right)$$

$$= 1.3 \text{ atm}$$

$$P_{He} = (1.06 \times 10^{-3} \text{ mol He}) \times \left(\frac{0.082\ 06 \text{ L}\cdot\text{atm}}{\text{K}\cdot\text{mol}}\right) \times (273 \text{ K}) \times \left(\frac{1}{500. \text{ mL}} \times \frac{1 \text{ mL}}{10^{-3} \text{ L}}\right)$$

$$= 0.0475 \text{ atm}$$

$$P_{total} = P_{H_2} + P_{NH_3} + P_{He} = 0.67 \text{ atm} + 1.3 \text{ atm} + 0.0475 \text{ atm}$$

$$= 2.02 \text{ atm} = 2.0 \text{ atm (2 sf)}$$

5.65 At $-10.0°C$, all water vapor is condensed and 607.1 Torr $= P_{air}$.

$$P_{air\ at\ 48.5°C} = (607.1 \text{ Torr}) \times \left(\frac{(273 + 48.5) \text{ K}}{(273 - 10) \text{ K}}\right) = 742.1 \text{ Torr}$$

$$P_{total} = P_{air} + P_{H_2O}$$

$$P_{H_2O} = P_{total} - P_{air} = 762.0 \text{ Torr} - 742.1 \text{ Torr} = 19.9 \text{ Torr}$$

$$n_{H_2O} = \frac{P_{H_2O}V}{RT} = \frac{19.9 \text{ Torr} \times 1.00 \text{ L}}{62.36 \text{ L}\cdot\text{Torr/K}\cdot\text{mol} \times 322 \text{ K}} = 9.91 \times 10^{-4} \text{ mol}$$

$$\text{mass} = 9.91 \times 10^{-4} \text{ mol} \times 18.02 \text{ g/mol} = 1.78 \times 10^{-2} \text{ g } H_2O$$

5.67 (a) $P_{N_2} = (803 \text{ kPa}) \times \left(\dfrac{4.0 \text{ L}}{14.0 \text{ L}}\right) = 22\bar{9} \text{ kPa} = 2.3 \times 10^2 \text{ kPa}$

$$P_{Ar} = (47.2 \text{ kPa}) \times \left(\frac{10.0 \text{ L}}{14.0 \text{ L}}\right) = 33.7 \text{ kPa}$$

(b) $P_T = P_{N_2} + P_{Ar} = 229 \text{ kPa} + 33.7 \text{ kPa} = 26\bar{3} \text{ kPa} = 2.63 \times 10^2 \text{ kPa}$

5.69 Total pressure $= P_{total}$

$$P_{total} = (97.6 \text{ kPa}) \times \left(\frac{1 \text{ atm}}{101.325 \text{ kPa}}\right) = 0.963 \text{ atm}$$

$$P_{H_2O} = (11.99 \text{ Torr}) \times \left(\frac{1 \text{ atm}}{760 \text{ Torr}}\right) = 0.015\ 78 \text{ atm}$$

If $P_{total} = P_{O_2} + P_{H_2O}$; $P_{O_2} = P_{total} - P_{H_2O}$

$$P_{O_2} = 0.963 \text{ atm} - 0.015\ 78 \text{ atm} = 0.947 \text{ atm}$$

$$n = \frac{PV}{RT} = (0.947 \text{ atm}) \times \left(25.7 \text{ mL} \times \frac{10^{-3} \text{ L}}{1 \text{ mL}}\right) \times \frac{1}{0.082\,06 \text{ L·atm/K·mol}} \times \left(\frac{1}{(273 + 14) \text{ K}}\right) = 1.03 \times 10^{-3} \text{ mol } O_2$$

$$(1.03 \times 10^{-3} \text{ mol } O_2) \times \left(\frac{2 \text{ mol } KClO_3}{3 \text{ mol } O_2}\right) \times \left(\frac{122.55 \text{ g } KClO_3}{1 \text{ mol } KClO_3}\right) = 8.42 \times 10^{-2} \text{ g } KClO_3$$

Molecules Motion in Gases

5.71 Rate of effusion $\propto \frac{1}{\sqrt{M}}$; M = molar mass

Therefore, $\frac{\text{rate (A)}}{\text{rate (B)}} = \sqrt{\frac{M_B}{M_A}}$

(a) $\frac{\text{rate (Ar)}}{\text{rate } (SO_2)} = \sqrt{\frac{64.06 \text{ g/mol}}{39.95 \text{ g/mol}}} = 1.266 > 1$, Ar is faster

(b) $\frac{\text{rate } (D_2)}{\text{rate } (H_2)} = \sqrt{\frac{2.016 \text{ g/mol}}{4.018 \text{ g/mol}}} = 0.7083 < 1$, H_2 is faster

5.73 We use Eq. 7 of Section 5.12. M = molar mass.

$$\frac{t(A)}{t(B)} = \sqrt{\frac{M_A}{M_B}} \text{ and } \frac{M_A}{M_B} = \left(\frac{t(A)}{t(B)}\right)^2$$

Let $M_B = M_{XeF_2}$

$$M_A = \left(\frac{t(A)}{t(XeF_2)}\right)^2$$

$$M_{XeF_2} = (2.7)^2 \times 169.3 \text{ g/mol} = 1.2 \times 10^3 \text{ g/mol}$$

5.75 Boyle's law states that pressure and volume are inversely proportional. As we decrease the volume, the distance that a gas molecule must travel prior to collision with the container wall decreases. It follows that more collisions will occur per unit time. Because pressure is the collective effect of these collisions, it should increase as the volume decreases.

5.77 Using Eq. 9 in section 5.13, we find $v = \sqrt{\frac{3 \text{ RT}}{\text{M}}}$

(a) $v = \sqrt{\frac{3 \times (8.314\,51 \text{ J/K·mol}) \times (273 \text{ K})}{2.016 \times 10^{-3} \text{ kg/mol}}} = 1.84 \times 10^3 \text{ m/s}$

(b) $v = \sqrt{\frac{3 \times (8.314\,51 \text{ J/K·mol}) \times (298 \text{ K})}{0.1313 \text{ kg/mol}}} = 238 \text{ m/s}$

(c) $v = \sqrt{\frac{3 \times (8.314\,51 \text{ J/K·mol}) \times (298 \text{ K})}{4.003 \times 10^{-3} \text{ kg/mol}}} = 1.36 \times 10^3 \text{ m/s}$

5.79 (a) $$\nu_{25^\circ C} = \sqrt{\frac{3 \times (8.314\ 51\ \text{J/K}\cdot\text{mol}) \times (298\ \text{K})}{0.3520\ \text{kg/mol}}} = 1.45 \times 10^2\ \text{m/s}$$

$$\nu_{100^\circ C} = \sqrt{\frac{3 \times (8.314\ 51\ \text{J/K}\cdot\text{mol}) \times (373\ \text{K})}{0.3520\ \text{kg/mol}}} = 1.62 \times 10^2\ \text{m/s}$$

The rms speed increases by a factor of 1.12.

(b) $$\nu_{26^\circ C} = \sqrt{\frac{3 \times (8.314\ 51\ \text{J}\cdot\text{K/mol}) \times (299\ \text{K})}{2.897 \times 10^{-2}\ \text{kg/mol}}} = 5.07 \times 10^2\ \text{m/s}$$

$$\nu_{-20^\circ C} = \sqrt{\frac{3 \times (8.314\ 51\ \text{J/K}\cdot\text{mol}) \times (253\ \text{K})}{2.897 \times 10^{-2}\ \text{kg/mol}}} = 4.67 \times 10^2\ \text{m/s}$$

Air at room temperature has a rms speed that is 1.08 times that of colder air.

Real Gases

5.81 Gases behave most ideally at high temperatures and low pressures. These conditions minimize the opportunity for interactions between molecules, that is, attractive and repulsive forces, which are the cause of deviations from ideality. The farther apart molecules are, the less interaction there will be.

5.83 Figure 5.30 shows this relationship. As two molecules approach, the strength of attraction increases until they are close enough to touch, then they repel each other strongly. An energy of interaction curve shows this effect as a decrease to a minimum value followed by a sharp increase in intermolecular energy.

5.85 (a) Intermolecular attractions in C_2H_4 result in its having a lower pressure than that of an ideal gas. Thus, the ideal gas vessel has the greater pressure.
(b) The free space is almost equal. Although the molecules of C_2H_4 take up some space, this space is a small percentage of the total volume.

5.87 The van der Waals equation is $\left(P + \frac{an^2}{V^2}\right)(V - nb) = nRT$

(a) $10.0\ \text{g}\ CO_2 \times \left(\frac{1\ \text{mol}\ CO_2}{44.01\ \text{g}\ CO_2}\right) = 0.227\ \text{mol}$

$a = 3.59\ \text{L}^2\cdot\text{atm/mol}^2$ and $b = 0.043\ \text{L/mol}$ for CO_2

$$\left[P + \left(\frac{3.59\ \text{L}^2\cdot\text{atm/mol}^2 \times (0.227\ \text{mol})^2}{(0.500\ \text{L})^2}\right)\right](0.500\ \text{L} - 0.227\ \text{mol} \times 0.043\ \text{L/mol})$$
$$= (0.227\ \text{mol}) \times (8.206 \times 10^{-2}\ \text{L}\cdot\text{atm/K}\cdot\text{mol})(283\ \text{K})$$

$(P + 0.740\ \text{atm})(0.490\ \text{L}) = 5.27\ \text{L}\cdot\text{atm}$

$$P = \frac{5.27\ \text{L·atm} - 0.363\ \text{L·atm}}{0.490\ \text{L}} = 10.0\ \text{atm}$$

(b) $10.0\ \text{g He} \times \left(\frac{1\ \text{mol He}}{4.003\ \text{g He}}\right) = 2.50\ \text{mol He}$

$a = 0.034\ \text{L}^2\cdot\text{atm/mol}^2$; $b = 0.024\ \text{L/mol}$

$$\left[P + \left(\frac{0.034\ \frac{\text{L}^2\cdot\text{atm}}{\text{mol}^2} \times (2.50\ \text{mol})^2}{(0.250\ \text{L})^2}\right)\right](0.250\ \text{L} - 2.50\ \text{mol} \times 0.024\ \text{L/mol})$$
$$= (2.50\ \text{mol}) \times (8.206 \times 10^{-2}\ \text{L·atm/K·mol}) \times (373\ \text{K})$$

$(P + 3.40\ \text{atm}) \times (0.190\ \text{L}) = 76.5\ \text{L·atm}$

$$P = \frac{76.5\ \text{L·atm} - 0.646\ \text{L·atm}}{0.190\ \text{L}} = 399\ \text{atm}$$

(c) $P_{CO_2} = \frac{nRT}{V}$

$$P_{CO_2} = \frac{(0.227\ \text{mol}) \times (8.206 \times 10^{-2}\ \text{L·atm/K·mol}) \times (283\ \text{K})}{0.500\ \text{L}} = 10.5\ \text{atm}\ CO_2$$

$$P_{He} = \frac{nRT}{V}$$

$$P_{He} = \frac{(2.50\ \text{mol}) \times (8.206 \times 10^{-2}\ \text{L·atm/K·mol}) \times (373\ \text{K})}{0.250\ \text{L}} = 306\ \text{atm He}$$

(d) Percentage error $\left|\frac{\text{actual} - \text{theoretical}}{\text{theoretical}}\right| \times 100$, where theoretical values are from the van der Waals equation.

For CO_2: $\frac{|10.5 - 10.0|}{10.0} \times 100 = 5\%$ error

For He: $\frac{|306 - 399|}{399} \times 100 = 23\%$ error

SUPPLEMENTARY EXERCISES

5.89 The ideal gas equation, $PV = nRT$, can be rewritten as $\frac{PV}{T} = nR =$ constant (at constant amount of gas, n).

Therefore, $\frac{P_1V_1}{T_1} = \frac{P_2V_2}{T_2}$

Solving for V_2, $V_2 = \left(\frac{T_2}{T_1}\right)\left(\frac{P_1}{P_2}\right) V_1$

For $T_1 = 273$ K and $P_1 = 1.00$ atm, this becomes $V_2 = \left(\frac{T_2}{273}\right)\left(\frac{1.00\ \text{atm}}{P_2}\right) V_1$

5.91 (a) The number of molecules is the same, but Cl_2 is diatomic.

(b) The temperature is the same, as stated in the problem.

(c) $n_{He} = \dfrac{(1.00 \text{ atm}) \times (1.000 \text{ L})}{(8.206 \times 10^{-2} \text{ L·atm/K·mol}) \times (298 \text{ K})}$

$= 4.09 \times 10^{-2} \text{ mol He} \times \left(\dfrac{4.003 \text{ g He}}{1 \text{ mol He}}\right) = 0.164 \text{ g He}$

$n_{Cl_2} = \dfrac{(1.00 \text{ atm}) \times (1.000 \text{ L})}{(8.206 \times 10^{-2} \text{ L·atm/K·mol}) \times (298 \text{ K})}$

$= 4.09 \times 10^{-2} \text{ mol } Cl_2 \times \left(\dfrac{70.90 \text{ g } Cl_2}{1 \text{ mol } Cl_2}\right) = 2.90 \text{ g } Cl_2$

The mass differs.

(d) average speed $\propto \sqrt{\dfrac{\text{temperature}}{\text{molar mass}}}$

$M_{Cl_2} > M_{He}$; therefore, average speed of He > average speed of Cl_2

(e) $d = \left(\dfrac{P}{RT}\right) M$ [See Exercise 5.53, M = molar mass]

$M_{Cl_2} > M_{He}$; therefore, $d_{Cl_2} > d_{He}$

(f) The average kinetic energy is proportional to temperature, so the average kinetic energy is the same for both gases.

5.93 $(1.5 \text{ in } H_2O) \times \left(\dfrac{1.0 \text{ g/cm}^3}{13.6 \text{ g/cm}^3}\right) = 0.11 \text{ in Hg}$

$P = (0.11 \text{ in Hg}) \times \left(\dfrac{2.54 \text{ cm}}{1 \text{ in.}}\right) \times \left(\dfrac{10 \text{ mm}}{1 \text{ cm}}\right) \times \left(\dfrac{1 \text{ Torr}}{1 \text{ mm Hg}}\right) = 2.8 \text{ Torr}$

5.95 $V = (2.0 \times 10^9 \text{ L}) \times \left(\dfrac{1.20 \text{ atm}}{6.00 \text{ atm}}\right) \times \left(\dfrac{[273 + (-25)] \text{ K}}{(273 + 40.) \text{ K}}\right) = 3.2 \times 10^8 \text{ L}$

5.97 $V = (10.0 \text{ L}) \times \left(\dfrac{2.00 \text{ atm}}{0.25 \text{ atm}}\right) \times \left(\dfrac{[273 + (-50)] \text{ K}}{(273 + 27) \text{ K}}\right) = 59 \text{ L}$

5.99 $P_{O_2} = \chi_{O_2}P$; $P_{O_2} = 0.21 \times 520. \text{ Torr} = 1.1 \times 10^2 \text{ Torr}$

5.101 (a) $PV = nRT$; $n = \dfrac{PV}{RT}$

$n = (5.00 \text{ atm}) \times \left(425 \text{ mL} \times \dfrac{10^{-3} \text{ L}}{1 \text{ mL}}\right) \times \left(\dfrac{\text{K·mol}}{0.082\,06 \text{ L·atm}}\right)$

$\times \left(\dfrac{1}{(273 + 23) \text{ K}}\right) = 8.75 \times 10^{-2} \text{ mol}$

$$\text{mass} = (8.75 \times 10^{-2}\ \text{mol}) \times \left(\frac{28.0\ \text{g}}{1\ \text{mol CO}}\right) = 2.45\ \text{g CO}$$

(b) $$\text{density} = \frac{\text{mass}}{\text{volume}} = \frac{2.45\ \text{g}}{425\ \text{mL} \times \left(\frac{10^{-3}\ \text{L}}{1\ \text{mL}}\right)} = 5.76\ \text{g/L}$$

(c) Because the mass and volume are fixed quantities in this experiment, the density does not change.

5.103 Rearrange Eg. 7 of section 5.12 to give $t_x = t_{Ar}\sqrt{\frac{\text{molar mass } x}{\text{molar mass Ar}}}$

(a) $$t_{CO_2} = (147\ \text{s})\sqrt{\frac{44.0\ \text{g/mol}}{40.0\ \text{g/mol}}} = 154\ \text{s}$$

(b) $$t_{C_2H_4} = (147\ \text{s})\sqrt{\frac{28.0\ \text{g/mol}}{40.0\ \text{g/mol}}} = 123\ \text{s}$$

(c) $$t_{H_2} = (147\ \text{s})\sqrt{\frac{2.02\ \text{g/mol}}{40.0\ \text{g/mol}}} = 33.0\ \text{s}$$

(d) $$t_{SO_2} = (147\ \text{s})\sqrt{\frac{64.0\ \text{g/mol}}{40.0\ \text{g/mol}}} = 186\ \text{s}$$

5.105 $$v_{-50°C} = \sqrt{\frac{3 \times (8.314\ 51\ \text{J/K·mol}) \times (223\ \text{K})}{2.897 \times 10^{-2}\text{kg/mol}}} = 4.38 \times 10^{2}\ \text{m/s}$$

$$v_{20°C} = \sqrt{\frac{3 \times (8.314\ 51\ \text{J/K·mol}) \times (293\ \text{K})}{2.897 \times 10^{-2}\ \text{kg/mol}}} = 5.02 \times 10^{2}\ \text{m/s}$$

5.107 (a) Hydrogen and helium molecules are very light. Consequently, they have speeds high enough to escape the Earth's gravitational pull.
(b) The CO_2 of the Earth's atmosphere may have been consumed by growing plants, which use CO_2 in the synthesis of glucose.

APPLIED EXERCISES

5.109 The lower the molecular mass, the larger the average molecular speed, so $O_3 < NO_2 < O_2 < NO$.

5.111 (a) $$4\ \text{molecules } SO_2 \times \left(\frac{1\ \text{mol } SO_2}{6.022 \times 10^{23}\ \text{molecules}}\right) = 4\ \text{mol}$$

(b) 4 molecules $SO_2 \times \left(\frac{1\text{ L}}{10^6\text{ molecules}}\right) = 4 \times 10^{-6}$ L or 4 μL

(c) 4 molecules $SO_2 \times \left(\frac{1\text{ m}^3}{10^6\text{ molecules}}\right) = 4 \times 10^{-6}\text{ m}^3 \times \left(\frac{100\text{ cm}}{1\text{ m}}\right)^3 = 4\text{ cm}^3$

(d) $P_{SO_2} = \chi_{SO_2}P \qquad P_{SO_2} = (7 \times 10^{-24}\text{ mol}) \times (760\text{ Torr}) = 5 \times 10^{-21}\text{ Torr}$

5.113 (a) $n = \frac{PV}{RT}$

$$n_{\text{butane}} = \frac{2.33\text{ atm} \times 250.\text{ L}}{0.082\ 06\text{ L·atm/K·mol} \times (273 + 150)\text{ K}} = 16.7\overline{8}\text{ mol}$$

$$\text{mass of CO}_2 = 16.7\overline{8}\text{ mol butane} \times \left(\frac{8\text{ mol CO}_2}{2\text{ mol butane}}\right) \times \left(\frac{44.01\text{ g CO}_2}{\text{mol CO}_2}\right)$$

$$= 2.95 \times 10^3\text{ g CO}_2 = 2.95\text{ kg CO}_2$$

(b) $n_{CO_2} = 16.7\overline{8}\text{ mol butane} \times \left(\frac{8\text{ mol CO}_2}{2\text{ mol butane}}\right) = 67.1\overline{2}\text{ mol CO}_2$

$$P = \frac{n_{CO_2}RT}{V} = \frac{67.1\overline{2}\text{ mol} \times 0.082\ 06\text{ L·atm/K·mol} \times 289\text{ K}}{4.000 \times 10^3\text{ L}}$$

$$= 0.398\text{ atm}$$

INTEGRATED EXERCISES

5.115 $n = PV/RT = (690.\text{ Torr}) \times (0.200\text{ L})/(62.36\text{ L·Torr/K·mol}) \times (293\text{ K})$

$= 7.55 \times 10^{-3}$ mol HCl dissolved in 100. mL water.

The resulting solution is $7.55 \times 10^{-3}\text{ mol}/0.100\text{ L} = 7.55 \times 10^{-2}$ M

$$\text{So } 0.100\text{ L} \times \left(\frac{7.55 \times 10^{-2}\text{ mol HCl}}{\text{L}}\right) \times \left(\frac{1\text{ mol NaOH}}{1\text{ mol HCl}}\right) \times \left(\frac{1}{0.0157\text{ L}}\right)$$

$= 0.481$ M

5.117 $85.7\text{ g C} \times \left(\frac{1\text{ mol C}}{12.01\text{ g}}\right) = 7.13\overline{6}\text{ mol C}$

$14.3\text{ g H} \times \left(\frac{1\text{ mol H}}{1.008\text{ g H}}\right) = 14.1\overline{9}\text{ mol H}$

Divide by 7.136 to get the ratio of C:H ≈ 1:2. So the empirical formula is CH_2.

Calculate the molecular mass from the remaining data.

$$PV = \left(\frac{m}{M}\right)RT \quad \text{or} \quad M = \frac{mRT}{PV}$$

$$M = \frac{1.77\text{ g} \times (62.364\text{ L·Torr/K·mol}) \times 290\text{ K}}{508\text{ Torr} \times 1.500\text{ L}} = 42.0\text{ g/mol}$$

The molecular formula is thus C_3H_6 $\left(\frac{42.0\text{ g/mol}}{14.03\text{ g/mol}} = 3\right)$.

5.119 We use Eq. 7 with HC = hydrocarbon, M = molar mass.

$$\frac{t(\text{Ar})}{t(\text{HC})} = \sqrt{\frac{M_{\text{Ar}}}{M_{\text{HC}}}}$$

Solve for M_{HC}

$$M_{\text{HC}} = \left(\frac{t(\text{HC})}{t(\text{Ar})}\right)^2 M_{\text{Ar}} = \left(\frac{349\ \text{s}}{210\ \text{s}}\right)^2 \times 39.95\ \text{g/mol} = 110\ \text{g/mol}$$

$$\frac{M_{\text{HC}}}{\text{molar mass } C_2H_3} = \frac{110\ \text{g/mol}}{27\ \text{g/mol}} \approx 4$$

Therefore, the molecular formula would be C_8H_{12}.

5.121 $$23.76\ \text{g S} \times \left(\frac{1\ \text{mol S}}{32.06\ \text{g S}}\right) = 0.7411\ \text{mol S}$$

$$23.71\ \text{g O} \times \left(\frac{1\ \text{mol O}}{16.00\ \text{g O}}\right) = 1.482\ \text{mol O}$$

$$52.54\ \text{g Cl} \times \left(\frac{1\ \text{mol Cl}}{35.45\ \text{g Cl}}\right) = 1.482\ \text{mol Cl}$$

Divide by 0.7411 to determine the ratio of S:O:Cl = 1:2:2. The empirical formula is SO_2Cl_2. Find the molecular weight as in Exercise 5.119.

$$M_{SO_2Cl_2} = \left(\frac{\text{t}(SO_2Cl_2)}{\text{t(Ar)}}\right)^2 M_{\text{Ar}} = \left(\frac{46.0}{25.0}\right)^2 \times 39.95\ \text{g/mol} = 135\ \text{g/mol}$$

$$\frac{135\ \text{g/mol}}{134.96\ \text{g/mol}} \approx 1$$ Therefore, the molecular formula would be SO_2Cl_2.

5.123 $$\text{number of moles of N} = 0.414\ \text{g N} \times \left(\frac{1\ \text{mol N}}{14.01\ \text{g N}}\right) = 0.029\ 5\bar{5}\ \text{mol N}$$

$$\text{number of moles of H} = 0.0591\ \text{g H} \times \left(\frac{1\ \text{mol H}}{1.0079\ \text{g H}}\right) = 0.058\ 6\bar{4}\ \text{mol H}$$

$$\frac{0.058\ 6\bar{4}\ \text{mol H}}{0.029\ 5\bar{5}\ \text{mol N}} = 1.98 \approx 2$$

Therefore, the empirical formula is NH_2.

The molar mass is calculated from $M = d\left(\frac{RT}{P}\right) = \left(\frac{m}{V}\right)\left(\frac{RT}{P}\right)$

$$M = \left(\frac{0.473\ \text{g}}{0.200\ \text{L}}\right) \times \left(\frac{0.082\ 06\ \text{L·atm/K·mol} \times 298\ \text{K}}{1.81\ \text{atm}}\right) = 32.0\ \text{g/mol}$$

Then, $$\frac{\text{molar mass of compound}}{\text{molar mass of empirical formula}} = \frac{32.0\ \text{g/mol}}{16.0\ \text{g/mol}} = 2$$

Therefore, the molecular formula is N_2H_4.

5.125 (a) $PV = nRT$; $n = \dfrac{PV}{RT}$

$$n_{O_2} = (2.33\ \text{atm}) \times (75.0\ \text{L}) \times \left(\frac{1}{0.082\ 06\ \text{L}\cdot\text{atm/K}\cdot\text{mol}}\right) \times \left(\frac{1}{(273 + 150)\ \text{K}}\right) = 5.03\ \text{mol}\ O_2$$

$$\text{mass of } Fe_2O_3 = (5.03\ \text{mol}\ O_2) \times \left(\frac{2\ \text{mol}\ Fe_2O_3}{11\ \text{mol}\ O_2}\right) \times \left(\frac{159.70\ \text{g}\ Fe_2O_3}{1\ \text{mol}\ Fe_2O_3}\right) = 146\ \text{g}\ Fe_2O_3$$

(b) $\#\ \text{moles of}\ SO_2 = (5.03\ \text{mol}\ O_2) \times \left(\dfrac{8\ \text{mol}\ SO_2}{11\ \text{mol}\ O_2}\right) = 3.66\ \text{mol}\ SO_2$, and

$$\text{molarity} = (3.66\ \text{mol}\ SO_2) \times \left(\frac{1\ \text{mol}\ H_2SO_3}{1\ \text{mol}\ SO_2}\right) \times \left(\frac{1}{5.00\ \text{L soln}}\right) = 0.732\ \text{M}\ H_2SO_3(aq)$$

CHAPTER 6
THERMOCHEMISTRY: THE FIRE WITHIN

EXERCISES

Internal Energy, Heat, and Work

6.1 (a) isolated (b) closed (c) closed

6.3 4.184 J = 1 cal, 4.184 kJ = 1 kcal

$$\text{energy content (kJ)} = 16 \text{ kcal} \times \left(\frac{4.184 \text{ kJ}}{1 \text{ kcal}}\right) = 67 \text{ kJ}$$

6.5 (a) Heat is absorbed by the system; q is positive; the system looks like it contracts, so work is done on the system; w is negative.
(b) Heat is released by the system; q is negative; the system looks like it contracts, so work is done on the system; w is negative.

6.7 $\Delta U = q + w = 550.\text{ kJ} + 700.\text{ kJ} = 1250 \text{ kJ or } 1.25 \times 10^3 \text{ kJ}$

6.9 $$q = \frac{100.\text{ J}}{5} \times 20.\text{ min} \times \left(\frac{60.5}{1 \text{ min}}\right) = 1.2 \times 10^5 \text{ J or } 1.2 \times 10^2 \text{ kJ}$$

$$w = -P_{ex}\Delta V = -1.0 \text{ atm } (2.5 \text{ L} - 2.0 \text{ L}) \times \left(\frac{101.325 \text{ J}}{1 \text{ L}\cdot\text{atm}}\right) = -51 \text{ J}$$

$\Delta U = q + w = 1.2 \times 10^5 \text{ J} + (-51 \text{ J}) = 1.2 \times 10^5 \text{ J}$
and $\Delta U = q - P\Delta V = 1.2 \times 10^5 \text{ J} - 51 \text{ J} = 1.2 \times 10^5 \text{ J}$

6.11 For this system, $\Delta U = -892.4$ kJ and $w = -492$ kJ (negative sign indicates expansion work). $\Delta U = q + w$; solve for q.
$q = \Delta U - w = -892.4 \text{ kJ} - (-492 \text{ kJ}) = -400.\text{ kJ}$
400. kJ of heat is lost from the system.

6.13 As the temperature of a system increases, the average velocity of the molecules in a system also increases.

Enthalpy Change

6.15 $\Delta U = q - P\Delta V$

At constant pressure, $\Delta H = \Delta U + P\Delta V$, so $\Delta H = -65 \text{ kJ} + 28 \text{ kJ} = -37 \text{ kJ}$

6.17 (a) endothermic ($\Delta H > 0$)

(b) exothermic ($\Delta H < 0$)

6.19 ΔH = heat lost by water

= mass × specific heat capacity × temperature change

$= 20.0 \text{ g} \times 4.184 \text{ J/g}\cdot(°\text{C}) \times (4.00°\text{C} - 20.0°\text{C})$

$= -1.3\bar{4} \times 10^3 \text{ J}$

$= -1.3$ kJ, exothermic, $\Delta H < 0$

Measuring Heat

6.21 (a) heat needed = temperature change × mass × specific heat capacity

$= (37.2 - 25.3)°\text{C} \times 50.0 \text{ g} \times 1.05 \text{ J/g}\cdot(°\text{C}) = 625 \text{ J}$

(b) ΔT = temperature change

heat supplied = mass × specific heat capacity × ΔT

Solving for ΔT,

$$\Delta T = \frac{\text{heat supplied}}{\text{mass} \times \text{specific heat capacity}}$$

$$= \frac{4.90 \times 10^5 \text{ J}}{1.0 \text{ kg} \times \left(\frac{10^3 \text{ g}}{1 \text{ kg}}\right) \times 0.90 \text{ J/g}\cdot(°\text{C})} = 5.4 \times 10^2 \text{ °C}$$

6.23 ss = stainless steel

heat = temperature change × mass × specific heat capacity

total heat needed = heat(ss) + heat(H_2O)

heat(ss) $= (100. - 25)°\text{C} \times 500.0 \text{ g} \times 0.51 \text{ J/g}\cdot(°\text{C}) = 1.9\bar{1} \times 10^4 \text{ J}$

heat(H_2O) $= 75°\text{C} \times 450.0 \text{ g} \times 4.18 \text{ J/g}\cdot(°\text{C}) = 1.4\bar{1} \times 10^5 \text{ J}$

total heat needed $= 1.9\bar{1} \times 10^4 + 1.4\bar{1} \times 10^5 \text{ J} = 1.6\bar{0} \times 10^5 \text{ J}$

$$\% \text{ heat}(H_2O) = \frac{\text{heat}(H_2O)}{\text{total heat}} \times 100\% = \frac{1.4\bar{1} \times 10^5 \text{ J}}{1.6\bar{0} \times 10^5 \text{ J}} \times 100\% = 88\%$$

6.25 (a) heat energy needed = temperature change × mass × specific heat capacity

heat energy needed $= (500. - 25)°\text{C} \times 10.0 \text{ g} \times 0.45 \text{ J/g}\cdot(°\text{C}) = 2.1 \times 10^3 \text{ J}$

(b) Solving the equation in part (a) for mass yields

$$\text{mass} = \frac{\text{heat energy}}{\text{specific heat capacity} \times \text{temperature change}}$$

$$\text{mass}_{Au} = \frac{2.1 \times 10^3\ \text{J}}{0.13\ \text{J/g}\cdot(°\text{C}) \times 475\ \text{K}} = 34\ \text{g}$$

6.27 Let ΔT = temperature change = T(final) − T(initial), c = specific heat capacity

Because all the energy lost by the metal is gained by the water, we can write

heat(metal) = − heat(H_2O)

heat(metal) = ΔT(metal) × mass(metal) × c(metal)

heat(H_2O) = ΔT(H_2O) × mass(H_2O) × c(H_2O)

Note that ΔT(metal) is negative and ΔT(H_2O) is positive. Then

ΔT(metal) × mass(metal) × c(metal) = − ΔT(H_2O) × mass(H_2O) × c(H_2O)

Solving for c(metal):

$$c(\text{metal}) = \frac{-\Delta T(H_2O) \times \text{mass}(H_2O) \times c(H_2O)}{\Delta T(\text{metal}) \times \text{mass}(\text{metal})}$$

$$= \frac{-(25.7 - 22.0)°\text{C} \times 50.7\ \text{g}\ H_2O \times 4.18\ \text{J/g}\cdot(°\text{C})}{(25.7 - 100.)°\text{C} \times 20.0\ \text{g metal}} = 0.53\ \text{J/g}\cdot(°\text{C})$$

6.29 ΔH_c = − 3227 kJ/mol. Because heat is given off by the reaction (exothermic), heat is absorbed by the calorimeter. We can write

$\Delta H_{cal} = -\Delta H_{rxn}$ (rxn means "reaction")

Let C_{cal} = heat capacity of the calorimeter and n = amount (moles) of benzoic acid:

$$n = 1.236\ \text{g} \times \left(\frac{1\ \text{mol}}{122.12\ \text{g}}\right) = 0.010\ 12\ \text{mol}$$

Then $\Delta H_{cal} = C_{cal} \times \Delta T$, ΔT = 2.345°C

$\Delta H_{rxn} = n \times \Delta H_c$ = 0.010 12 mol × (− 3227 kJ/mol) = − 32.66 kJ

Then $\Delta H_{cal} = -\Delta H_{rxn}$, C_{cal} × 2.345°C = + 32.66 kJ

$$C_{cal} = \frac{32.66\ \text{kJ}}{2.345°\text{C}} = 13.93\ \text{kJ/(°C)}$$

6.31 Let C_{cal} = heat capacity of calorimeter

$\Delta H_{comb} = -\Delta H_{cal}$ (comb means "combustion")

$\Delta H_{cal} = C_{cal} \times \Delta T$ = 5.24 kJ/(°C) × (23.17 − 22.45)°C = 3.8 kJ

ΔH_{comb} = − 3.8 kJ

6.33 Let C_{cal} = heat capacity of calorimeter

(a) $\Delta H_{neut} = -\Delta H_{cal} = -C_{cal} \times \Delta T$ (neut means "neutralization")

$= -525.0\ \text{J/(°C)} \times (21.3 - 18.6)°\text{C} = -1.4\bar{2} \times 10^3\ \text{J}$

(b) number of moles HNO_3 = 0.0500 L × 0.500 mol/L = 0.0250 mol

$$\Delta H = \frac{\Delta H_{neut}}{n_{HNO_3}} = \frac{-1.4\bar{2} \times 10^3 \text{ J}}{0.0250 \text{ mol}} = -57 \text{ kJ/mol}$$

Enthalpy of Physical Change

6.35 ΔH = heat required, n = number of moles

(a) $\Delta H_{vap} = \frac{\Delta H}{n} = \frac{1.93 \text{ kJ}}{0.235 \text{ mol}} = 8.21 \text{ kJ/mol}$

(b) $\Delta H_{vap} = \frac{\Delta H}{n} = \left(\frac{21.2 \text{ kJ}}{22.45 \text{ g } C_2H_5OH}\right) \times \left(\frac{46.07 \text{ g } C_2H_5OH}{1 \text{ mol } C_2H_5OH}\right) = 43.5 \text{ kJ/mol}$

6.37 (a) number of moles of H_2O = 100.0 g H_2O × $\left(\frac{1 \text{ mol } H_2O}{18.02 \text{ g } H_2O}\right)$ = 5.55 mol H_2O

$\Delta H° = 5.55 \text{ mol } H_2O \times 40.7 \text{ kJ/mol } H_2O = +226 \text{ kJ}$

(b) number of moles of NH_3 = 600. g NH_3 × $\left(\frac{1 \text{ mol } NH_3}{17.03 \text{ g } NH_3}\right)$ = 35.2 mol NH_3

$\Delta H° = 35.2 \text{ mol } NH_3 \times 5.65 \text{ kJ/mol } NH_3 = +199 \text{ kJ}$

6.39 heat needed = $6.01 \times 10^3 \text{ J/mol} \times \left(\frac{1 \text{ mol}}{18.02 \text{ g } H_2O}\right) \times 50.0 \text{ g } H_2O$

$+4.18 \text{ J/g}\cdot(°C) \times 50.0 \text{ g} \times 25°C = 16.\bar{7} \times 10^3 \text{ J} + 5.2\bar{2} \times 10^3 \text{ J}$

$= 21.\bar{9} \times 10^3 \text{ J} = 22 \text{ kJ}$

6.41 Heating curve for bromine:

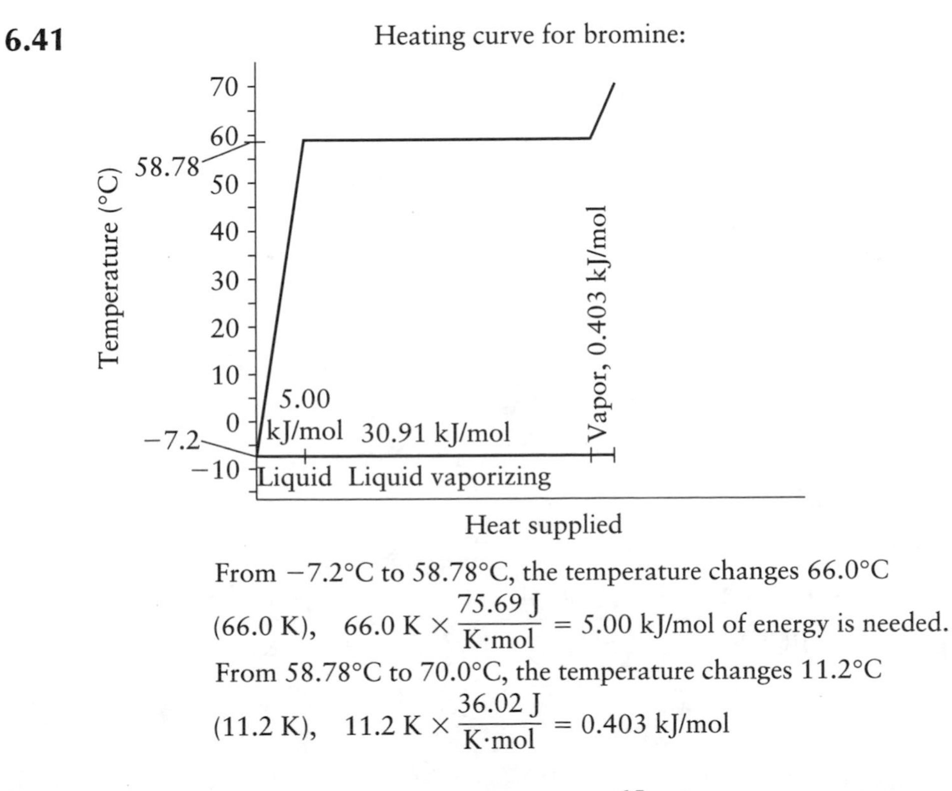

From −7.2°C to 58.78°C, the temperature changes 66.0°C

(66.0 K), $66.0 \text{ K} \times \frac{75.69 \text{ J}}{\text{K}\cdot\text{mol}} = 5.00$ kJ/mol of energy is needed.

From 58.78°C to 70.0°C, the temperature changes 11.2°C

(11.2 K), $11.2 \text{ K} \times \frac{36.02 \text{ J}}{\text{K}\cdot\text{mol}} = 0.403$ kJ/mol

Enthalpy of Chemical Change

6.43 The standard state of a substance is its pure form at 1 atmosphere pressure. Standard state data are usually reported at 25°C, but can be reported for any temperature.

6.45 (a) heat absorbed $= 0.20 \text{ mol S}_8 \times \left(\dfrac{+358.8 \text{ kJ}}{1 \text{ mol S}_8}\right) = +72 \text{ kJ}$

(b) heat absorbed $= 20.0 \text{ g C} \times \left(\dfrac{1 \text{ mol C}}{12.01 \text{ g C}}\right) \times \left(\dfrac{+358.8 \text{ kJ}}{4 \text{ mol C}}\right) = +149 \text{ kJ}$

(c) mass of CS_2 produced $= 217 \text{ kJ} \times \left(\dfrac{4 \text{ mol CS}_2}{358.8 \text{ kJ}}\right) \times \left(\dfrac{76.13 \text{ g CS}_2}{1 \text{ mol CS}_2}\right) = 184 \text{ g}$

6.47 (a) mass of octane $= -12 \times 10^6 \text{ J} \times \left(\dfrac{2 \text{ mol octane}}{-10.942 \times 10^6 \text{ J}}\right) \times \left(\dfrac{114.22 \text{ g octane}}{1 \text{ mol octane}}\right)$

$= 2.5 \times 10^2 \text{ g octane}$

(b) 1 gal = 3.785 L

heat evolved $= 1.0 \text{ gal} \times \left(\dfrac{3.785 \text{ L}}{1 \text{ gal}}\right) \times \left(\dfrac{1 \text{ mL}}{10^{-3} \text{ L}}\right) \times \left(\dfrac{0.70 \text{ g octane}}{1 \text{ mL}}\right)$

$\times \left(\dfrac{1 \text{ mol octane}}{114.22 \text{ g octane}}\right) \times \left(\dfrac{-10.942 \times 10^6 \text{ J}}{2 \text{ mol octane}}\right) = -1.3 \times 10^8 \text{ J}$

Hess's Law

6.49 (1) $S(s) + O_2(g) \longrightarrow SO_2(g)$ $\quad \Delta H° = -296.83$ kJ

(2) $2\,S(s) + 3\,O_2(g) \longrightarrow 2\,SO_3(g)$ $\quad \Delta H° = -791.44$ kJ

Reverse (1) and multiply by 2, then add to (2).

$2\,SO_2(g) \longrightarrow 2\,S(s) + 2\,O_2(g)$ $\quad \Delta H° = -2(-296.83 \text{ kJ}) = 593.66$ kJ

$2\,S(s) + 3\,O_2(g) \longrightarrow 2\,SO_3(g)$ $\quad \Delta H° = -791.44$ kJ

$2\,SO_2(g) + O_2(g) \longrightarrow 2\,SO_3(g)$ $\quad \Delta H° = (593.66 - 791.44) \text{ kJ} = -197.78$ kJ

6.51 (1) $P_4(s) + 6\,Cl_2(g) \longrightarrow 4\,PCl_3(l)$ $\quad \Delta H° = -1278.8$ kJ

(2) $PCl_3(l) + Cl_2(g) \longrightarrow PCl_5(s)$ $\quad \Delta H° = -124$ kJ

Multiply (2) by 4, then add to (1).

$P_4(s) + 6\,Cl_2(g) \longrightarrow 4\,PCl_3(l)$ $\quad \Delta H° = -1278.8$ kJ

$4\,PCl_3(l) + 4\,Cl_2(g) \longrightarrow 4\,PCl_5(s)$ $\quad \Delta H° = 4(-124 \text{ kJ}) = -496$ kJ

$P_4(s) + 10\,Cl_2(g) \longrightarrow 4\,PCl_5(s)$ $\quad \Delta H° = (-1278.8 - 496)$ kJ

$= -1775$ kJ

6.53 (1) $2\ C_2H_2(g) + 5\ O_2(g) \longrightarrow 4\ CO_2(g) + 2\ H_2O(l)$ $\quad \Delta H° = -2600.\ kJ$

(2) $2\ C_2H_6(g) + 7\ O_2(g) \longrightarrow 4\ CO_2(g) + 6\ H_2O(l)$ $\quad \Delta H° = -3120.\ kJ$

(3) $H_2(g) + \frac{1}{2}\ O_2(g) \longrightarrow H_2O(l)$ $\quad \Delta H° = -286\ kJ$

Multiply (1) by $\frac{1}{2}$, reverse (2) and multiply by $\frac{1}{2}$, multiply (3) by 2, and then add all together.

$C_2H_2(g) + \frac{5}{2}\ O_2(g) \longrightarrow 2\ CO_2(g) + H_2O(l)$	$\Delta H° = \frac{1}{2} \times (-2600.\ kJ) = -1300.\ kJ$
$2\ CO_2(g) + 3\ H_2O(l) \longrightarrow C_2H_6(g) + \frac{7}{2}\ O_2(g)$	$\Delta H° = -\frac{1}{2} \times (-3120.\ kJ) = 1560.\ kJ$
$2\ H_2(g) + O_2(g) \longrightarrow 2\ H_2O(l)$	$\Delta H° = 2 \times (-286\ kJ) = -572\ kJ$
$C_2H_2(g) + 2\ H_2(g) \longrightarrow C_2H_6(g)$	$\Delta H° = (-1300. + 1560. - 572)\ kJ = -312\ kJ$

6.55 (1) $NH_3(g) + HCl(g) \longrightarrow NH_4Cl(s)$ $\quad \Delta H° = -176.0\ kJ$

(2) $N_2(g) + 3\ H_2(g) \longrightarrow 2\ NH_3(g)$ $\quad \Delta H° = -92.22\ kJ$

(3) $N_2(g) + 4\ H_2(g) + Cl_2(g) \longrightarrow 2\ NH_4Cl(s)$ $\quad \Delta H° = -628.86\ kJ$

Reverse (1) and multiply by 2, reverse (2), then add to (3).

$2\ NH_4Cl(s) \longrightarrow 2\ NH_3(g) + 2\ HCl(g)$	$\Delta H° = -2(-176.0\ kJ) = 352.0\ kJ$
$2\ NH_3(g) \longrightarrow N_2(g) + 3\ H_2(g)$	$\Delta H° = -(-92.22\ kJ) = 92.22\ kJ$
$N_2(g) + 4\ H_2(g) + Cl_2(g) \longrightarrow 2\ NH_4Cl(s)$	$\Delta H° = -628.86\ kJ$
$H_2(g) + Cl_2(g) \longrightarrow 2\ HCl(g)$	$\Delta H° = (352.0 + 92.22 - 628.86)\ kJ = -184.7\ kJ$

6.57 (1) $C(graphite) + O_2(g) \longrightarrow CO_2(g)$ $\quad \Delta H° = -393.51\ kJ/mol$

(2) $C(diamond) + O_2(g) \longrightarrow CO_2(g)$ $\quad \Delta H° = -395.41\ kJ/mol$

Reverse (2) and then add to (1).

$C(graphite) + O_2(g) \longrightarrow CO_2(g)$	$\Delta H° = -393.51\ kJ/mol$
$CO_2(g) \longrightarrow C(diamond) + O_2(g)$	$\Delta H° = -(-395.41\ kJ/mol) = 395.41\ kJ/mol$
$C(graphite) \longrightarrow C(diamond)$	$\Delta H° = (-393.51 + 395.41)\ kJ/mol = +1.90\ kJ/mol$

Heat Output

6.59 specific enthalpy $= 4854\ kJ/mol \times \left(\frac{1\ mol}{100.20\ g}\right) = 48.44\ kJ/g$

enthalpy density $= 48.44\ kJ/g \times 0.68\ g/mL \times \left(\frac{1\ mL}{10^{-3}\ L}\right) = 3.3 \times 10^4\ kJ/L$

6.61 The reactions are

$Mg(s) + \frac{1}{2} O_2(g) \longrightarrow MgO(s)$

$2\ Al(s) + \frac{3}{2} O_2(g) \longrightarrow Al_2O_3(s)$

For Mg: $\dfrac{601.70\ \text{kJ/mol}}{24.31\ \text{g/mol}} = 24.75\ \text{kJ/g}$

For Al: $\dfrac{837.85\ \text{kJ/mol}}{26.98\ \text{g/mol}} = 31.05\ \text{kJ/g}$, so Al is better. Note that there are two moles of Al in one mole of Al_2O_3.

Enthalpy of Formation

6.63 (a) $K(s) + \frac{1}{2} Cl_2(g) + \frac{3}{2} O_2(g) \longrightarrow KClO_3(s)\quad \Delta H_f^\circ = -397.73\ \text{kJ/mol}$

(b) $\frac{5}{2} H_2(g) + \frac{1}{2} N_2(g) + 2\ C(\text{graphite}) + O_2(g) \longrightarrow H_2NCH_2COOH(s)$

$\Delta H_f^\circ = -532.9\ \text{kJ/mol}$

(c) $2\ Al(s) + \frac{3}{2} O_2(g) \longrightarrow Al_2O_3(s)\quad \Delta H_f^\circ = -1675.7\ \text{kJ/mol}$

6.65 (a) $\Delta H^\circ = (2\ \text{mol} \times \Delta H_f^\circ[SO_3(g)]) - (2\ \text{mol} \times \Delta H_f^\circ[SO_2(g)])$

$\Delta H^\circ = 2\ \text{mol} \times (-395.72\ \text{kJ/mol}) - 2\ \text{mol} \times (-296.83\ \text{kJ/mol})$

$= -197.78\ \text{kJ}$

rxn means "reaction."

$\Delta H_{rxn}^\circ(10.0\ \text{g}\ SO_2) = 10.0\ \text{g}\ SO_2 \times \left(\dfrac{1\ \text{mol}\ SO_2}{64.06\ \text{g}\ SO_2}\right) \times \left(\dfrac{-197.78\ \text{kJ}}{2\ \text{mol}\ SO_2}\right)$

$= -15.4\ \text{kJ}$

(b) $\Delta H^\circ = 1\ \text{mol} \times \Delta H_f^\circ[H_2O(l)] - 1\ \text{mol} \times \Delta H_f^\circ[CuO(s)]$

$\Delta H^\circ = -285.83\ \text{kJ} - (-157.3\ \text{kJ}) = -128.5\ \text{kJ}$

6.67 (a) $\Delta H^\circ = 1\ \text{mol} \times \Delta H_f^\circ[H_2O(l)] - 1\ \text{mol} \times \Delta H_f^\circ[D_2O(l)]$

$\Delta H^\circ = 1\ \text{mol} \times (-285.83\ \text{kJ/mol}) - 1\ \text{mol} \times (-294.60\ \text{kJ/mol})$

$= +8.77\ \text{kJ}$

(b) $\Delta H^\circ = 2\ \text{mol} \times \Delta H_f^\circ[H_2O(l)] - 2\ \text{mol} \times \Delta H_f^\circ[H_2S(g)] - 1\ \text{mol} \times \Delta H_f^\circ[SO_2(g)]$

$\Delta H^\circ = 2\ \text{mol} \times (-285.83\ \text{kJ/mol}) - 2\ \text{mol} \times (-20.63\ \text{kJ/mol}) - 1\ \text{mol} \times (-296.83\ \text{kJ/mol}) = -233.57\ \text{kJ}$

(c) $\Delta H^\circ = 4\ \text{mol} \times \Delta H_f^\circ[NO(g)] + 6\ \text{mol} \times \Delta H_f^\circ[H_2O(g)] - 4\ \text{mol} \times \Delta H_f^\circ[NH_3(g)]$

$\Delta H^\circ = 4\ \text{mol} \times (90.25\ \text{kJ/mol}) + 6\ \text{mol} \times (-241.82\ \text{kJ/mol}) - 4\ \text{mol} \times (-46.11\ \text{kJ/mol}) = -905.48\ \text{kJ}$

6.69 $\Delta H_f^\circ(NO)$ is 90.25 kJ/mol, available in Appendix 2A.

(1) $2\ NO(g) + O_2(g) \longrightarrow 2\ NO_2(g) \quad \Delta H^\circ = -114.1\ kJ$

(2) $4\ NO_2(g) + O_2(g) \longrightarrow 2\ N_2O_5(g) \quad \Delta H^\circ = -110.2\ kJ$

(3) $\frac{1}{2}\ N_2(g) + \frac{1}{2}\ O_2(g) \longrightarrow NO(g) \quad \Delta H^\circ = +90.25\ kJ$

Multiply (1) by 2, (3) by 4, and add to (2).

$$4\ NO(g) + 2\ O_2(g) \longrightarrow 4\ NO_2(g) \quad \Delta H^\circ = 2 \times (-114.1\ kJ) = -228.2\ kJ$$
$$4\ NO_2(g) + O_2(g) \longrightarrow 2\ N_2O_5(g) \quad \Delta H^\circ = -110.2\ kJ$$
$$2\ N_2(g) + 2\ O_2(g) \longrightarrow 4\ NO(g) \quad \Delta H^\circ = 4 \times (90.25\ kJ) = 361.0\ kJ$$
$$2\ N_2(g) + 5\ O_2(g) \longrightarrow 2\ N_2O_5(g) \quad \Delta H^\circ = (-228.2 - 110.2 + 361.0)\ kJ$$
$$= +22.6\ kJ$$
$$\Delta H_f^\circ = \frac{+22.6\ kJ}{2\ mol} = +11.3\ kJ/mol$$

6.71 (1) $P(s) + \frac{3}{2}\ Cl_2(g) \longrightarrow PCl_3(l) \quad \Delta H^\circ = -319.7\ kJ$

(2) $PCl_3(l) + Cl_2(g) \longrightarrow PCl_5(s) \quad \Delta H^\circ = -124\ kJ$

(1) + (2) = $P(s) + \frac{5}{2}\ Cl_2(g) \longrightarrow PCl_5(s)$

$\Delta H^\circ = (-319.7\ kJ) + (-124\ kJ) = -444\ kJ$

$\Delta H_f^\circ = -444\ kJ/mol$

SUPPLEMENTARY EXERCISES

6.73 (a) The internal energy of an open system can be increased by adding or removing matter, heating or cooling it, or by doing work on the system or letting the system do work.

(b) The internal energy of a closed system can be increased by heating or cooling it, or by doing work on the system.

(c) None of these methods can be used to increase the internal energy of an isolated system, because neither matter nor energy can be exchanged in an isolated system.

6.75 (a) ΔU, internal energy, is the sum of all kinetic and potential energies of all atoms and molecules in a sample. Internal energy is identified with the heat supplied at constant volume. ΔH, enthalpy, is identified with the heat supplied at constant pressure.

(b) ΔH equals ΔU when $P\Delta V$ work is zero; in other words, $\Delta H = \Delta U$ when there is no volume change at constant pressure.

6.77 (a) True under the condition when $P\Delta V$ work is zero

(b) Always true

(c) Always false

(d) True when $P\Delta V$ work is zero; when this is true, however, $q = 0$ as well, so $\Delta U = 0$.

(e) True, because $\Delta U = q + w$ and $q = 0$.

6.79 (a) $2.00 \text{ mol } C_6H_6 \times \left(\dfrac{15 \text{ mol } O_2}{2 \text{ mol } C_6H_6}\right) = 15 \text{ mol } O_2$

$V = \dfrac{nRT}{P}$, $T = 273$ K, $P = 1.00$ atm

$$V_{O_2} = \frac{(15 \text{ mol } O_2)(0.082\ 06 \text{ L}\cdot\text{atm/K}\cdot\text{mol})(273 \text{ K})}{1.00 \text{ atm}} = 336.\bar{0} \text{ L } O_2$$

Find the volume of gaseous products, then find the volume change, assuming that the reactant volume is due to the oxygen gas.

$$V_{CO_2} = \frac{(12 \text{ mol } CO_2)(0.082\ 06 \text{ L}\cdot\text{atm/K}\cdot\text{mol})(273 \text{ K})}{1.00 \text{ atm}} = 268.\bar{8} \text{ L } CO_2$$

$$V_{H_2O} = \frac{(6 \text{ mol } H_2O)(0.082\ 06 \text{ L}\cdot\text{atm/K}\cdot\text{mol})(273 \text{ K})}{1.00 \text{ atm}} = 134.\bar{4} \text{ L } H_2O$$

The volume change is $(269 \text{ L} + 134 \text{ L}) - 336 \text{ L} = 67.\bar{2}$ L

$$-P\Delta V \text{ work} = -1.00 \text{ atm} \times 67 \text{ L} \times \left(\frac{101.325 \text{ J}}{1 \text{ L}\cdot\text{atm}}\right) = -6.81 \times 10^3 = -6.81 \text{ kJ}$$

(b) $\Delta H^\circ_{rxn} 12 \times \Delta H^\circ_f[CO_2(g)] + 6 \times \Delta H^\circ_f[H_2O(g)] - 2 \times \Delta H^\circ_f[C_6H_6(l)] - 15 \times \Delta H^\circ_f[O_2(g)]$

$\Delta H^\circ_{rxn} = 12(-393.51 \text{ kJ/mol}) + 6(-241.82 \text{ kJ/mol}) - 2(49.0 \text{ kJ/mol}) - 15(0)$

$\Delta H^\circ_{rxn} = -4722.12 \text{ kJ/mol} + (-1450.92 \text{ kJ/mol}) - 98.0 \text{ kJ/mol})$

$= -6271.0 \text{ kJ/mol}$

(c) $\Delta H^\circ = \Delta U^\circ + P\Delta V$

$\Delta U^\circ = -6271.0 \text{ kJ/mol} + (-6.81 \text{ kJ}) = -6277.8 \text{ kJ/mol}$

6.81 For the calorimeter, heat = heat capacity × temperature rise, so $\Delta H_{cal} = C_{cal} \times \Delta T$

For the reaction, $\Delta H = -\Delta H_{cal} = -C_{cal} \times \Delta T = -8.92 \text{ kJ/(°C)} \times 2.37\text{°C} = -21.1\bar{4} \text{ kJ}$

$$\text{mass} = -21.1\bar{4} \text{ kJ} \times \left(\frac{30.\text{ g}}{-460.\text{ kJ}}\right) = 1.4 \text{ g}$$

6.83 Let t_f = final temperature

Because no heat is lost, we may write

total heat = heat(ice) + heat(water) = 0

$$\text{heat(ice)} = 50.0 \text{ g} \times \left(\frac{1 \text{ mol}}{18.02 \text{ g}}\right) \times \left(\frac{6.01 \times 10^3 \text{ J}}{1 \text{ mol}}\right) + 50.0 \text{ g} \times 4.184 \text{ J/(°C)}\cdot\text{g} \times (t_f - 0)$$

heat(water) = 400. g × 4.184 J/(°C)·g × (t_f − 45)

or $166\,\overline{76}$ J + 209 J/(°C) × t_f = −$167\overline{4}$ J/(°C) × t_f + $753\,\overline{12}$ J

Solving for t_f

$$t_f = \frac{5.85\overline{6} \times 10^4 \text{ J}}{1883 \text{ J/(°C)}} = 31.1\text{°C}$$

6.85 $Pb(s) + O_2(g) \longrightarrow PbO_2(s)$

$$\Delta H_f^\circ[PbO_2(s)] = \left(\frac{-3.76 \text{ kJ}}{3.245 \text{ g } PbO_2}\right) \times \left(\frac{239.19 \text{ g } PbO_2}{1 \text{ mol } PbO_2}\right) = -277 \text{ kJ/mol}$$

6.87 (1) $3\,Fe_2O_3(s) + CO(g) \longrightarrow 2\,Fe_3O_4 + CO_2(g)$ $\Delta H^\circ = -47.2$ kJ

(2) $Fe_2O_3(s) + 3\,CO(g) \longrightarrow 2\,Fe(s) + 3\,CO_2(g)$ $\Delta H^\circ = -24.7$ kJ

(3) $Fe_3O_4(s) + CO(g) \longrightarrow 3\,FeO(s) + CO_2(g)$ $\Delta H^\circ = +35.9$ kJ

Reverse (1), multiply (2) by 3, reverse (3) and multiply by 2, then add.

$2\,Fe_3O_4 + CO_2(g) \longrightarrow 3\,Fe_2O_3(s) + CO(g)$ $\Delta H^\circ = +47.2$ kJ

$3\,Fe_2O_3(s) + 9\,CO(g) \longrightarrow 6\,Fe(s) + 9\,CO_2(s)$ $\Delta H^\circ = 3 \times (-24.7 \text{ kJ})$
$= -74.1$ kJ

$6\,FeO(s) + 2\,CO_2(g) \longrightarrow 2\,Fe_3O_4(s) + 2\,CO(g)$ $\Delta H^\circ = -2 \times (35.9 \text{ kJ})$
$= -71.8$ kJ

$6\,FeO(s) + 6\,CO(g) \longrightarrow 6\,Fe(s) + 6\,CO_2(g)$

$\Delta H^\circ = (+47.2 - 74.1 - 71.8) \text{ kJ} = -98.7$ kJ

Dividing the reaction and ΔH° by 6 gives

$FeO(s) + CO(g) \longrightarrow Fe(s) + CO_2(s)$ $\Delta H^\circ = -16.4$ kJ

6.89 (1) $2\,C_2H_2(s) + 5\,O_2(g) \longrightarrow 4\,CO_2(g) + 2\,H_2O(l)$ $\Delta H^\circ = -2600.$ kJ

(2) $2\,C_2H_4(s) + 6\,O_2(g) \longrightarrow 4\,CO_2(g) + 4\,H_2O(l)$ $\Delta H^\circ = -2822$ kJ

(3) $2\,H_2(g) + O_2(g) \longrightarrow 2\,H_2O(l)$ $\Delta H^\circ = -572$ kJ

Multiply (1) by $\frac{1}{2}$, reverse (2) and multiply by $\frac{1}{2}$, multiply (3) by $\frac{1}{2}$, and then add.

$C_2H_2(g) + \frac{5}{2}\,O_2(g) \longrightarrow 2\,CO_2(g) + H_2O(l)$ $\Delta H^\circ = \frac{1}{2} \times (-2600. \text{ kJ})$
$= -1300.$ kJ

$2\,CO_2(g) + 2\,H_2O(l) \longrightarrow C_2H_4(g) + 3\,O_2(g)$ $\Delta H^\circ = -\frac{1}{2} \times (-2822 \text{ kJ})$
$= +1411$ kJ

$H_2(g) + \frac{1}{2}\,O_2(g) \longrightarrow H_2O(l)$ $\Delta H^\circ = \frac{1}{2} \times (-572 \text{ kJ})$
$= -286$ kJ

$C_2H_2(g) + H_2(g) \longrightarrow C_2H_4(g)$

$\Delta H^\circ = (-1300. + 1411 - 286) \text{ kJ} = -175$ kJ

6.91 The heat evolved is ΔH°_{sol}, see Chapter 12.

$$\Delta H^\circ_{sol} = 1 \text{ mol} \times \Delta H^\circ_f[\text{Na}^+(\text{aq})] + 1 \text{ mol} \times \Delta H^\circ_f[\text{OH}^-(\text{aq})] - 1 \text{ mol} \times \Delta H^\circ_f[\text{NaOH(s)}]$$

$$\Delta H^\circ_{sol} = 1 \text{ mol} \times (-240.12 \text{ kJ/mol}) + 1 \text{ mol} \times (-229.99 \text{ kJ/mol}) - 1 \text{ mol} \times (-425.61 \text{ kJ/mol}) = -44.50 \text{ kJ}$$

$$\Delta H^\circ_{rxn} = 20.0 \text{ g NaOH} \times \left(\frac{1 \text{ mol NaOH}}{40.00 \text{ g NaOH}}\right) \times \left(\frac{-44.50 \text{ kJ}}{1 \text{ mol NaOH}}\right) = -22.2\overline{5} \text{ kJ}$$

6.93 $\Delta H^\circ_c(C_8H_{18}) = -5471$ kJ/mol

$$\text{heat produced} = 0.700 \times 0.100 \text{ g } C_8H_{18} \times \left(\frac{1 \text{ mol } C_8H_{18}}{114.23 \text{ g } C_8H_{18}}\right) \times \left(\frac{-5471 \text{ kJ}}{1 \text{ mol } C_8H_{18}}\right)$$

$$= -3.35 \text{ kJ}$$

heat added = +3.35 kJ

$$\Delta T = \frac{+3.35 \times 10^3 \text{ J}}{2.42 \text{ J/g}\cdot(^\circ\text{C}) \times 250.0 \text{ g}} = 5.54^\circ\text{C}$$

APPLIED EXERCISES

6.95 specific enthalpy of cheese = 17.0 kJ/g (Table in Case Study)

$$\text{mass in grams} = 2 \text{ oz} \times \left(\frac{28.35 \text{ g}}{1 \text{ oz}}\right) = 56.7 \text{ g}$$

$$\text{energy (= enthalpy change)} = 56.7 \text{ g} \times 17.0 \text{ kJ/g} = 963.\overline{9} \text{ kJ}$$

$$\text{time spent} = \frac{963.\overline{9} \text{ kJ}}{30 \text{ kJ/min}} = 30 \text{ min}$$

6.97 (a) $\left(\frac{1250 \text{ kJ}}{\text{h}}\right) \times (1 \text{ h total time}) \times (150 \text{ days}) = 1.9 \times 10^5 \text{ kJ}$

(b) $0.40 \text{ gal} \left(\frac{3.785 \text{ L}}{\text{gal}}\right)\left(\frac{\text{mL}}{10^{-3} \text{ L}}\right)\left(\frac{0.702 \text{ g}}{\text{mL}}\right)\left(\frac{1 \text{ mol } C_8H_{18}}{114.23 \text{ g } C_8H_{18}}\right)\left(\frac{5471 \text{ kJ}}{\text{mol}}\right)(150 \text{ days}) = 7.6 \times 10^6 \text{ kJ}$

6.99 (a) Heat lost by cereal = heat gained by bomb calorimeter

$$\text{Brand X} = \frac{600.\text{ J}}{^\circ\text{C}} \times (309.0 - 300.2)\text{K} = 5.3 \times 10^3 \text{ J per 1.00 g cereal}$$

$$\text{Brand ABC} = \frac{600.\text{ J}}{^\circ\text{C}} \times (307.5 - 299.0)\text{K} = 5.1 \times 10^3 \text{ J per 1.00 g cereal}$$

(b) and (c) $\text{Brand X} = \frac{5.3 \times 10^3 \text{ J}}{1.00 \text{ g}} \times 100.\text{ g} = 5.3 \times 10^2 \text{ kJ}$

$$\text{Brand ABC} = \frac{5.1 \times 10^3\ \text{J}}{1.00\ \text{g}} \times 100.\ \text{g} = 5.1 \times 10^2\ \text{kJ}$$

(c) $\text{Brand X} = 5.3 \times 10^5\ \text{J}\left(\frac{\text{cal}}{4.184\ \text{J}}\right)\left(\frac{\text{Cal}}{1000\ \text{cal}}\right) = 127\ \text{Cal, round to } 130.\ \text{Cal}$

$$\text{Brand ABC} = 5.1 \times 10^5\ \text{J}\left(\frac{\text{cal}}{4.184\ \text{J}}\right)\left(\frac{\text{Cal}}{1000\ \text{cal}}\right) = 122\ \text{Cal, round to } 120.\ \text{Cal}$$

6.101 $\Delta H° = 28\ \text{mol} \times \Delta H_f°[CO_2(g)] + 10\ \text{mol} \times \Delta H_f°[H_2O(g)] - 4\ \text{mol} \times \Delta H_f°[C_7H_5N_3O_6(s)]$

$\Delta H° = 28\ \text{mol} \times (-393.51\ \text{kJ/mol}) + 10\ \text{mol} \times (-241.82\ \text{kJ/mol}) - 4\ \text{mol} \times (-67\ \text{kJ/mol}) = -13\ \overline{168}\ \text{kJ}$

$\text{enthalpy density} = \left(\frac{+13\ \overline{168}\ \text{kJ}}{4\ \text{mol TNT}}\right) \times \left(\frac{1\ \text{mol TNT}}{227.14\ \text{g TNT}}\right) \times \left(\frac{1.65\ \text{g TNT}}{1\ \text{cm}^3\ \text{TNT}}\right)$

$= +23.9\ \text{kJ/cm}^3 = +23.9 \times 10^3\ \text{kJ/L}$ (higher than the enthalpy density of hydrogen or methanol, but not as high as that of octane)

INTEGRATED EXERCISES

6.103 (a) $n = \frac{PV}{RT}$ $\quad n = \frac{(12.0\ \text{atm})(30.0\ \text{L})}{(0.082\ 06\ \text{L}\cdot\text{atm/K}\cdot\text{mol})(298\ \text{K})} = 14.7$ moles of H_2

(b) $H_2(g) + \frac{1}{2}O_2(g) \longrightarrow H_2O(g)$

$\Delta H° = \frac{-285.8\ \text{kJ}}{\text{mol}}(14.7\ \text{mol}) = 4.20 \times 10^3$ kJ of heat was released.

6.105 (a) $Cu(s) + 2\ AgNO_3(aq) \longrightarrow Cu(NO_3)_2(aq) + 2\ Aq(s)$

This is an oxidation-reduction reaction.

(b) $\Delta H° = 1\ \text{mol} \times \Delta H_f°[Cu^{2+}(aq)] - 2\ \text{mol} \times \Delta H_f°[Ag^+(aq)]$

$\Delta H° = 1\ \text{mol} \times (64.77\ \text{kJ/mol}) - 2\ \text{mol} \times (105.58\ \text{kJ/mol}) = -146.39\ \text{kJ}$

$\Delta H°_{rxn} = 1.88\ \text{g Ag} \times \left(\frac{1\ \text{mol Ag}}{107.87\ \text{g Ag}}\right) \times \left(\frac{-146.39\ \text{kJ}}{2\ \text{mol Ag}}\right) = -1.28\ \text{kJ}$

(c) Reaction is exothermic.

6.107 (a) (1) $CO_2(g) + 2\ H_2(g) \longrightarrow C(s) + 2\ H_2O(l)$

(2) $2\ H_2O(l) \longrightarrow 2\ H_2(g) + O_2(g)$

Add reactions (1) and (2) to get

$CO_2(g) \longrightarrow C(s) + O_2(g)$

(b) Determine the moles of gas produced. $n = \frac{PV}{RT}$

$$n = \frac{(0.80\ \text{atm})(45.0\ \text{L})}{(0.082\ 06\ \text{L}\cdot\text{atm/K}\cdot\text{mol})(300.\ \text{K})} = 1.4\bar{6}\ \text{mol}$$

$\Delta H° = +393.51$ kJ/mol. The production of $1.4\overline{6}$ mol of gas requires
$1.4\overline{6}$ mol (393.51 kJ/mol) $= 5.7 \times 10^2$ kJ

(c) The reaction is endothermic, so the reactor will need to be heated to maintain constant temperature.

(d) CO_2

CHAPTER 7
INSIDE THE ATOM

EXERCISES

The Characteristics of Light

7.1 (a) $400\text{ nm} \times \left(\dfrac{10^{-9}\text{ m}}{1\text{ nm}}\right) = 4 \times 10^{-7}\text{ m}$ and $\nu = \dfrac{c}{\lambda} = \dfrac{3.00 \times 10^{8}\text{ m/s}}{4 \times 10^{-7}\text{ m}}$

$$= 7 \times 10^{14}/\text{s}$$

Because $1\text{ s}^{-1} = 1\text{ Hz}$, the frequency is 7×10^{14} Hz.

$700\text{ nm} \times \left(\dfrac{10^{-9}\text{ m}}{1\text{ nm}}\right) = 7 \times 10^{-7}\text{ m}$ and $\nu = \dfrac{c}{\lambda} = \dfrac{3.00 \times 10^{8}\text{ m/s}}{7 \times 10^{-7}\text{ m}}$

$$= 4 \times 10^{14}/\text{s}$$

Because $1\text{ s}^{-1} = 1\text{ Hz}$, the frequency is 4×10^{14} Hz. Therefore, the wavelength range is 4×10^{14} Hz to 7×10^{14} Hz.

(b) $\nu = \dfrac{c}{\lambda} = \dfrac{3.00 \times 10^{8}\text{ m/s}}{250\text{ m}} = 1.2 \times 10^{6}/\text{s} = 1.2 \times 10^{6}\text{ Hz}$

7.3 (a) $\lambda = \dfrac{c}{\nu} = \left(\dfrac{3.00 \times 10^{8}\text{ m/s}}{7.1 \times 10^{14}/\text{s}}\right) \times \left(\dfrac{1\text{ nm}}{10^{-9}\text{ m}}\right) = 420\text{ nm}$

(b) $\lambda = \dfrac{c}{\nu} = \left(\dfrac{3.00 \times 10^{8}\text{ m/s}}{2.0 \times 10^{18}/\text{s}}\right) \times \left(\dfrac{1\text{ pm}}{10^{-12}\text{ m}}\right) = 150\text{ pm}$

Quanta and Photons

7.5 (a) $589\text{ nm} \times \left(\dfrac{10^{-9}\text{ m}}{1\text{ nm}}\right) = 5.89 \times 10^{-7}\text{ m}$

$$E = \frac{hc}{\lambda} = \frac{(6.63 \times 10^{-34}\text{ J}\cdot\text{s}) \times (3.00 \times 10^{8}\text{ m/s})}{5.89 \times 10^{-7}\text{ m}} = 3.38 \times 10^{-19}\text{ J}$$

(b) $3.38 \times 10^{-19}\text{ J/photon} \times \left(\dfrac{1\text{ photon}}{1\text{ Na atom}}\right)$

$$\times \left(\frac{6.022 \times 10^{23}\text{ Na atoms}}{1.00\text{ mol Na atoms}}\right) = 2.03 \times 10^{5}\text{ J}$$

7.7 (a) $86\text{ pm} \times \left(\dfrac{10^{-12}\text{ m}}{1\text{ pm}}\right) = 8.6 \times 10^{-11}\text{ m}$

$$E = \frac{hc}{\lambda} = \frac{(6.63 \times 10^{-34}\ \text{J}\cdot\text{s}) \times (3.00 \times 10^{8}\ \text{m/s})}{8.6 \times 10^{-11}\ \text{m}} = 2.3 \times 10^{-15}\ \text{J}$$

(b) $470.\ \text{nm} \times \left(\frac{10^{-9}\ \text{m}}{1\ \text{nm}}\right) = 4.70 \times 10^{-7}\ \text{m}$

$$E = \frac{hc}{\lambda} = \frac{(6.63 \times 10^{-34}\ \text{J}\cdot\text{s}) \times (3.00 \times 10^{8}\ \text{m/s})}{4.70 \times 10^{-7}\ \text{m}} = 4.23 \times 10^{-19}\ \text{J}$$

and $4.23 \times 10^{-19}\ \text{J/photon} \times \left(\frac{6.022 \times 10^{23}\ \text{photons}}{1.0\ \text{mol photons}}\right) = 2.5 \times 10^{5}\ \text{J}$

7.9 (a) Electrons are most easily ejected from rubidium, because the lowest frequency light is required.
(b) The kinetic energy drops to zero at different frequencies because each metal requires that light of some threshold frequency or higher be present for an electron to be ejected. The frequency at which electrons can be ejected is characteristic of the metal; different metals have different frequencies.
(c) The fastest electrons are produced by rubidium, because the kinetic energy of the electron emitted is the difference between energy of radiation and the ionization energy; rubidium also has the lowest ionization energy.

Atomic Spectra

7.11 (a) Let $n_1 = 2$, $n_u = 6$, then $E(\text{photon}) = -\Delta E(\text{atom}) = h\mathcal{R} \times \left(\frac{1}{{n_1}^2} - \frac{1}{{n_u}^2}\right)$

and because $E(\text{photon}) = h\nu$

$$\nu = \mathcal{R} \times \left[\frac{1}{{n_1}^2} - \frac{1}{{n_u}^2}\right] = 3.29 \times 10^{15}\ \text{Hz} \times \left(\frac{1}{2^2} - \frac{1}{6^2}\right) = 3.29 \times 10^{15}\ \text{Hz}\left(\frac{2}{9}\right)$$

$\nu = 7.31 \times 10^{14}\ \text{Hz}$

and $\lambda = \frac{c}{\nu} = \frac{3.00 \times 10^{8}\ \text{m/s}}{7.31 \times 10^{14}\ \text{Hz}} = 4.10 \times 10^{-7}\ \text{m (or 410 nm)}$

(b) violet

7.13 (a) See the solution to Exercise 7.11(a).

$\nu = \mathcal{R} \times \left(\frac{1}{{n_1}^2} - \frac{1}{{n_u}^2}\right)$ The highest frequency photon corresponds to the largest value of $\left(\frac{1}{{n_1}^2} - \frac{1}{{n_u}^2}\right)$. This frequency occurs when $n_1 = 1$ and $n_u = \infty$.

$$\nu = \mathcal{R} \times \left(\frac{1}{1^2} - \frac{1}{\infty^2}\right) = \mathcal{R} = 3.29 \times 10^{15}\ \text{Hz}$$

(b) $\lambda = \frac{c}{\nu} = \frac{3.00 \times 10^8 \text{ m/s}}{3.29 \times 10^{15}\text{/s}} = 9.12 \times 10^{-8}$ m; x-ray or gamma ray

Particles and Waves

7.15 (a) electron mass $= 9.11 \times 10^{-31}$ kg

$h = 6.63 \times 10^{-34} \text{ J·s} \equiv 6.63 \times 10^{-34} \text{ kg·m}^2\text{/s}$

$$\lambda = \frac{h}{mv} = \frac{6.63 \times 10^{-34} \text{ kg·m}^2\text{/s}}{(9.11 \times 10^{-31} \text{ kg}) \times (1.5 \times 10^7 \text{ m·s})} = 4.8 \times 10^{-11} \text{ m}$$

(b) neutron mass $= 1.67 \times 10^{-27}$ kg

$h = 6.63 \times 10^{-34} \text{ kg·m}^2\text{/s}$

$$\lambda = \frac{h}{mv} = \frac{6.63 \times 10^{-34} \text{ kg·m}^2\text{/s}}{(1.67 \times 10^{-27} \text{ kg}) \times (1.5 \times 10^7 \text{ m/s})} = 2.6 \times 10^{-14} \text{ m}$$

7.17 $\lambda = \frac{h}{mv}$. If v = velocity, and each person runs at the same speed, $\lambda \propto \frac{1}{m}$, and the larger the mass, the shorter the wavelength. Therefore, the person weighing 80 kg has the shorter wavelength.

Atomic Orbitals and Quantum Numbers

7.19 A hard sphere, such as a billiard ball, has a uniform density and its mass is confined within a fixed and well-defined region of space. In atoms, the electron density spreads out from the central nucleus in a nonuniform manner, producing an electron cloud that is neither hard nor confined to a limited region. The electron density varies with distance in the manner shown in Figure 7.17.

7.21 Subshells are characterized by the allowed l values, which range from 0 to $l_{max} = n - 1$. The number of subshells is simply given by n.

(a) $n = 2$; therefore, $l_{max} = 1$ and subshells with $l = 0$ and 1; so two subshells

(b) $n = 3$; therefore, $l_{max} = 2$ and subshells with $l = 0, 1, 2$; so three subshells

(c) $n = 3$; therefore, l can be 0, 1, and 2

7.23 In each case, determine $2l + 1$.

(a) $l = 0$, $2l + 1 = 2 \times 0 + 1 = 1$

(b) $l = 2$, $2l + 1 = 2 \times 2 + 1 = 5$

(c) $l = 1$, $2l + 1 = 2 \times 1 + 1 = 3$

(d) $l = 3$, $2l + 1 = 2 \times 3 + 1 = 7$

7.25 (a) number of subshells $= n = 2$; $l = 0$ and $l = 1$; so two subshells
(b) number of orbitals $= 2l + 1 = 2 \times 2 + 1 = 5$
(c) number of orbitals $= n^2 = 2^2 = 4$; one *s*- and three *p*-orbitals
(d) for $3d$, $l = 2$; number of orbitals $= 2l + 1 = 2 \times 2 + 1 = 5$

7.27 In each case, calculate $2l + 1 =$ number of orbitals.
(a) $3d$, 5 (b) $1s$, 1 (c) $6f$, 7 (d) $2p$, 3

7.29 In each case, determine the number of orbitals and multiply by 2.
(a) $2 \times (2l + 1) = 2 \times (2 + 1) = 6$ electrons
(b) $2 \times 1 = 2$ electrons
(c) $2 \times n^2 = 2 \times 2^2 = 8$ electrons
(d) $2 \times 1 = 2$ electrons

7.31 (a) $n = 2$, $l = 0, 1$; *s* and *p* only, so $2d$ not allowed
(b) $n = 4$, $l = 0, 1, 2, 3$; *s*, *p*, *d*, *f* allowed, so $4d$ allowed
(c) $n = 4$, $l = 0, 1, 2, 3$; *s*, *p*, *d*, *f* only, so $4g$ not allowed
(d) $n = 6$, $l = 0, 1, 2, 3, 4, 5$; *s*, *p*, *d*, *f*, *g*, *h* allowed, so $6f$ allowed

7.33 (a) allowed
(b) not allowed; if $l = 0$, then m_l can only be 0. n and l cannot be equal:
$l = 0 \ldots, n - 1$
(c) not allowed; n can't equal l

7.35 (a) $n = 2$, $l = 1$, $m_l = 0$, $m_s = -\frac{1}{2}$
(b) $n = 5$, $l = 2$, $m_l = +1$, $m_s = +\frac{1}{2}$

Orbital Energies

7.37 The average distance from the nucleus for a 2*s*-electron is much less than that for a 3*s*-electron; consequently, the electrostatic force of attraction between the nucleus and the 2*s*-electron is much greater than that for a 3*s*-electron. The electron cloud of the 2*s*-electron is bunched more densely around the nucleus than is the 3*s*-electron cloud.

7.39 (a) $n = 1, 2, 3, 4$; energy increases from 1 to 4
(b) $l = 0, 1, 2, 3$; energy increases from 1 to 3

Electron Configurations and the Building-Up Principle

7.41 $1s$, $2p$, $3s$, $3d$, $5d$; energy increases left to right

7.43 (a) Ca: $[Ar]4s^2$
(b) N: $1s^22s^22p^3$
(c) Br: $[Ar]3d^{10}4s^24p^5$
(d) U: $[Rn]5f^36d^17s^2$
For U, the *Aufbau* rules would give $[Rn]5f^47s^2$. Uranium is an exception.

7.45 (a) Ni: $[Ar]3d^84s^2$
(b) Cd: $[Kr]4d^{10}5s^2$
(c) Pb: $[Xe]4f^{14}5d^{10}6s^26p^2$
(d) Ag: $[Kr]4d^{10}5s^1$

7.47 (a) Fe^{2+}: $[Ar]3d^6$
(b) Cl^-: $[Ne]3s^23p^6$
(c) Tl^+: $[Xe]4f^{14}5d^{10}6s^2$

7.49 (a) Si: $[Ne]3s^23p^2$
(b) Ne: $[He]2s^22p^6$
(c) Cs: $[Xe]6s^1$
(d) S: $[Ne]3s^23p^4$

7.51 (a) Group: ns^2
(b) Group 18: ns^2np^6

Atomic and Ionic Radius

7.53 The outermost electrons of an atom determine the atomic radius. Proceeding down a group, the outermost electrons occupy shells that lie farther and farther from the nucleus, so the size (radius) increases down a group.

7.55 (a) Group 1 (alkali metals)
(b) The radius of the cation is less than the radius of the neutral atom.
(c) Among a set of cations with the same number of electrons, cations with the largest number of protons will have the smallest radius.

7.57 (a) S (size decreases from left to right)
(b) S^{2-} (same number of electrons; Cl has more protons, so it is smaller)
(c) Na (size decreases from left to right)
(d) Mg^{2+} (same number of electrons; Al has more protons, so it is smaller)

Ionization Energy

7.59 Ionization energies decrease down a group because the outermost electron occupies a shell that is farther from the nucleus and is therefore less tightly bound. Ionization energies increase across a period because the effective nuclear charge increases from left to right across a given period. As a result, the outermost electron is gripped more tightly and the ionization energies increase.

7.61 (a) Mg; consistent with the trend
(b) N; consistent with the trend
(c) P; inconsistent with the trend; S might have been predicted.

7.63 The ionizations energies for Group 16 elements are less than those for Group 15 elements. The group configurations are Group 16, ns^2np^4, and Group 15, ns^2np^3. The half-filled subshell of Group 15 is more stable than simple theory suggests. This stability makes removal of the electron more difficult; therefore, it has a higher ionization energy. In Group 16, the fourth *p*-electron pairs with another electron, producing stronger electron repulsion, which makes it easier to remove this electron. Although there is the competing effect of increasing effective nuclear charge, the electron repulsion effect predominates.

7.65 Both the first and second electrons lost from Mg are 3*s*, Mg: [Ne]$3s^2$. The second electron for sodium, Na: [Ne]$3s^1$ must be removed from a 2*p* level that not only is filled but also is considerably lower in energy than the 3*s* level. The second ionization energy of sodium is very high; sodium exists only as Na^+, not Na^{2+}.

Electron Affinity

7.67 (a) Group 17 (halogens) (b) Electron affinities generally increase (left to right) across a period. However, anomalies are common.

7.69 (a) Cl (b) O (c) Cl

7.71 Atomic radii increase and ionization energies decrease from top to bottom within a group. Atomic radii decrease and ionization energies increase (left to right) across a period. In both cases, there is an inverse correlation.

Trends in Chemical Properties

7.73 A diagonal relationship is a similarity between diagonal neighbors in the periodic table, especially for elements at the left of the table. Two examples are (1) Li and Mg, both of which burn in nitrogen to form the nitride; (2) Be and Al, which are both amphoteric, reacting with both acids and bases. Other examples are the metalloids, which fall in a diagonal band across the periodic table.

7.75 (a) The reactivity of metals decreases, going from left to right across a period.
(b) The reactivity of metals increases, going from top to bottom of a main group.

7.77 (a) aluminum, metal (b) carbon, nonmetal (c) germanium, metalloid
(d) arsenic, metalloid

7.79 See Fig. 7.49.
(a) do (b) do (c) do, but not strongly (d) do, but not strongly

SUPPLEMENTARY EXERCISES

7.81 $\lambda = (3 \text{ cm}) \times \left(\frac{10^{-2} \text{ m}}{1 \text{ cm}}\right) = 3 \times 10^{-2} \text{ m}$

$$\nu = \frac{c}{\lambda} = \frac{3.00 \times 10^{8} \text{ m/s}}{3 \times 10^{-2} \text{ m}} = 1 \times 10^{10}\text{/s} = 1 \times 10^{10} \text{ Hz}$$

7.83 mass of one H_2 molecule $= \left(\frac{2.016 \text{ g } H_2}{1 \text{ mol } H_2}\right) \times \left(\frac{1 \text{ mol } H_2}{6.022 \times 10^{23} \text{ molecules}}\right)$

$$\times \left(\frac{1 \text{ kg}}{10^{3} \text{ g}}\right) = 3.348 \times 10^{-27} \text{ kg per } H_2 \text{ molecule}$$

and $h = 6.626 \times 10^{-34} \text{ J}\cdot\text{s} \equiv 6.626 \times 10^{-34} \text{ kg}\cdot\text{m}^2\text{/s}$

$$\lambda = \frac{h}{mv} = \frac{6.626 \times 10^{-34} \text{ kg}\cdot\text{m}^2\text{/s}}{(3.348 \times 10^{-27} \text{ kg per } H_2 \text{ molecule})(1930. \text{ m/s})} = 1.025 \times 10^{-10} \text{ m}$$

7.85 (a) Because $l = 5$ for an h-subshell, $2l + 1 = 2 \times 5 + 1 = 11$ orbitals
(b) Because $l = 3$ for a $5f$-subshell, $2l + 1 = 2 \times 3 + 1 = 7$ orbitals

(c) number of orbitals = $2l + 1 = 2 \times 2 + 1 = 5$ orbitals

maximum number of electrons = 2 × number of orbitals = $2 \times 5 = 10$ electrons

(d) Sb: [Kr]$4d^{10}5s^2 5p^3$; there are three $5p$-electrons.

7.87 (b) is a p_y orbital.

7.89 (a) Negative values of l are not acceptable.

(b) l values greater than $n - 1$ are not acceptable.

(c) and (d) are acceptable.

7.91 (a) Group 15 (b) Group 12 (c) Group 1 (d) Group 18 (e) Group 8 (f) Group 2

7.93 (a) Zr: [Kr]$4d^2 5s^2$

(b) Se: [Ar]$3d^{10}4s^2 4p^4$

(c) Rb: [Kr]$5s^1$

(d) Cl: [Ne]$3s^2 3p^5$

(e) Sb: [Kr]$4d^{10}5s^2 5p^3$

(f) Pu: [Rn]$5f^6 7s^2$

(g) Si: [Ne]$3s^2 3p^2$

(h) Ar: [Ne]$3s^2 3p^6$

7.95 (a) $n = 3; l = 1; m_l = +1; m_s = +\frac{1}{2}$

(b) $n = 3; l = 0; m_l = 0; m_s = -\frac{1}{2}$

7.97 (a) S (b) F^- (c) Cs (d) F (e) Ca (f) Fe (g) Ba^{2+} (h) S^{2-} (i) N (j) Cl (k) I (l) K (m) 4g (n) 4 (o) 7

7.99 (a) Ag: [Kr]$4d^{10}5s^1$

Cu: [Ar]$3d^{10}4s^1$

Au: [Xe]$4f^{14}5d^{10}6s^1$

Silver, copper, and gold each have one electron in a high s-orbital. This electron can be removed easily, making these metals good conductors.

(b) Group I metals are very reactive, especially with water (see Chapter 19), forming salts that do not conduct as solids.

7.101 (a) d-block (b) s-block (c) p-block (d) d-block (e) p-block (f) d-block

7.103 (a) Zn, Cd, and Hg lose two outermost, highest energy *s*-electrons to form Zn^{2+}, Cd^{2+}, and Hg^{2+} as do the alkaline earth metals, such as Mg, which also lose two outermost, highest energy *s*-electrons to form Mg^{2+}. For Zn, Cd, and Hg, the next highest level, *d*, is completely filled, and a significant energy difference exists between the highest energy *s*- and *d*-electrons.

(b) Differences between Group IIA and IIB may be explained by the effect of the filled *d*-orbitals in IIB elements. As we go across a transition row, the *d*-orbitals fill, and the elements become more similar to the *p*-block elements. Group IIB is at the end of the row, closest to the *p*-block.

IIA metals are more reactive than IIB metals because their ionization energies are lower; their outermost electrons can be lost more easily and so they are more reactive.

7.105

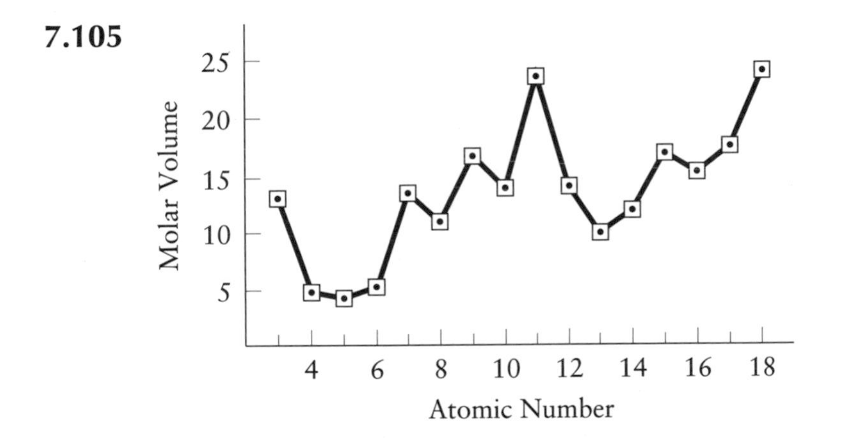

The molar volume roughly parallels atomic size (volume) which increases as the *s*-sublevel begins to fill and subsequently decreases as the *p*-sublevel fills (refer to textbook discussion of periodic variation of atomic radii). In the above plot, this effect is most clearly seen in passing from Ne (10) to Na (11) and Mg (12), then to Al (13) and Si (14). Ne has a filled 2*p*-sublevel; the 3*s*-sublevel fills with Na and Mg; and the 3*p*-sublevel begins to fill with Al.

7.107 (a) The energy of the incident radiation is $\Delta E = h\nu$ and the energy of the emitted electron is in the form $E = \frac{1}{2} mv^2$. The energy difference between the two is the energy required to ionize the electron.

(b) $h\nu = I + \frac{1}{2} m_e v^2$

$$\frac{(6.63 \times 10^{-34}\ \text{J·s}) \times (3.00 \times 10^{8}\ \text{m/s})}{5.84 \times 10^{-8}\ \text{m}}$$

$$= I + \tfrac{1}{2}(9.11 \times 10^{-31}\ \text{kg}) \times (2.450 \times 10^{6}\ \text{m/s})^2$$

$I = 6.72 \times 10^{-19}$ J/atom (404 kJ/mol)

APPLIED EXERCISES

7.109 $\Delta E = -\frac{hc}{\lambda} = -(6.63 \times 10^{-34}\ \text{J}\cdot\text{s}) \times \left(\frac{3.00 \times 10^{8}\ \text{m/s}}{\lambda}\right)$

$= -\frac{1.99 \times 10^{-25}\ \text{J}\cdot\text{m}}{\lambda}$

Substitute each λ in turn into this expression to get the energy per atom, then multiply each result by $N_A = 6.022 \times 10^{23}$ atoms/mol to get the energy per mole.

$$-\frac{1.99 \times 10^{-25}\ \text{J}\cdot\text{m}}{4.87 \times 10^{-7}\ \text{m}} = -4.09 \times 10^{-19}\ \text{J/atom} = -2.46 \times 10^{5}\ \text{J/mol}$$

$$-\frac{1.99 \times 10^{-25}\ \text{J}\cdot\text{m}}{5.24 \times 10^{-7}\ \text{m}} = -3.80 \times 10^{-19}\ \text{J/atom} = -2.29 \times 10^{5}\ \text{J/mol}$$

$$-\frac{1.99 \times 10^{-25}\ \text{J}\cdot\text{m}}{5.43 \times 10^{-7}\ \text{m}} = -3.66 \times 10^{-19}\ \text{J/atom} = -2.20 \times 10^{5}\ \text{J/mol}$$

$$-\frac{1.99 \times 10^{-25}\ \text{J}\cdot\text{m}}{5.53 \times 10^{-7}\ \text{m}} = -3.60 \times 10^{-19}\ \text{J/atom} = -2.17 \times 10^{5}\ \text{J/mol}$$

$$-\frac{1.99 \times 10^{-25}\ \text{J}\cdot\text{m}}{5.78 \times 10^{-7}\ \text{m}} = -3.44 \times 10^{-19}\ \text{J/atom} = -2.07 \times 10^{5}\ \text{J/mol}$$

7.111 (a) From exercise 7.86, we know $\lambda = \frac{c}{\mathcal{R}\left(\frac{1}{n_1^2} - \frac{1}{n_2^2}\right)}$

$$\lambda = \frac{3.00 \times 10^{8}\ \text{m/s}}{3.29 \times 10^{15}/\text{s}\left(\frac{1}{2^2} - \frac{1}{100^2}\right)} = 3.65 \times 10^{-7}\ \text{m or 365 nm}$$

(b) Balmer series

(c) They would be longer, because

$$\lambda = \frac{3.00 \times 10^{8}\ \text{m/s}}{3.29 \times 10^{15}/\text{s}\left(\frac{1}{100^2} - \frac{1}{110^2}\right)} = 5.25 \times 10^{-3}\ \text{m}$$

7.113 (a) Cesium is desirable because cesium has a low ionization energy, so harmless infrared radiation can be used.

(b) Cesium reacts very vigorously (explosively) with water and moist air.

INTEGRATED EXERCISES

7.115 The elements that are gases at 298 K and 1 atm are
H_2, He, N_2, O_2, F_2, Ne, Cl_2, Ar, Kr, Xe, Rn

Because density = molar mass/molar volume, density $\propto$ molar volume so $H_2 < He < Ne < N_2 < O_2 < F_2 < Ar < Cl_2 < Kr < Xe < Rn$

7.117 $Ba(s) + 2\ H_2O(l) \longrightarrow Ba(OH)_2(aq) + H_2(g)$
$Ca(s) + 2\ H_2O(l) \longrightarrow Ca(OH)_2(aq) + H_2(g)$

7.119 Given the following equations:

(1) $K(s) \longrightarrow K(g)$
(2) $K(g) \longrightarrow K^+(g) + e^-$
(3) $Br_2(l) \longrightarrow Br_2(g)$
(4) $Br_2(g) \longrightarrow 2\ Br(g)$
(5) $Br(g) + e^- \longrightarrow Br^-(g)$
(6) $K(s) + \frac{1}{2}\ Br_2(l) \longrightarrow KBr(s)$

Multiple (3) and (4) by $\frac{1}{2}$, reverse (6), then add reactions (1), (2), (3), (4), (5), and (6) together.

(1) $K(s) \longrightarrow K(g)$	$+89.2$ kJ
(2) $K(g) \longrightarrow K^+(g) + e^-$	$+425.0$ kJ
(3) $\frac{1}{2}\ Br_2(l) \longrightarrow \frac{1}{2}\ Br_2(g)$	$\frac{1}{2}(+30.9\ kJ) = +15.4\overline{5}\ kJ$
(4) $\frac{1}{2}\ Br_2(g) \longrightarrow Br(g)$	$\frac{1}{2}(+192.9\ kJ) = +96.45\ kJ$
(5) $Br(g) + e^- \longrightarrow Br^-(g)$	-331.0 kJ
(6) $KBr(s) \longrightarrow K(s) + \frac{1}{2}\ Br_2(l)$	$+393.80$ kJ/mol
$KBr(s) \longrightarrow K^+(g) + Br^-(g)$	$\Delta H° = (+89.2 + 425.0 + 15.4\overline{5} + 96.45 - 331.0 + 393.80)kJ = 688.9\ kJ$

CHAPTER 8
CHEMICAL BONDS

EXERCISES

Ionic Bonds

8.1 Refer to Section 1.10 and Example 1.4 for the rules of assigning charges to monatomic ions formed from main-group elements.
(a) $+1$ (b) -2 (c) $+2$ (d) $+3$

8.3 Na^+ has the configuration $1s^2 2s^2 2p^6$, which is a very stable electronic configuration, identical to that of Ne, which has an octet of electrons in the $n = 2$ shell. These electrons are all core electrons, with high ionization energies. To lose one, as required to form Na^{2+}, would take a great deal of energy, because these core electrons are so tightly held. Detailed consideration of the energetics of all the processes resulting in the formation of ionic compounds of Na^{2+} shows that the existence of such compounds is energetically unfavorable. See the discussion of the Born-Haber cycle in Section 8.3. The ΔH_f° of ionic compounds of Na^{2+} would be predicted to be positive and is not consistent with the existence of stable ionic compounds.

8.5 (a) $\mathrm{Ca}:$ (b) $\cdot\ddot{\underset{\cdot\cdot}{\mathrm{S}}}\cdot$ (c) $[:\ddot{\underset{\cdot\cdot}{\mathrm{O}}}:]^{2-}$ (d) $[:\ddot{\underset{\cdot\cdot}{\mathrm{N}}}:]^{3-}$

8.7 (a) $\mathrm{K}^+[:\ddot{\underset{\cdot\cdot}{\mathrm{F}}}:]^-$

(b) $[:\ddot{\underset{\cdot\cdot}{\mathrm{S}}}:]^{2-}\ \mathrm{Al}^{3+}\ [:\ddot{\underset{\cdot\cdot}{\mathrm{S}}}:]^{2-}\ \mathrm{Al}^{3+}\ [:\ddot{\underset{\cdot\cdot}{\mathrm{S}}}:]^{2-}$

(c) $\mathrm{Ca}^{2+}\ [:\ddot{\underset{\cdot\cdot}{\mathrm{N}}}:]^{3-}\ \mathrm{Ca}^{2+}\ [:\ddot{\underset{\cdot\cdot}{\mathrm{N}}}:]^{3-}\ \mathrm{Ca}^{2+}$

8.9 The inert pair effect, or the tendency to form cations two units lower in charge than expected from the group number, can be observed in indium. Indium forms both In^{3+} and In^+.

Lattice Enthalpies

8.11 This difference is a result of the difference in ionic radii between Mg^{2+} and Ba^{2+}. See Fig. 7.36. The radius of Mg^{2+} (72 pm) is smaller than the radius of Ba^{2+} (136 pm), so the distance between Mg^{2+} and O^{2-} ions is less than that between Ba^{2+} and O^{2-} ions in the crystal lattice. Thus, the lattice enthalpy of MgO exceeds that of BaO. In MgO, O^{2-} is so much bigger than Mg^{2+} that O^{2-} almost touches other O^{2-} anions. This is not the case in BaO. In other words, O^{2-}—O^{2-} repulsions are greater in MgO than in BaO because of the shorter distance between them.

8.13 The lattice enthalpy corresponds to the $\Delta H°$ of the process

$AgF(s) \longrightarrow Ag^{+}(g) + F^{-}(g)$

This process can be broken down into the following steps:

	$\Delta H°$, kJ/mol
(1) $Ag(s) \longrightarrow Ag(g)$	+284
(2) $Ag(g) \longrightarrow Ag^{+}(g) + e^{-}$	+731
(3) $\frac{1}{2}F_2(g) \longrightarrow F(g)$	+179
(4) $F(g) + e^{-} \longrightarrow F^{-}(g)$	−(+328) [Figure 7.43]
(5) $AgF(s) \longrightarrow Ag(s) + \frac{1}{2}F_2(g)$	−(−205)
$AgF(s) \longrightarrow Ag^{+}(g) + F^{-}(g)$	+1071 = $\Delta H_L°$ kJ/mol

Note that the enthalpy change of step 4 is the negative of the electron affinity of Fig. 7.43 and that step 5 is the reverse of the formation of AgF(s) from the elements.

8.15 (a) endothermic; all ionization energies of the elements correspond to positive ΔH values. See Section 7.17.

(b) exothermic

(c) endothermic; sublimation always corresponds to a positive ΔH value.

(d) exothermic; the ΔH for this process is negative, because it is the negative of the lattice enthalpy, which is always positive.

(e) exothermic; Appendix 2A shows that the enthalpy of formation of CuO(s) is −157.3 kJ/mol.

Lewis Structures

8.17 (a) $\mathrm{H{-}\ddot{\underset{..}{F}}{:}}$ (b) $\mathrm{{:}\ddot{\underset{..}{Cl}}{-}\ddot{\underset{..}{O}}{-}\ddot{\underset{..}{Cl}}{:}}$ (c) $\mathrm{CF_4}$: central C bonded to four $\mathrm{{:}\ddot{\underset{..}{F}}{:}}$ atoms

8.19 (a) $\left[\mathrm{NH_4}\right]^+$: central N bonded to four H (b) $\left[\mathrm{{:}\ddot{\underset{..}{Cl}}{-}\ddot{\underset{..}{O}}{:}}\right]^-$ (c) $\left[\mathrm{BF_4}\right]^-$: central B bonded to four $\mathrm{{:}\ddot{\underset{..}{F}}{:}}$ atoms

8.21 (a) $\left[\mathrm{NH_4}\right]^+$ (central N bonded to four H) $\left[\mathrm{{:}\ddot{\underset{..}{Cl}}{:}}\right]^-$ (b) $\mathrm{K^+}$ $\mathrm{K^+}$ $\mathrm{K^+}$ $\left[\mathrm{PO_4}\right]^{3-}$: central P bonded to four $\mathrm{{:}\ddot{\underset{..}{O}}{:}}$ atoms (c) $\mathrm{Na^+}$ $\left[\mathrm{{:}\ddot{\underset{..}{Cl}}{-}\ddot{\underset{..}{O}}{:}}\right]^-$

8.23 (a) X is silicon, Y is sulfur
(b) X is carbon, Y is boron

8.25 (a) $\mathrm{H{-}C({=}\ddot{O}{:}){-}H}$ (b) $\mathrm{H{-}CH_2{-}\ddot{\underset{..}{O}}{-}H}$ (central C bonded to H above and below)

Resonance

8.27 An example is the benzene molecule, which can be represented as a resonance hybrid of two principal Lewis structures, although other structures contribute slightly.

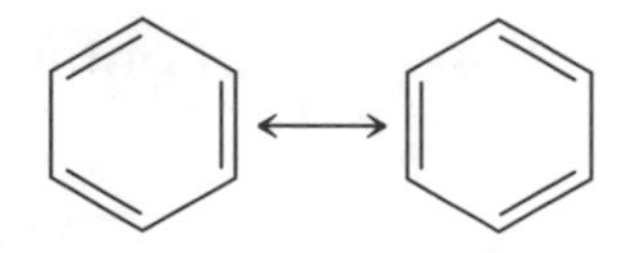

Resonance is a blending of Lewis structures, none of which by themselves are capable of adequately describing the properties of the molecule. Two consequences of resonance are that (1) the blended structure, or resonance hybrid, has a significantly lower energy than any of the individual structures. Thus, resonance plays a significant role in the chemical properties of the compound. The compound is less reactive than would have been predicted for a molecule existing in the form of just one of the structures. (2) Bond lengths are significantly different from what would be expected without resonance. In benzene, the C—C bond distance is intermediate between that expected for a C—C (single) bond and a C=C (double) bond, but the distance is not merely an average of the two.

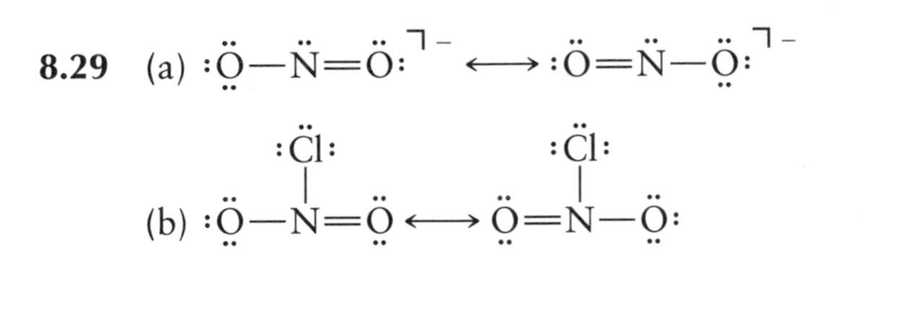

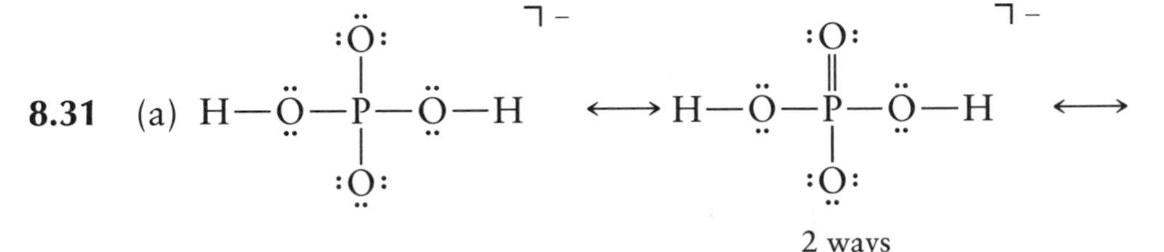

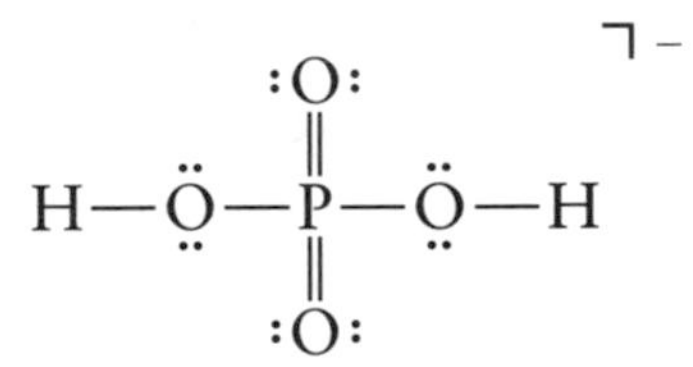

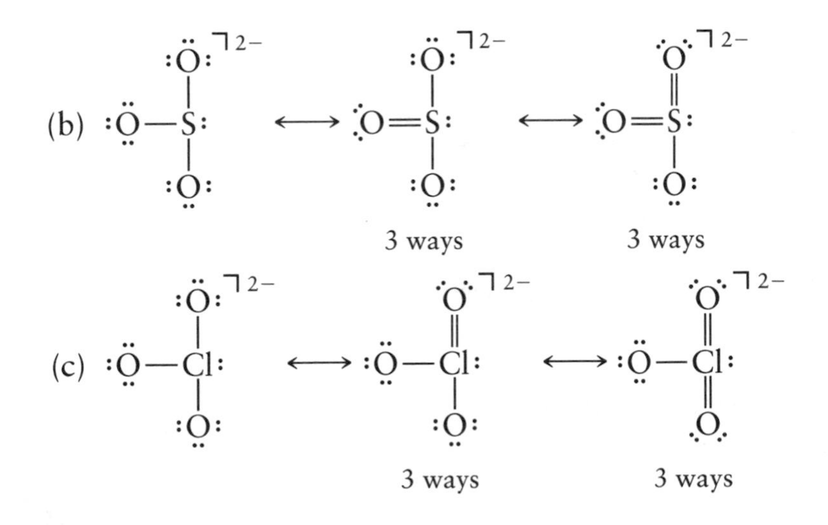

Formal Charge

8.33 Formal charge is a good "bookkeeping" system for electrons and is based on the assumption that the molecule is "perfectly covalent." However, as with any model, we must be careful not to follow it too rigorously. Molecules do not have such a clearly defined distribution of charge. Formal charge is determined by dividing the electrons in a bond equally between the atoms connected by the bond. The concept is used primarily for deciding between otherwise equally plausible Lewis structures.

8.35 We use the general expression FC = $V - (L + \frac{1}{2}S)$. See Section 8.9 for the meaning of these symbols.

(a) $H_2\ddot{N}-\ddot{N}H_2$ (each N bonded to two H)

$FC(N) = 5 - 2 - \frac{1}{2}(6) = 0$

$FC(H) = 1 - 0 - \frac{1}{2}(2) = 0$

(b) $[:\ddot{N}=N=\ddot{N}:]^-$

$FC(N), \text{ends} = 5 - 4 - \frac{1}{2}(4) = -1$

$FC(N), \text{center} = 5 - 0 - \frac{1}{2}(8) = +1$

8.37 In each case, we use the formula FC = $V - (L + \frac{1}{2}S)$. The best structure is the one with formal charges closest to 0.

(a)

	First structure	Second structure
	$-\ddot{S}=$	$=\underset{..}{S}=$
FC:	+1	0
	$:\ddot{O}=$	$:\ddot{O}=$
FC:	0	0
	$-\ddot{\underset{..}{O}}:$	$=\ddot{O}:$
FC:	−1	0

The second structure has formal charges closest to 0; it is the more plausible structure.

(b)

	First structure	Second structure
	$=S=$ (with $\Vert$ above)	$=S=$ (with $\vert$ above)
FC:	0	+1
	$:\ddot{O}=$	$:\ddot{O}=$
FC:	0	0
	$\ddot{O}$ (with $\Vert$ below)	$:\ddot{O}:$ (with $\vert$ below)
FC:	0	−1

The first structure is more plausible.

8.39 The structures differ only about the S—O bond. Therefore, look at the formal charge for these two atoms.

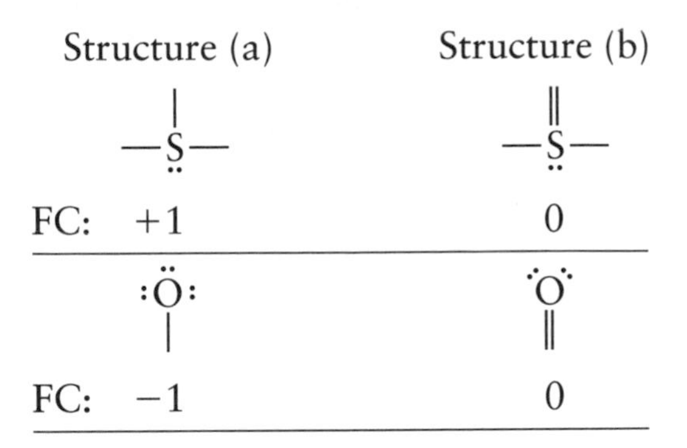

Structure (b) has formal charges closest to 0; it is the more plausible.

8.41 (a)

	First structure	Second structure
	:F̤̈—	:F̤̈—
FC:	0	+1
	—Xe—	—Xe=
FC:	0	−1
	—F̤̈:	=F:
FC:	0	+1

The first structure is dominant.

(b)

	First structure	Second structure
	:O=	:Ö—
FC:	0	−1
	=C=	—C≡
FC:	0	0
	=O:	≡O:
FC:	0	+1

The first structure is dominant.

Exceptions to the Octet Rule

8.43 A radical is any species with an unpaired electron. Examples are

(1) $\cdot CH_3$ or $\cdot$C(H)(H)—H

(2) NO or :Ṅ=Ö:

(3) NO_2 or Ö=Ṅ—Ö: ⟷ :Ö—Ṅ=Ö

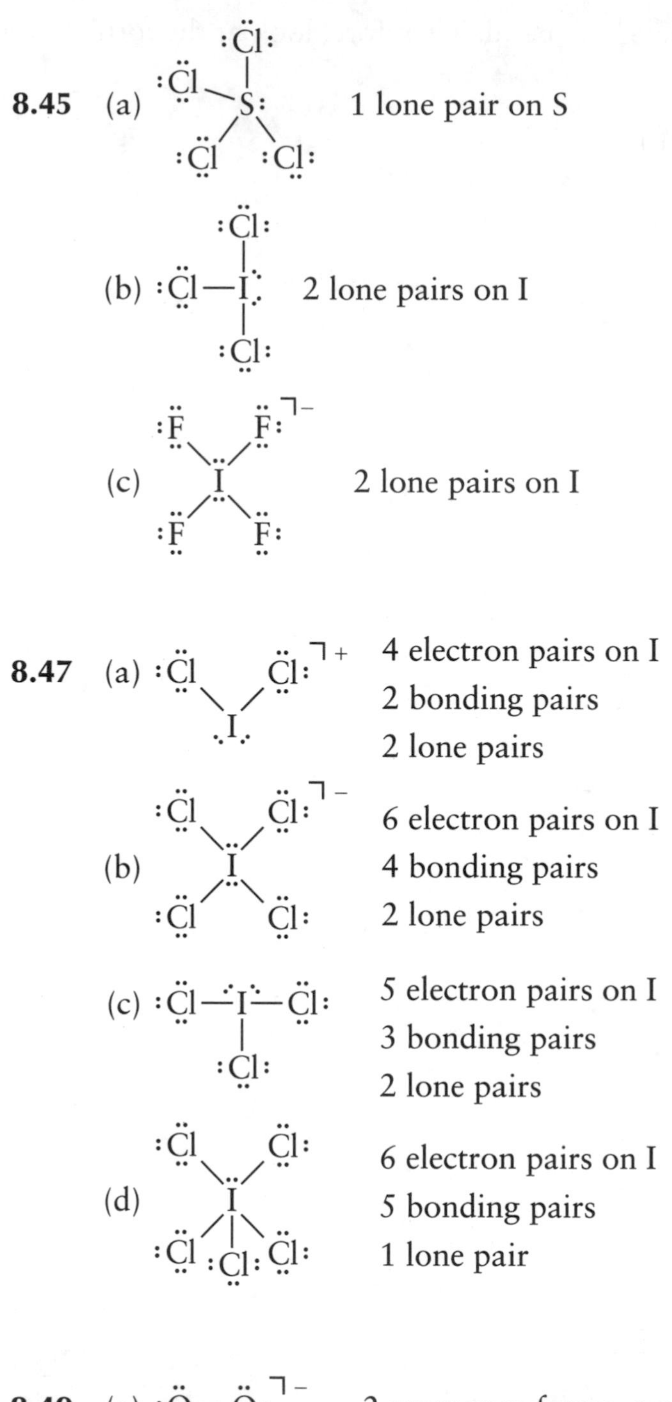

8.45 (a) 1 lone pair on S

(b) 2 lone pairs on I

(c) 2 lone pairs on I

8.47 (a) 4 electron pairs on I
2 bonding pairs
2 lone pairs

(b) 6 electron pairs on I
4 bonding pairs
2 lone pairs

(c) 5 electron pairs on I
3 bonding pairs
2 lone pairs

(d) 6 electron pairs on I
5 bonding pairs
1 lone pair

8.49 (a) $[:\ddot{\underset{..}{O}}-\ddot{\underset{..}{O}}\cdot]^-$ 2 resonance forms; a radical

(b) $:\ddot{\underset{..}{F}}-\dot{\underset{..}{N}}-\ddot{\underset{..}{F}}:$ a radical

(c) $\dot{\underset{.}{N}}=\ddot{O}-\ddot{\underset{..}{Cl}}:$ one resonance form shown; not a radical

8.51 (a) $:\ddot{\underset{..}{F}}-\ddot{Xe}(=\ddot{O}:)-\ddot{\underset{..}{F}}:$ 2 lone pairs on Xe

(b) $:\ddot{\underset{..}{F}}-\underset{..}{\ddot{Xe}}-\ddot{\underset{..}{F}}:$ 3 lone pairs on Xe

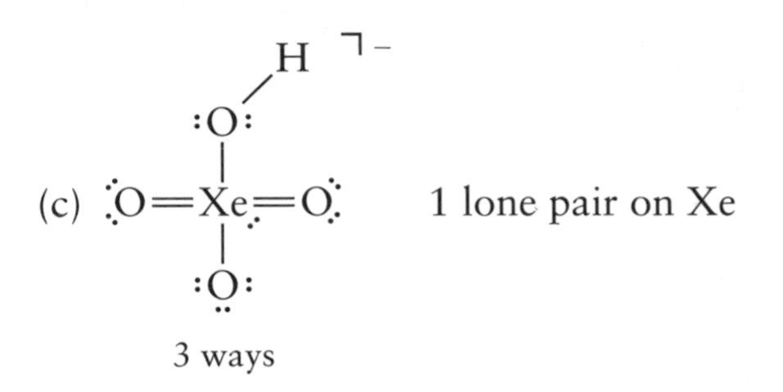

Lewis Acids and Bases

8.53 A *Lewis acid* is an electron pair acceptor; therefore, its electronic structure must allow an additional electron pair to become attached to it. A *Lewis base* is an electron pair donor; therefore, it must contain a lone pair of electrons that it can donate.

Lewis acids: H^+, Al^{3+}, BF_3

Lewis bases: OH^-, NH_3, H_2O

8.55 The net ionic reaction for an acid-base neutralization is

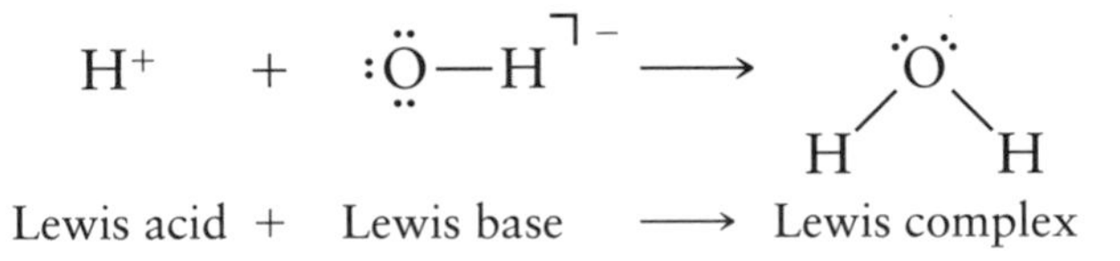

8.57 (a) NH_3 There is a lone pair that can be donated; NH_3 is a Lewis base.

(b) BF_3 There is an incomplete octet on B; there is room to accept a pair. Therefore, BF_3 is a Lewis acid.

(c) Ag^+ There are a number of empty orbitals close in energy to the highest occupied orbital that could accept an electron pair; Ag^+ is a Lewis acid.

(d) F^- There are four lone pairs; F^- is a Lewis base.

8.59

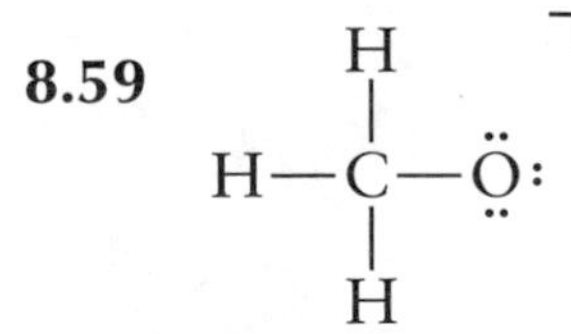

There are available electron pairs on the O; thus CH_3O^- would be expected to be a Lewis base.

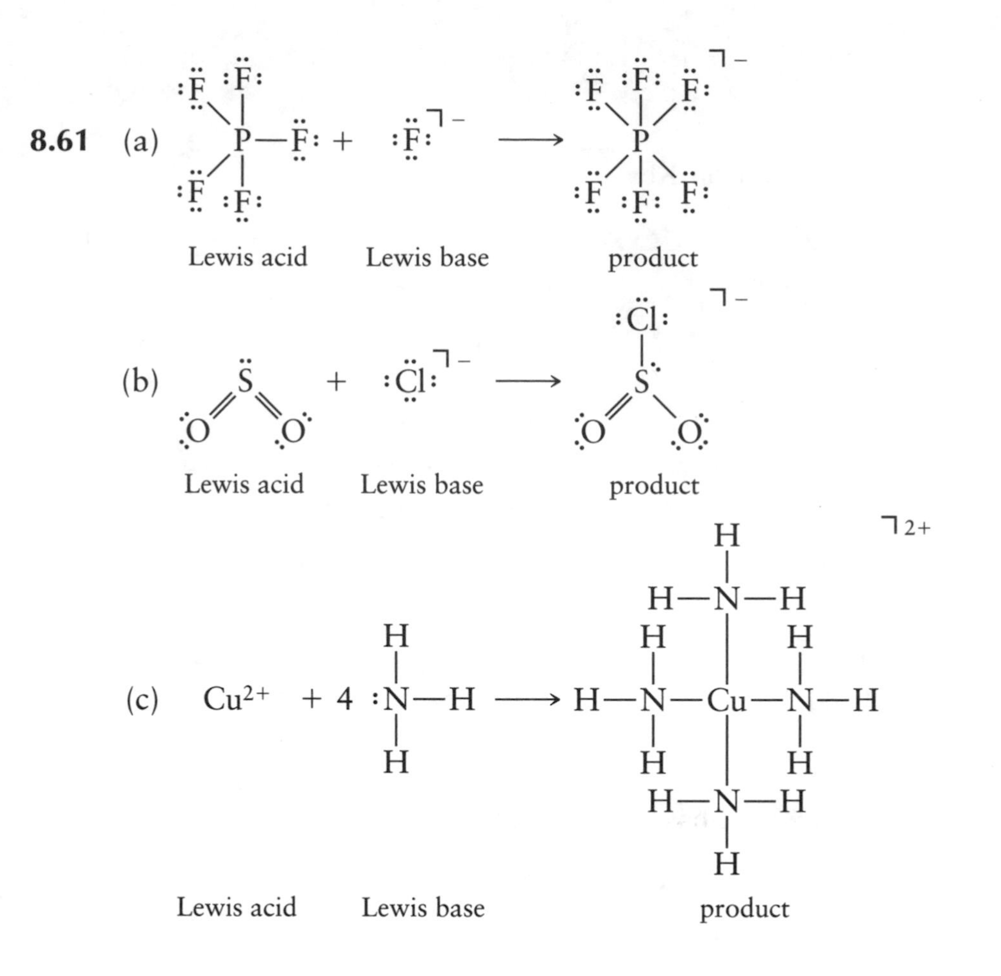

Ionic and Covalent Bonding

8.63 A compound is likely to be ionic if it is a binary compound of a metal and nonmetal between which there is a large difference in electronegativity. Binary compounds of nonmetals are likely to be nonionic.

(a) ionic, made of a metal and a nonmetal

(b) nonionic, made of a nonmetal and a nonmetal

(c) ionic, made of a metal and a nonmetal

8.65 We use the rule of thumb that an electronegativity difference of $\Delta\chi > 2.0$ corresponds to an ionic bond; $\Delta\chi < 1.5$ corresponds to a covalent bond; and $1.5 \leq \Delta\chi \leq 2.0$ corresponds to a significant mix of both ionic and covalent bonding.

(a) Ba—Cl, $\Delta\chi = 3.2 - 0.89 = 2.3$; therefore, ionic

(b) Bi—I, $\Delta\chi = 2.7 - 2.0 = 0.7$; therefore, covalent

(c) Si—H, $\Delta\chi = 2.2 - 1.9 = 0.3$; therefore, covalent

8.67 The order is most likely $Rb^{+} < Sr^{2+} < Be^{2+}$, which corresponds to the order of ionic size in reverse. Higher charge predominates over lower charge, and for the same charge, smaller size predominates over larger size.

8.69 $O^{2-} < N^{3-} < Cl^{-} < Br^{-}$. The order parallels the ionic size: Br^{-} is largest, O^{2-} is smallest. The bigger the ion, the more electrons present and the more spread out the charge, thus making the ion more susceptible to charge distortion by another charge.

8.71 In each case, calculate the difference in electronegativity, $\Delta\chi$, between the atoms in the bonds. Larger $\Delta\chi$ corresponds to greater ionic character in the bond. Use Fig. 8.21.

(a) $\Delta\chi$ (H—Cl) = 3.2 − 2.2 = 1.0; HCl is more ionic.
$\Delta\chi$ (H—I) = 2.7 − 2.2 = 0.5

(b) $\Delta\chi$ (C—H) = 2.6 − 2.2 = 0.4
$\Delta\chi$ (C—F) = 4.0 − 2.6 = 1.4; CF_4 is more ionic.

(c) $\Delta\chi$ (C—O) = 3.4 − 2.6 = 0.8; CO_2 is more ionic.
$\Delta\chi$ (C—S) = 2.6 − 2.6 = 0

8.73 In each case, calculate the electronegativity difference of the bond, $\Delta\chi$, and then use the rule of thumb given in the last paragraph of Section 8.14. See also Exercise 8.65.

(a) $\Delta\chi$ (Ag—F) = 4.0 − 1.9 = 2.1; $\Delta\chi > 2.0$, therefore ionic
(b) $\Delta\chi$ (Ag—I) = 2.7 − 1.9 = 0.8; $\Delta\chi < 1.5$, therefore significantly covalent
(c) $\Delta\chi$ (Al—Cl) = 3.2 − 1.6 = 1.6; $\Delta\chi > 1.5$, therefore mixed ionic and covalent
(d) $\Delta\chi$ (Al—F) = 4.0 − 1.6 = 2.4; $\Delta\chi > 2.0$, therefore ionic

SUPPLEMENTARY EXERCISES

8.75 Boron is in Group 13 and Period 2 of the periodic table and has only 3 valence electrons. In order to have an octet of electrons, boron would have to add 5 electrons by sharing with other atoms. That is improbable, especially because the other atoms involved are likely to be highly electronegative nonmetals. Other elements in Period 2 or later periods, that form covalent bonds have to acquire no more than 4 additional electrons to complete their octet and thus are not likely to form electron-deficient compounds.

8.77 In each case, we calculate the electronegativity difference, $\Delta\chi$. The greater $\Delta\chi$, the greater the ionic character.

$\Delta\chi$ (Na—Cl) = 3.2 − 0.93 = 2.3, most ionic
$\Delta\chi$ (Al—Cl) = 3.2 − 1.6 = 1.6

$\Delta\chi$ (C—Cl) = 3.2 − 2.6 = 0.6
$\Delta\chi$ (Br—Cl) = 3.2 − 3.0 = 0.2, least ionic

8.79 (a) $Li^+\ [:H]^-$

(b) $[:\ddot{\underset{..}{Cl}}:]^-\ Cu^{2+}\ [:\ddot{\underset{..}{Cl}}:]^-$

(c) $Ba^{2+}\ [:\ddot{\underset{..}{N}}:]^{3-}\ Ba^{2+}\ [:\ddot{\underset{..}{N}}:]^{3-}\ Ba^{2+}$

(d) $[:\ddot{\underset{..}{O}}:]^{2-}\ Ga^{3+}\ [:\ddot{\underset{..}{O}}:]^{2-}\ Ga^{3+}\ [:\ddot{\underset{..}{O}}:]^{2-}$

8.81 (a) The lattice enthalpy corresponds to the $\Delta H°$ of the following process:
$Na_2O(s) \longrightarrow 2\ Na^+(g) + O^{2-}(g)$
This process can be broken down into the following steps:

Step	$\Delta H°$, kJ/mol
(1) $2\ Na(s) \longrightarrow 2\ Na(g)$	$2 \times (+107.32)$
(2) $2\ Na(g) \longrightarrow 2\ Na^+(g) + 2e^-$	$2 \times (+494)$
(3) $\frac{1}{2}\ O_2(g) \longrightarrow O(g)$	+249
(4) $O(g) + 2e^- \longrightarrow O^{2-}(g)$	+703
(5) $Na_2O(s) \longrightarrow 2\ Na(s) + \frac{1}{2}\ O_2(g)$	−(−409)
$Na_2O(s) \longrightarrow 2\ Na^+(g) + O^{2-}(g)$	+2564 kJ/mol = $\Delta H_L°$

The numerical value for $\Delta H°$ in step 4 is obtained from the electron affinity of oxygen, Appendix 2D, as shown:
$\Delta H°$ (electron gain enthalpy) = [−(−844) + 141] kJ/mol = 703 kJ/mol
(b) The lattice enthalpy corresponds to the $\Delta H°$ of the following process:
$AlCl_3(s) \longrightarrow Al^{3+}(g) + 3\ Cl^-(g)$
This process can be broken down into the following steps:

Step	$\Delta H°$, kJ/mol
(1) $Al(s) \longrightarrow Al(g)$	+326
(2) $Al(g) \longrightarrow Al^+(g) + e^-$	+577
$Al^+(g) \longrightarrow Al^{2+}(g) + e^-$	+1820
$Al^{2+}(g) \longrightarrow Al^{3+}(g) + e^-$	+2740
(3) $\frac{3}{2}\ Cl_2(g) \longrightarrow 3\ Cl(g)$	$3 \times (+121.68)$
(4) $3\ Cl(g) + 3e^- \longrightarrow 3\ Cl^-(g)$	$3 \times (-349)$
(5) $AlCl_3(s) \longrightarrow Al(s) + \frac{3}{2}\ Cl_2(g)$	−(−704.2)
$AlCl_3(s) \longrightarrow Al^{3+}(g) + 3\ Cl^-(g)$	+5485 kJ/mol = $\Delta H_L°$

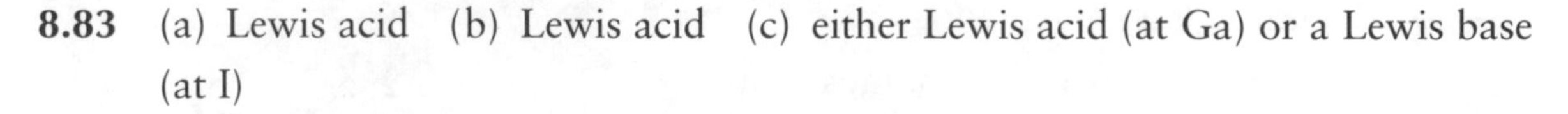

8.83 (a) Lewis acid (b) Lewis acid (c) either Lewis acid (at Ga) or a Lewis base (at I)

8.85 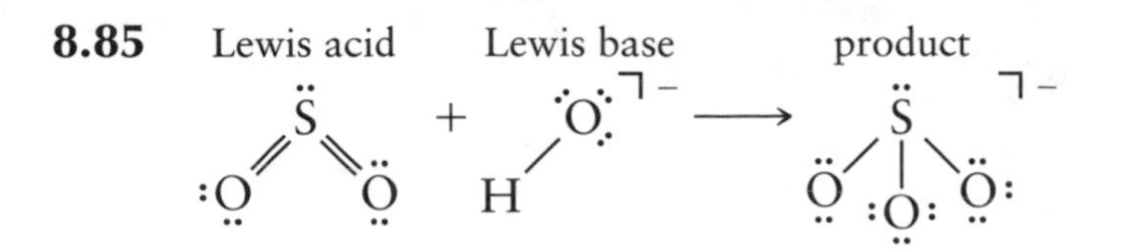

8.87 CH_3^-, because the Lewis structure of CH_3^- shows a lone pair on the C atom; there is no lone pair in CH_4.

8.89 (a)

	First structure	Second structure
	:O=	:Ö—
FC:	0	−1
	:O: ‖	:Ö: |
FC:	0	−1
	—H	—H
FC:	0	0
	—Ö—	—Ö—
FC:	0	0
	‖ =Cl—	| —Cl—
FC:	0	+2

The first structure has the lower energy.

(b)

O=	=C=	=S:
FC: 0	FC: 0	FC: 0

(c)

H—	—C≡	≡N:
FC: 0	FC: 0	FC: 0

(d)

	First structure	Second structure
	:N=	:N≡
FC:	−1	0
	=C=	≡C—
FC:	0	0
	=N:	—N̈:
FC:	−1	−2

The first structure has the lower energy.

(e) First structure Second structure

	First structure	Second structure
	:Ö: (single bond)	Ö (double bond)
FC:	−1	0
	:Ö—	:Ö—
FC:	−1	−1
	:Ö:	:Ö:
FC:	−1	−1
	—Ö:	—Ö:
FC:	−1	−1
	—As—	—As—
FC:	+1	0

Structure 2 has the lower energy.

8.91 In a very broad sense, yes. All bonds involve a pair of electrons influenced by two nuclear centers. It becomes a question of degree as to where the electron pair more closely resides. Nonpolar covalent bonds share the electron pair equally, polar covalent bonds less so, and ionic bonds hardly at all. The octet rule is useful in the understanding of all the bonding situations mentioned to the extent that it applies. Recall that there are exceptions. The coordinate covalent bond that results from complex formation is like any other covalent bond, once it has formed.

8.93 The processes can be broken down into the following steps:

	$\Delta H°$, kJ/mol	
$Mg^+(g) + Cl^-(g) \longrightarrow MgCl(s)$	−717	(i.e., one assumes that ΔH°_L
$Mg(s) \longrightarrow Mg(g)$	+148	is the same as that of KCl)
$Mg(g) \longrightarrow Mg^+(g) + e^-$	+736	
$\frac{1}{2} Cl_2(g) \longrightarrow Cl(g)$	+121.68	(from Exercise 8.81)
$Cl(g) + e^- \longrightarrow Cl^-(g)$	−349	
$Mg(s) + \frac{1}{2} Cl_2(g) \longrightarrow MgCl(s)$	−60. kJ/mol $= \Delta H^\circ_f$	

	$\Delta H°$, kJ/mol
$Mg^{2+}(g) + 2\ Cl^-(g) \longrightarrow MgCl_2(s)$	-2524
$Mg(s) \longrightarrow Mg(g)$	$+148$
$Mg(g) \longrightarrow Mg^{2+}(g) + 2e^-$	$+2186$
$Cl_2(g) \longrightarrow 2\ Cl(g)$	$+243.36$
$2\ Cl(g) + 2e^- \longrightarrow 2\ Cl^-(g)$	-698
$Mg(s) + Cl_2(g) \longrightarrow MgCl_2(s)$	-645 kJ/mol $= \Delta H_f°$

(Furthermore, for $2\ MgCl(s) \longrightarrow MgCl_2(s) + Mg(s)$; the energy released is thus $\Delta H° = +645 - (2)(-60.) = +525$ kJ/mol)

Thus, $MgCl_2$ is the far more likely chloride of magnesium.

APPLIED EXERCISES

8.95 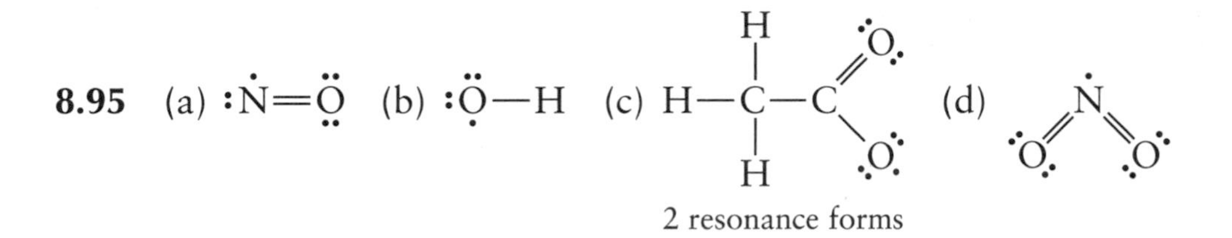

2 resonance forms

8.97

H :O:
| ‖
H—C—C—H
|
H

Resonance is unlikely because it would result in one trivalent carbon atom and one pentavalent carbon atom.

8.99 $CO_3^{2-}(s) \longrightarrow CO_2(g) + O^{2-}(s)$

complex — Lewis acid — Lewis base

$Ca^{2+}O^{2-}(s) + SiO_2(s) \longrightarrow CaSiO_3(s)$

Lewis base — Lewis acid — complex

INTEGRATED EXERCISES

8.101 Large anions such as chromate and phosphate (which also have larger charges) are more polarizable than smaller anions like carbonate, chloride, and nitrate. In general, it appears that the more covalent character, and thus the more polarizable the ion, the more insoluble the salts of the anion tend to be. Because solubility of ionic compounds requires the formation of ions in solution, ions that have more covalent character will probably not form salts as readily in solution. As such, these compounds will tend to be insoluble.

8.103 Electronegativity difference is not the only factor that determines the physical properties of a compound. The rule based on $\Delta\chi$ for determining the type of bond is at best only a qualitative one. Other factors related to the electronic configurations of the atoms involved, such as ionization energy, electron affinity, as well as size and polarizability, all play a role.

(a) The electronic configurations of Mg and I are such that the octet rule may be easily satisfied by an electron transfer process. After the transfer, the small Mg^{2+} ion, with its highly concentrated positive charge, has a strong electrostatic attraction to the large iodide ions. The bonding energy is then primarily electrostatic or ionic.

(b) In SiF_4, for the octet rule to be satisfied on Si by electron transfer to F, the Si^{4+} ion would have to generate such a highly charged ion that, given the bonding energy in SiF_4, one should conclude that the bond is primarily a result of electron sharing or covalent bonding. The radius of Si^{4+} is also quite small (26 pm), leading to significant covalent character.

8.105 (a) Cd: [Kr]$4d^{10}5s^2$ In^+: [Kr]$4d^{10}5s^2$ Sn^{2+}: [Kr]$4d^{10}5s^2$

All are the same, because the ions form by the loss of $5p$ electrons, and cadmium itself has no $5p$ electrons.

(b) There are no unpaired electrons in any of the species.

(c) In^{3+}: [Kr]$4d^{10}$, which is isoelectronic with palladium.

CHAPTER 9
MOLECULAR STRUCTURE

EXERCISE

The Shapes of Molecules and Ions

9.1 (a) trigonal bipyramidal, 120° bond angles in equatorial plane, 90° between axial bonds and equatorial plane
(b) linear, 180°
(c) trigonal pyramidal, slightly less than 109°
(d) angular, slightly less than 109°
(e) tetrahedral, 109.5°

9.3 (a) must be (b) may be, AX_2E_3

9.5 (a) H—O—Cl (angular Lewis structure) AX_2E_2, angular

(b) PF_3 (P with three F) AX_3E, trigonal pyramidal

(c) N=N=O AX_2, linear

(d) O=O—O two resonance forms AX_2E, angular

9.7 (a) $[H_3O]^+$ (O bonded to three H) AX_3E, trigonal pyramidal

(b) $[SO_4]^{2-}$ (S singly bonded to four O) AX_4, tetrahedral

Lewis structures with one and two S=O double bonds also contribute to the overall resonance hybrid that represents SO_4^{2-}. All structures are AX_4 and have a tetrahedral shape.

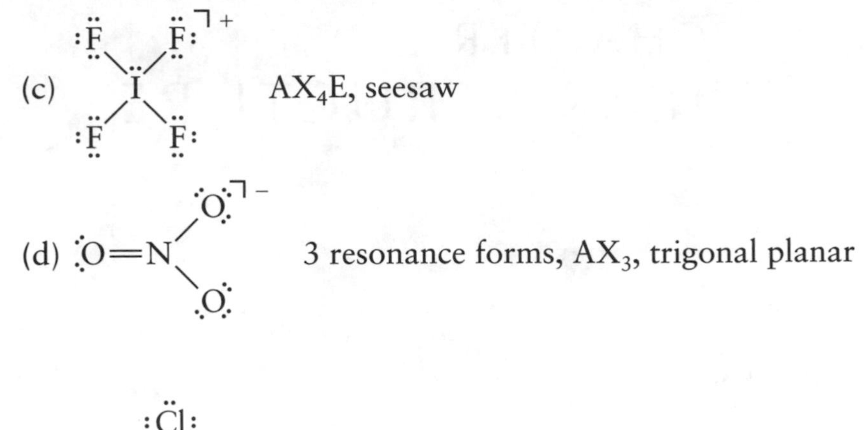


```
          :Cl:
            |
     :Cl—S:
          /  \
      :Cl    :Cl:
```

9.9 (a) AX$_4$E, seesaw

```
       :Cl:
         |
 :Cl—I:
         |
       :Cl:
```

(b) AX$_3$E$_2$, T-shaped

```
  :F      F:  ]−
     \   /
       I
     /   \
  :F      F:
```

(c) AX$_4$E$_2$, square planar

```
        Xe
      /  |  \
   :O:  :O:  :O:
```

(d) AX$_3$E, trigonal pyramidal

Resonance forms with Xe=O double bonds also contribute to the overall structure. All are AX$_3$E and have the trigonal-pyramidal shape.

9.11 (a) :I—I—I: central I is AX$_2$E$_3$, linear, 180°

(b) :F—I—F: AX$_3$E$_2$, T-shaped, slightly less than 90°

```
:F—I—F:
    |
   :F:
```

```
  :O:    :O:  ]−
     \   /
       I
     /   \
  :O:    :O:
```

(c) AX$_4$, tetrahedral, 109.5°

```
         :F:
   :F     |     F:
      \   |   /
         Te
      /   |   \
   :F     |     F:
         :F:
```

(d) AX$_6$, octahedral, 90°

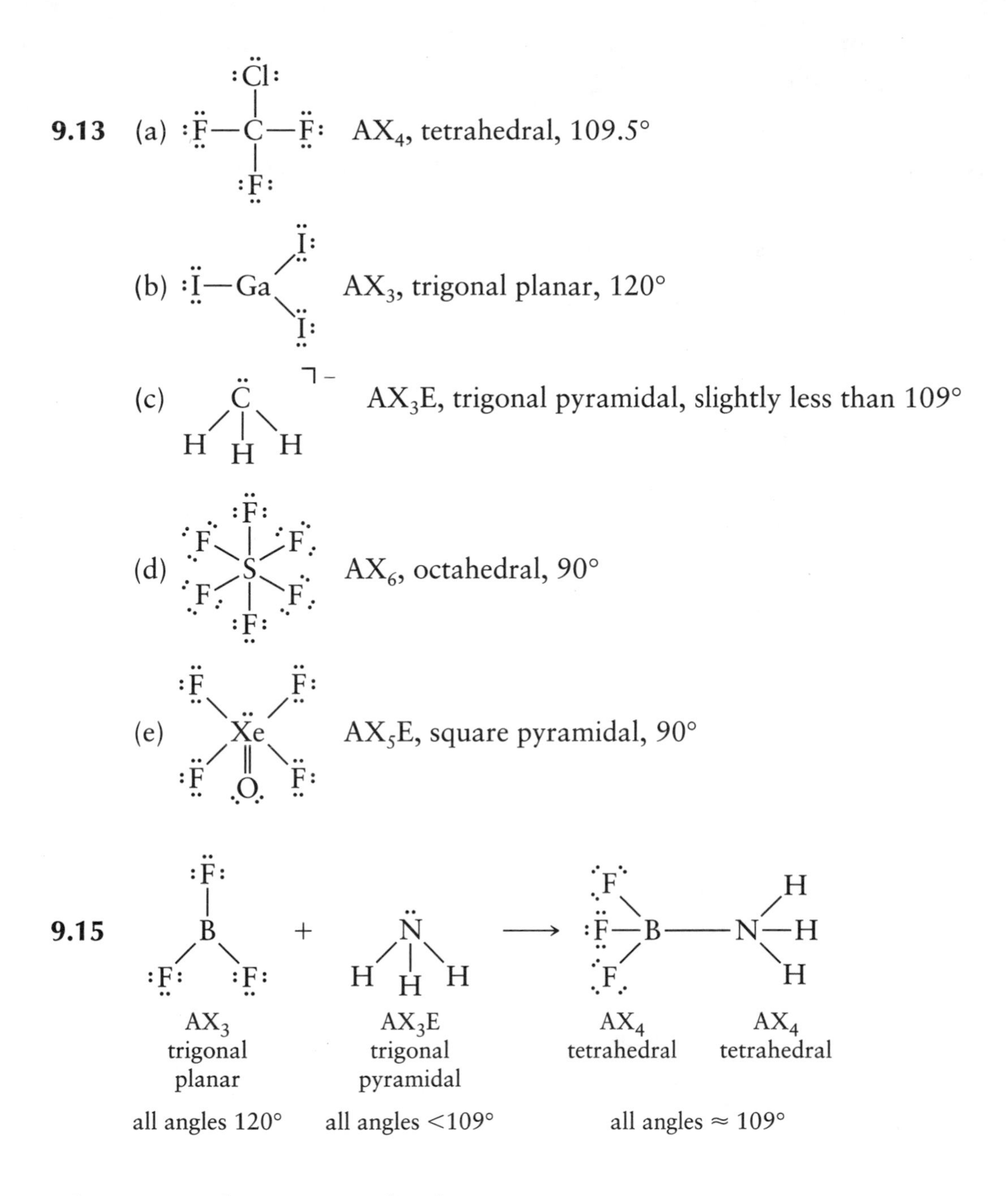

Charge Distribution in Molecules

9.17 An electric dipole ("two poles") is a positive charge next to an equal but opposite negative charge. An electric dipole moment is a measure of the magnitude of the electric dipole in debye (D) units.

9.19 The direction of the dipole moment is toward the element in the bond with the larger electronegativity, χ.

(a) O—H; $\chi_O = 3.4$, $\chi_H = 2.2$; therefore, toward O
(b) O—F; $\chi_O = 3.4$, $\chi_F = 4.0$; therefore, toward F
(c) F—Cl; $\chi_F = 4.0$, $\chi_{Cl} = 3.2$; therefore, toward F
(d) O—S; $\chi_O = 3.4$, $\chi_S = 2.6$; therefore, toward O

9.21 (a) Br_2; one nonpolar bond

(b) H_2NNH_2; $\mathrm{H_2\ddot{N}{-}\ddot{N}H_2}$, four polar N—H bonds, one nonpolar N—N bond

(c) CH_4; all C—H bonds are slightly polar

(d) O_3; $\mathrm{O{=}\ddot{O}{-}O}$, AX_2E, both bonds are polar

9.23 See Fig. 9.15 in the text.

(a) AX_3, trigonal planar, nonpolar

(b) AX_4E_2, square planar, nonpolar

(c) AX_3E, trigonal pyramidal, polar

(d) AX_2E_2, angular, polar

9.25 See Fig. 9.15 in the text.

(a) CCl_4 (C bonded to four :Cl:) AX_4, tetrahedral, nonpolar

(b) $\mathrm{:\!S{=}C{=}S\!:}$ AX_2, linear, nonpolar

(c) PCl_5 (P bonded to five :Cl:) AX_5, trigonal bipyramidal, nonpolar

(d) 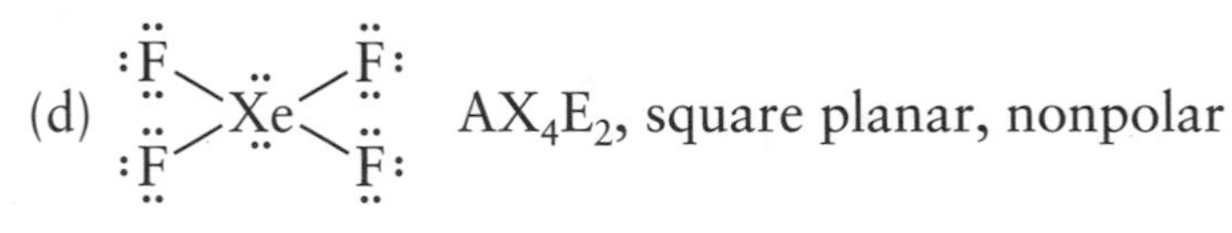

AX_4E_2, square planar, nonpolar

9.27 (a)

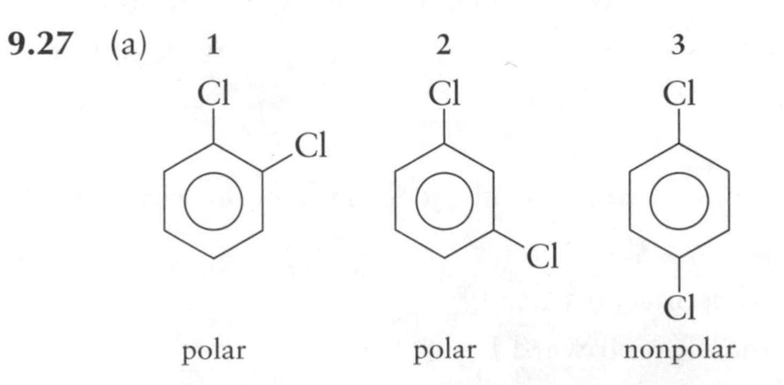

(b) Structure **1** is the most unsymmetrical and has the largest separation of charge; it is the most polar and has the largest dipole moment.

Bond Strengths and Bond Lengths

9.29 (a) $H_2O \longrightarrow 2\,H + O$

$\Delta H = 2 \times \Delta H_B\,(O{-}H) = (2 \times 463)$ kJ/mol $= +926$ kJ/mol

(b) $CO_2 \longrightarrow C + 2\,O$ $\quad O{=}C{=}O$

$\Delta H = 2 \times \Delta H_B(C{=}O) = (2 \times 743)$ kJ/mol $= +1486$ kJ/mol

(c) $CH_3COOH \longrightarrow 2\,C + 4\,H + 2\,O$

H H—C—C(=O)—O—H, with a third H on the first C

$\Delta H = 3 \times \Delta H_B(C{-}H) + 1 \times \Delta H_B(C{-}C) + 1 \times \Delta H_B(C{=}O) + 1 \times \Delta H_B(C{-}O) + 1 \times \Delta H_B(O{-}H)$

$= [3(412) + 1(348) + 1(743) + 1(360) + 1(463)]$ kJ/mol

$= +3150$ kJ/mol

(d) $CH_3NH_2 \longrightarrow C + 5\,H + N$

H₃C—N̈H₂ (H—C—N with three H on C, two H and a lone pair on N)

$\Delta H = 3 \times \Delta H_B(C{-}H) + 1 \times \Delta H_B(C{-}N) + 2 \times \Delta H_B(N{-}H)$

$= [3(412) + 1(305) + 2(388)]$ kJ/mol $= +2317$ kJ/mol

9.31 Use Tables 9.2 and 9.3.

(a) $\frac{1}{2}\,H_2 + \frac{1}{2}\,Cl_2 \longrightarrow HCl$

$\Delta H_f^\circ = -\Delta H_B(H{-}Cl) + \frac{1}{2}\Delta H_B(H{-}H) + \frac{1}{2}\Delta H_B(Cl{-}Cl)$

$= [-(431) + \frac{1}{2}(436) + \frac{1}{2}(242)]$ kJ/mol

$= -92$ kJ/mol

(b) $H_2 + O_2 \longrightarrow H_2O_2$ $\quad H{-}O{-}O{-}H$

$\Delta H_f^\circ = -\Delta H_B(O{-}O) - 2 \times \Delta H_B(O{-}H) + \Delta H_B(O{=}O) + \Delta H_B(H{-}H)$

$= [-157 - 2(463) + 496 + 436]$ kJ/mol $= -151$ kJ/mol

(c) $C + 2\,Cl_2 \longrightarrow CCl_4$

$\Delta H_f^\circ = -4 \times \Delta H_B(C{-}Cl) + 2 \times \Delta H_B(Cl{-}Cl)$

$= [-4(338) + 2(242)]$ kJ/mol $= -868$ kJ/mol

(d) $\frac{1}{2}\,N_2 + \frac{3}{2}\,H_2 \longrightarrow NH_3$

$\Delta H_f^\circ = -3 \times \Delta H_B(N{-}H) + \frac{1}{2} \times \Delta H_B(N{\equiv}N) + \frac{3}{2} \times \Delta H_B(H{-}H)$

$= [-3(388) + \frac{1}{2}(944) + \frac{3}{2}(436)]$ kJ/mol $= -38$ kJ/mol

9.33 (a) $HCl(g) + F_2(g) \longrightarrow HF(g) + ClF(g)$ or

$H{-}Cl + F{-}F \longrightarrow H{-}F + Cl{-}F$

The enthalpy required to break the bonds in the reactants is

$\Delta H^\circ = 1 \times \Delta H_B(H{-}Cl) + 1 \times \Delta H_B(F{-}F) = [1(431) + 1(158)]$ kJ $= 589$ kJ

The enthalpy released when the products form is

$\Delta H° = -[1 \times \Delta H_B(\text{H—F}) + 1 \times \Delta H_B(\text{Cl—F})] = -[1(565) + 1(256)]$ kJ

$= -821$ kJ

The sum of these two enthalpy changes is the reaction enthalpy:

$\Delta H° = 589 \text{ kJ} - 821 \text{ kJ} = -232$ kJ

(b) $C_2H_4(g) + HCl(g) \longrightarrow CH_3CH_2Cl(g)$ or

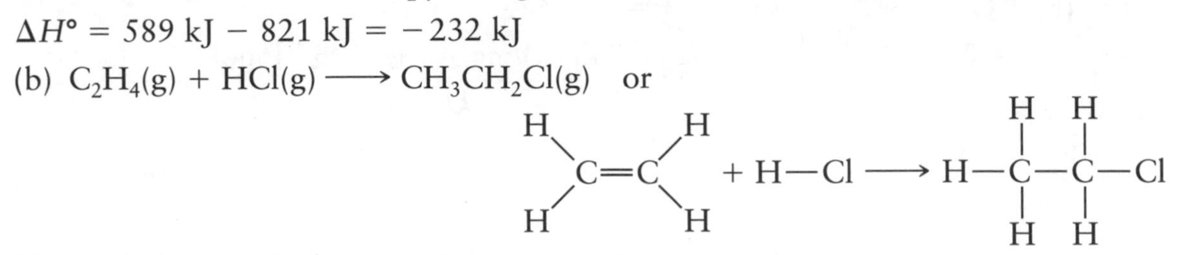

The enthalpy required to atomize reactants is

$\Delta H° = 1 \times \Delta H_B(\text{C=C}) + 4 \times \Delta H_B(\text{C—H}) + 1 \times \Delta H_B(\text{H—Cl})$

$= [1(612) + 4(412) + 1(431)] \text{ kJ} = 2691$ kJ

The enthalpy released when the products form is

$\Delta H° = -[1 \times \Delta H_B(\text{C—C}) + 1 \times \Delta H_B(\text{C—Cl}) + 5 \times \Delta H_B(\text{C—H})]$

$= -[1(348) + 1(338) + 5(412)] \text{ kJ} = -2746$ kJ

The sum of these two enthalpy changes is the reaction enthalpy:

$\Delta H° = 2691 \text{ kJ} - 2746 \text{ kJ} = -55$ kJ

(c) $C_2H_2(g) + 2\,H_2(g) \longrightarrow CH_3CH_3$ or

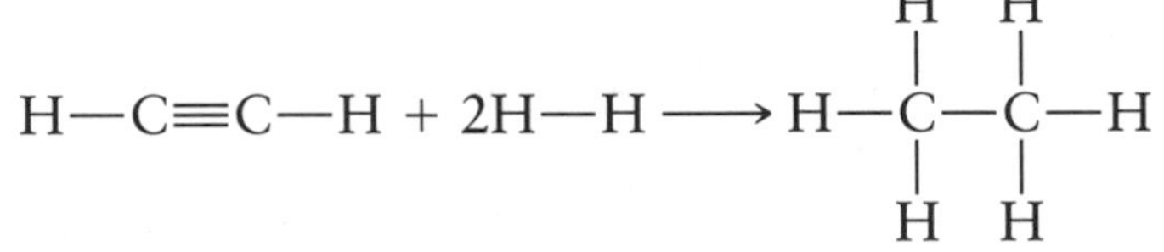

The enthalpy required to atomize reactants is

$\Delta H° = 1 \times \Delta H_B(\text{C≡C}) + 2 \times \Delta H_B(\text{C—H}) + 2 \times \Delta H_B(\text{H—H})$

$= [1(837) + 2(412) + 2(436)] \text{ kJ} = 2533$ kJ

The enthalpy released when the products form is

$\Delta H° = -[1 \times \Delta H_B(\text{C—C}) + 6 \times \Delta H_B(\text{C—H})]$

$= -[1(348) + 6(412)] \text{ kJ} = -2820$ kJ

The sum is the reaction enthalpy: $\Delta H° = 2533 \text{ kJ} - 2820 \text{ kJ} = -287$ kJ

9.35

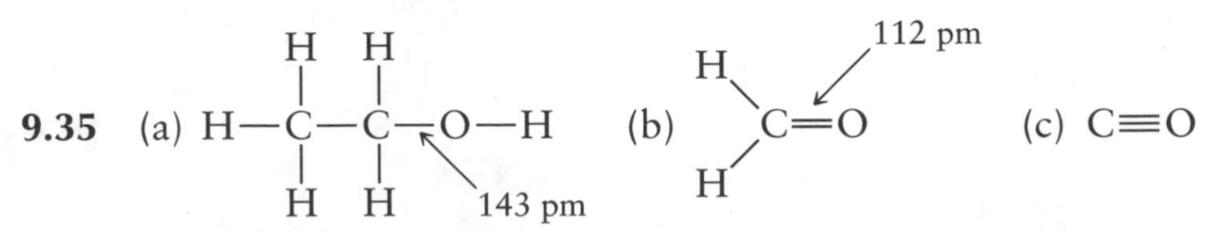

Data are not available for C≡O in Table 9.4, but it is expected to be the shortest of the three, because largest bond order corresponds to shortest bond length. Therefore, (c) < (b) < (a).

9.37 (a)

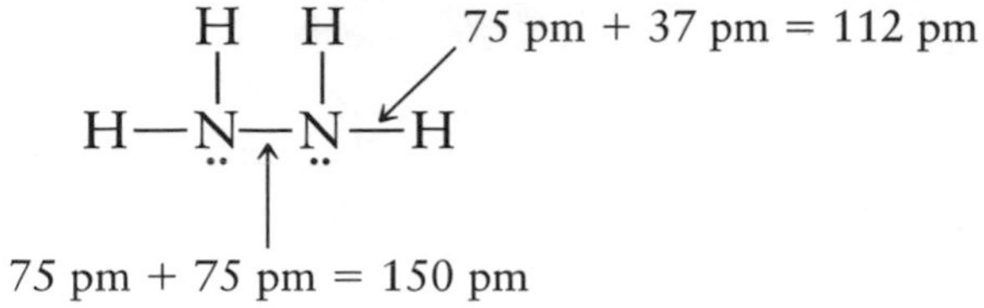

H H
H—N—N—H
75 pm + 37 pm = 112 pm
75 pm + 75 pm = 150 pm

(b)

O=C=O
60 pm + 67 pm = 127 pm

(c)

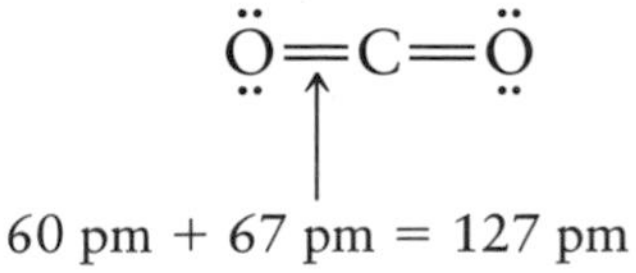

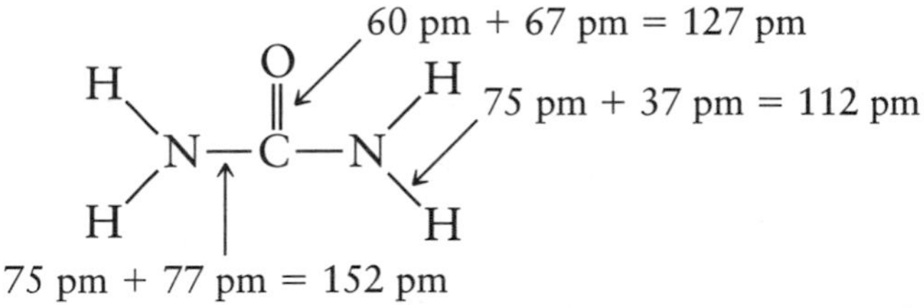

O
H
N—C—N
H
60 pm + 67 pm = 127 pm
75 pm + 37 pm = 112 pm
75 pm + 77 pm = 152 pm

(d)

H—N=N—H
75 pm + 37 pm = 112 pm
60 pm + 60 pm = 120 pm

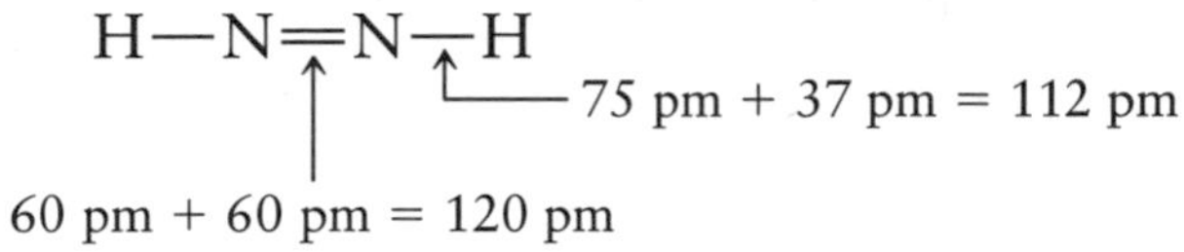

Orbitals and Bonds

9.39 Cl: $[Ne]3s^2 3p_x^{\ 2}\ 3p_y^{\ 2}\ 3p_z^{\ 1}$; so, in the diatomic molecule, $3p_z$—$3p_z$ overlap occurs to form the bond, which is a σ bond.

9.41 See Table 9.5.
(a) tetrahedral (b) linear (c) octahedral (d) trigonal planar

9.43 (a) SF_4 is AX_4E; therefore, the electron pairs adopt a trigonal bipyramidal arrangement that corresponds to dsp^3 hybridization on S.
(b) BCl_3 is AX_3; trigonal planar electron pairs; sp^2 hybridization on B
(c) NH_3 is AX_3E; tetrahedral electron pairs; sp^3 hybridization on N
(d) $(CH_3)_2$ Be is AX_2; linear electron pairs; *sp* hybridization on Be

9.45 (a)

sp2
sp2 sp2 sp2
p

(b)

sp3
sp3 sp3 sp3 sp3

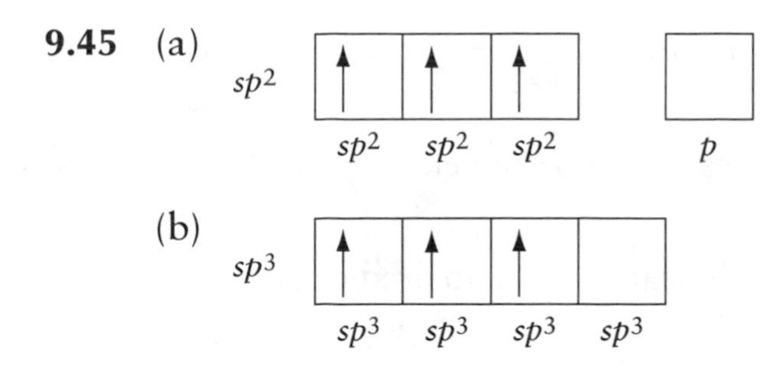

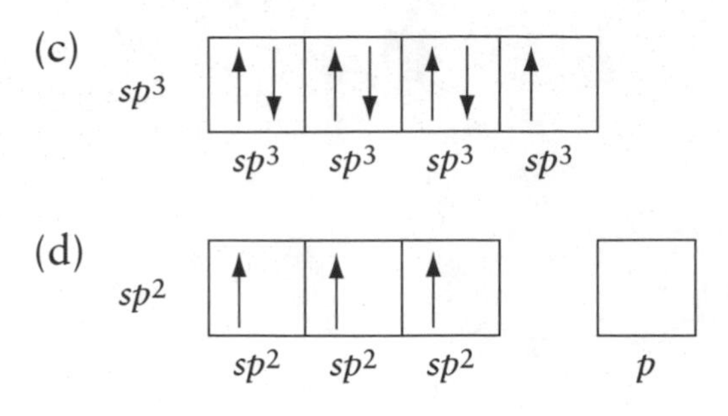

9.47 (a) each C is A of AX_4; tetrahedral electron pairs; sp^3 hybrid orbitals
(b) each C is A of AX_2; linear electron pairs; *sp* orbitals
(c) PCl_5 is AX_5; trigonal bipyramidal pairs; dsp^3 hybrid orbitals
(d) HOCl is AX_2E_2; tetrahedral pairs; sp^3 hybrid orbitals

Molecular Orbitals

9.49 (a) There are 4 valence electrons in the Be_2 molecule. The electrons will fill the σ_{2s} and $\sigma_{2s}{}^*$ orbitals, so the valence-shell electron configuration is $\sigma_{2s}{}^2\ \sigma_{2s}{}^{*2}$.
The bond order is equal to $\frac{1}{2}$ the number of electrons in bonding orbitals minus $\frac{1}{2}$ the number of electrons in antibonding orbitals.
$BO = \frac{1}{2} \times (2 - 2) = 0.$
(b) There are 6 valence electrons in the B_2 molecule. Using the energy diagram from Figure 9.56, we find $\sigma_{2s}{}^2\sigma_{2s}{}^{*2}\pi_{2p}{}^2$.
$BO = \frac{1}{2} \times (4 - 2) = 1$
(c) There are 16 valence electrons in the Ne_2 molecule. Assuming Ne_2 follows the pattern for O_2 and N_2, in which the σ_{2p} orbital is of lower energy than the π_{2p} orbital (Figure 9.57), we find $\sigma_{2s}{}^2\sigma_{2s}{}^{*2}\sigma_{2p}{}^2\pi_{2p}{}^4\pi_{2p}{}^{*4}\sigma_{2p}{}^{*2}$.
$BO = \frac{1}{2}(8 - 8) = 0$

9.51 The criteria for paramagnetism include unpaired electrons. By examining the valence-shell electron configurations for the species of interest, we can determine which are paramagnetic.
(a) 12 valence electrons, so $\sigma_{2s}{}^2\sigma_{2s}{}^{*2}\sigma_{2p}{}^2\pi_{2p}{}^4\pi_{2p}{}^{*2}$. The $\pi_{2p}{}^{*2}$ electrons are unpaired, so O_2 is paramagnetic.
(b) Adding one electron to form $O_2{}^-$ leaves one unpaired electron, so $O_2{}^-$ is paramagnetic.
(c) Removing one electron to form $O_2{}^+$ still leaves one unpaired electron, so $O_2{}^+$ is paramagnetic.

9.53 (a) F_2; $\sigma_{2s}{}^2\sigma_{2s}{}^{*2}\sigma_{2p}{}^2\pi_{2p}{}^4\pi_{2p}{}^{*4}$ BO $= \frac{1}{2}(8 - 6) = 1$

$F_2{}^{2-}$; $\sigma_{2s}{}^2\sigma_{2s}{}^{*2}\sigma_{2p}{}^2\pi_{2p}{}^4\pi_{2p}{}^{*4}\sigma_{2p}{}^{*2}$ BO $= \frac{1}{2}(8 - 8) = 0$

F_2 has the stronger bond.

(b) B_2; $\sigma_{2s}{}^2\sigma_{2s}{}^{*2}\pi_{2p}{}^2$ BO $= \frac{1}{2}(4 - 2) = 1$

$B_2{}^+$; $\sigma_{2s}{}^2\sigma_{2s}{}^{*2}\pi_{2p}{}^1$ BO $= \frac{1}{2}(3 - 2) = \frac{1}{2}$

B_2 has the stronger bond.

9.55 First, determine the valence-electron configuration of the ion. Then, determine the bond order. HeH^-; $\sigma_{1s}{}^2\sigma_{1s}{}^{*2}$ BO $= \frac{1}{2}(2 - 2) = 0$

Because the bond order of HeH^- is zero, the ion probably does not exist. For HeH^+; $\sigma_{1s}{}^2$ BO $= \frac{1}{2}(2 - 0) = 1$. This species is more stable because it has a higher bond order.

SUPPLEMENTARY EXERCISES

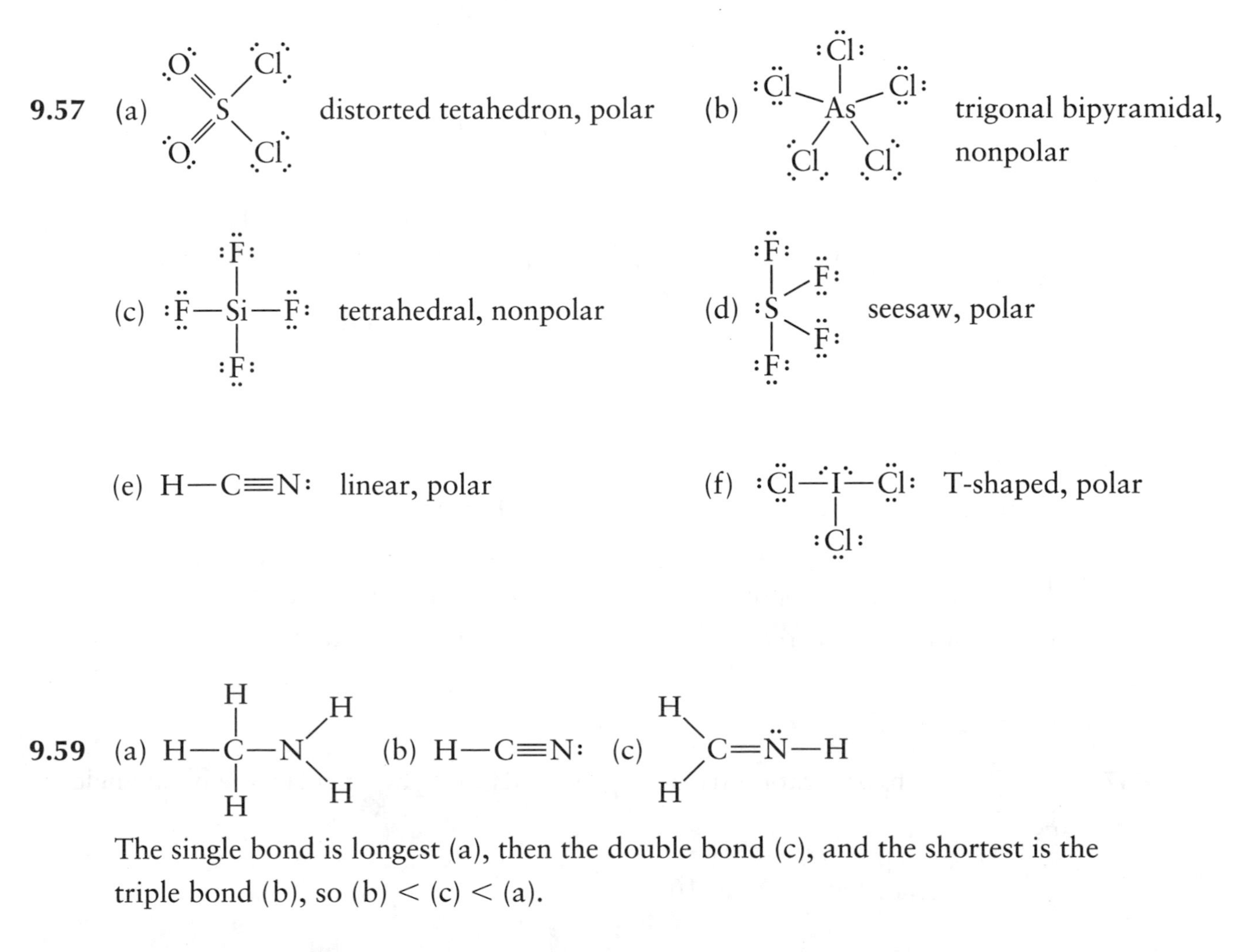

The single bond is longest (a), then the double bond (c), and the shortest is the triple bond (b), so (b) < (c) < (a).

9.61 (a) :F̤—T̈e—F̤: AX_4E, seesaw
:F̤: :F̤:

(b) $[H—\ddot{N}—H]^-$ AX_2E_2, angular

(c) $\ddot{S}=C=\ddot{S}$ AX_2, linear

(d) $[NH_4]^+$ (N bonded to four H) AX_4, tetrahedral

(e) GeH_4 (Ge bonded to four H) AX_4, tetrahedral

(f) $\ddot{O}=C=\ddot{S}$ AX_2, about C, linear

9.63 (a) 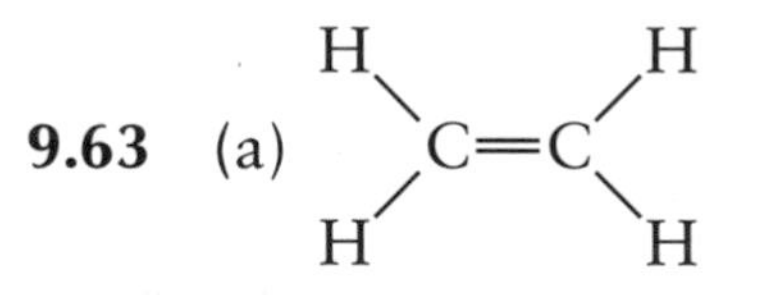

Each C is A of AX_3; therefore, the molecule is trigonal planar, with bond angles of about 120°.

(b) :C̤l—C≡N: 180° (c) $POCl_3$ (:Ö=P bonded to three :C̤l:) 109° (d) $H_2\ddot{N}—\ddot{N}H_2$ 109°

9.65 (a) CF_4, 77 pm (for C) + 72 pm (for F) = 149 pm
(b) SiF_4, 111 pm (for Si) + 72 pm (for F) = 183 pm
(c) SnF_4, 141 pm (for Sn) + 72 pm (for F) = 213 pm
The bond length is proportional to the increasing size of the atom attached to the fluorine atom; size increases down a group in the periodic table.

9.67 :Ö—F̤: (O also bonded to :F̤:) sp^3 hybridization (two sp^3—p bonds), AX_2E_2, angular; the bond angle is somewhat less than 109°

9.69 The Kekulé structures are

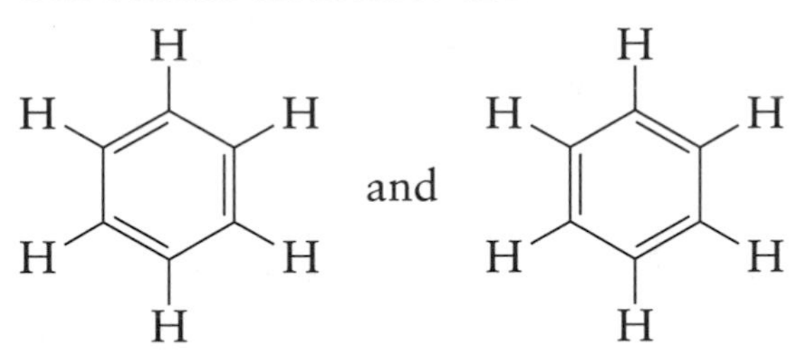

Each has 6 C—H single bonds, 3 C=C double bonds, and 3 C—C single bonds. The resonance structure can be written

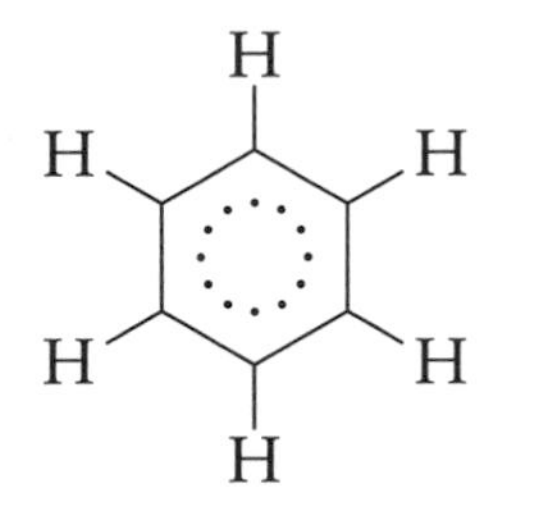

This structure has 6 C—H single bonds and 6 $\mathrm{C}\overset{\cdots}{-}\mathrm{C}$ resonance bonds. The difference between the resonance structure and the Kekulé structure is in the C—C bonds, not in the C—H bonds.

For the Kekulé structure,

$$\text{total bond enthalpy} = 3 \times \Delta H_B(\mathrm{C{=}C}) + 3 \times \Delta H_B(\mathrm{C{-}C})$$
$$= 3 \times 612\ \text{kJ/mol} + 3 \times 348\ \text{kJ/mol} = 2880\ \text{kJ/mol}$$

For the resonance structure,

$$\text{total bond enthalpy} = 6 \times \Delta H_B(\mathrm{C}\overset{\cdots}{-}\mathrm{C}) = 6 \times 518\ \text{kJ/mol} = 3108\ \text{kJ/mol}$$

The lowering in energy (enthalpy) is then 2880 kJ/mol − 3108 kJ/mol = −228 kJ/mol, which is very close to the accepted value of −206 kJ/mol.

9.71

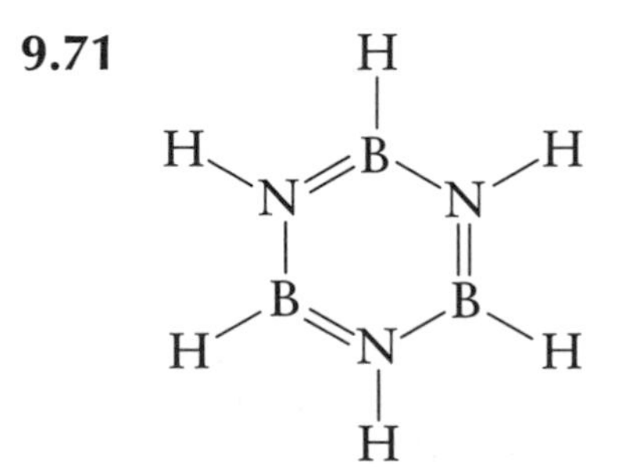

The hexagonal structure of the molecule dictates that all bond angles be 120°, so we expect that each B and N atom is sp^2 hybridized.

9.73 (a) 10 valence electrons, so NO^+; $\sigma_{2s}{}^2\sigma_{2s}{}^{*2}\pi_{2p}{}^4\sigma_{2p}{}^2$;

$BO = \frac{1}{2}(8 - 2) = 3$

(b) 9 valence electrons, so $N_2{}^+$; $\sigma_{2s}{}^2\sigma_{2s}{}^{*2}\pi_{2p}{}^4\sigma_{2p}{}^1$;

$BO = \frac{1}{2}(7 - 2) = \frac{5}{2}$

(c) 13 valence electrons, so O_2^-; $\sigma_{2s}^2\sigma_{2s}^{*2}\pi_{2p}^4\sigma_{2p}^2\pi_{2p}^{*3}$;
$BO = \frac{1}{2}(8 - 5) = \frac{3}{2}$

APPLIED EXERCISES

9.75 (a) Phenolphthalein, acid form | phenolphthalein, base form

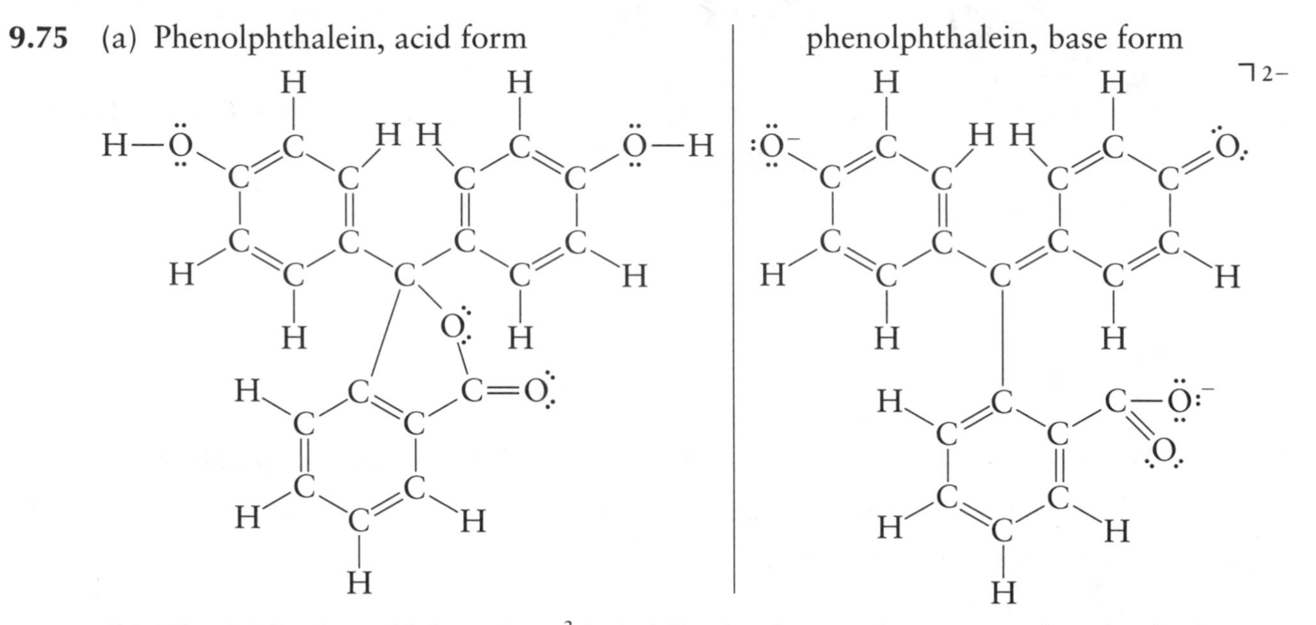

(b) The *C in the acid form is sp^3-hybridized, whereas the same carbon in the base form is sp^2-hybridized.

(c) The change of hybridization on the asterisk-labeled carbon changes the electronic structure of phenolphthalein. The base form has a more extensive conjugated system (see Investigating Matter 9.2), which may also account for the highly colored base form.

9.77 Assume that Figure 9.56 is appropriate.

(a) CO has 10 valence electrons; $\sigma_{2s}^2\sigma_{2s}^{*2}\pi_{2p}^4\sigma_{2p}^2$

(b) NO has 11 valence electrons; $\sigma_{2s}^2\sigma_{2s}^{*2}\pi_{2p}^4\sigma_{2p}^2\pi_{2p}^{*1}$

(c) CN^- has 10 valence electrons; $\sigma_{2s}^2\sigma_{2s}^{*2}\pi_{2p}^4\sigma_{2p}^2$

9.79 (a) 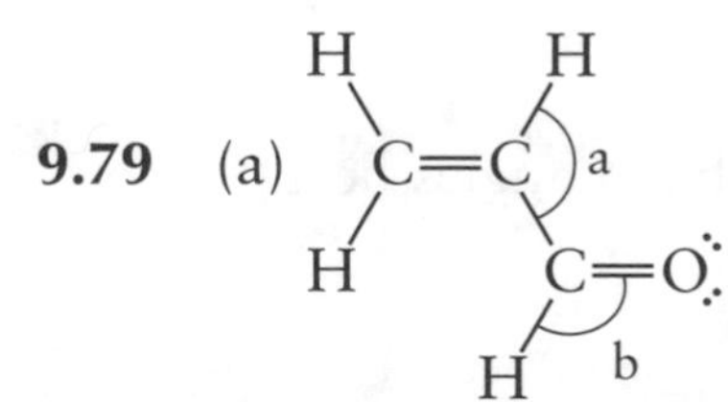

All C atoms can be considered as the central A atom of AX_3 structures; therefore, angles a and b are approximately 120°.

(b) Each carbon is sp^2-hybridized, the O is sp^2-hybridized.

(c) There are 7 σ bonds and 2 π bonds in the molecule.

INTEGRATED EXERCISES

9.81 The formula of the compound can be determined by the standard techniques fully described in the examples and exercises of Chapter 2. Here, it can be done more directly by noting that there must be at least one C atom and one O atom in the compound. Then, the number of H atoms is

$$\frac{(32.04 - 12.01 - 16.00)\ \text{g/mol}}{1.008\ \text{g/mol}} = 4$$

The formula is CH_4O.

(a) The only possible atomic arrangement is

```
      H
      |     ..
  H — C  —  O — H
      |     ..
      H
```

C is A of AX_4; therefore, the bond angles are 109.5°. O is A of AX_2E_2; therefore, the bond angles are slightly less than 109°.

(b) The electron-pair geometry about both C and O is tetrahedral; therefore, both have sp^3 hybridization.

(c) O is A of AX_2E_2; therefore, the molecule is polar.

9.83 (a), (b), and (c) are shown below:

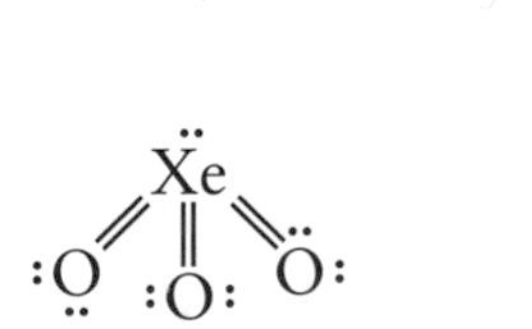

sp^3 hybridized
trigonal pyramidal
Bonds are slightly
less than 109°.

dsp^3 hybridized
or seesaw
Bonds are slightly
less than 120°.

d^2sp^3 hybridized
octahedral
All bonds are 90°.

(d) Xe is +6 in XeO_3, +6 in XeO_4^{2-}, and +8 in XeO_6^{4-}.

9.85 Assume the Lewis structures drawn.

$N_2(g) + \frac{1}{2} O_2(g) \longrightarrow N_2O(g)$ $\quad :\ddot{N}{=}N{=}\ddot{O}:$

$$\Delta H_f^\circ = -\Delta H_B(N{=}N) - \Delta H_B(N{=}O) + \Delta H_B(N{\equiv}N) + \tfrac{1}{2}\Delta H_B(O{=}O)$$
$$= -(-409 - 630 + 944 + \tfrac{1}{2}(496))\ \text{kJ/mol} = 153\ \text{kJ/mol}$$

$\frac{1}{2} N_2(g) + O_2(g) \longrightarrow NO_2(g)$ $\quad :\ddot{O}{-}\dot{N}{=}\ddot{O}:$

$$\Delta H_f^\circ = -\Delta H_B(N{=}O) - \Delta H_B(N{-}O) + \Delta H_B(O{=}O) + \tfrac{1}{2}\Delta H_B(N{\equiv}N)$$
$$= [-630 - 210 + 496 + \tfrac{1}{2}(944)]\ \text{kJ/mol} = 128\ \text{kJ/mol}$$

From Appendix 2A, $\Delta H_f^\circ = +82.05$ kJ/mol for N_2O and $\Delta H_f^\circ = +33.18$ kJ/mol for NO_2.

(a) Resonance stabilization energy can be determined by finding the difference between the standard enthalpy of formation in Appendix 2A and the enthalpy of formation calculated from bond enthalpy.
(b) For N_2O, we find (153 − 82.05) kJ/mol = 71 kJ/mol
For NO_2, we find (128 − 33.18) kJ/mol = 95 kJ/mol
In each case, the enthalpy of formation is lowered from the value predicted by bond enthalpies by the amount calculated above. This quantity is the resonance stabilization energy.
(c) NO_2 must have more resonance forms that contribute to its actual structure than has N_2O.

9.87 (a) O_2 has 12 valence electrons, as we know it. However, if three electrons occupied an orbital, then each oxygen would have 3 electrons in a full orbital (what we call 1*s*), and 5 valence electrons. O_2 would therefore have 10 valence electrons and a configuration of $\sigma_{2s}{}^{3}\sigma_{2s}{}^{*3}\sigma_{2p}{}^{3}\pi_{2p}{}^{1}$ (based on the orbital arrangement found in Figure 9.57). Oxygen atoms would then have 5 valence electrons.
(b) Ar has 18 electrons. The 1*s* orbital equivalent would contain 3 electrons, the 2*s* would contain 3 electrons, and the 2*p* orbitals would contain 9 electrons, leaving three valence electrons in the 3*s* orbitals. Ar_2 would contain 6 valence electrons and an electron configuration of $\sigma_{3s}{}^{3}\sigma_{3s}{}^{*3}$. The bond order is zero, so Ar_2 would not likely exist.
(c) The "octet rule" would be the "dozen rule," because each atom would be most stable with 12 valence electrons. Single bonds would require 3 shared electrons; a "lone triplet" would have 3 electrons as well. Double bonds would contain 6 electrons and triple bonds would contain 9 electrons.

CHAPTER 10
LIQUIDS AND SOLIDS

EXERCISES

Intermolecular Forces

10.1 (a) London forces
(b) dipole-dipole, London forces
(c) London forces
(d) dipole-dipole, London forces

10.3 (a) :C≡O: London forces; very weak dipole-dipole forces also exist.

(b) H—Ö—H (bent, with two lone pairs on O) Hydrogen bonding; dipole-dipole and London forces also occur.

(c) NF_3 (N with one lone pair, bonded to three :F: atoms) London forces; very weak dipole-dipole forces

(d) H—C—H with H above and below C (CH_4) London forces

10.5 A hydrogen bond may be described as A—H···B with the dotted bond, ···, representing the hydrogen bond. A must be N, O, or F. They are the only elements sufficiently electronegative to produce the strongly polar A—H bond necessary to attract the lone pair of electrons on the neighboring B atoms. In like manner, the B atoms should also be N, O, or F, because they are the smallest of the highly electronegative elements. Their small size results in the strongest interaction with the positive side of the polar A—H bond.

10.7 (a) HF, (c) NH_3, and (d) CH_3OH can form hydrogen bonds. Hydrogen bonding is especially strong in HF. All three have higher normal boiling points, higher enthalpies of vaporization, higher surface tension, and lower vapor pressures at a given temperature than similar compounds of their congeners.

10.9 (a)

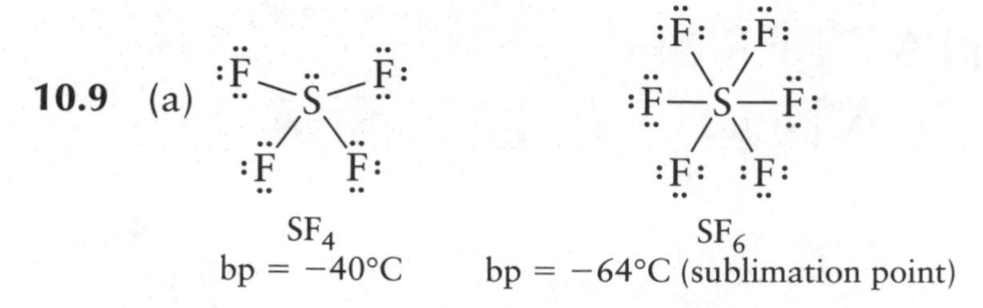

VSEPR predicts that SF_4 has the seesaw shape; whereas SF_6 is octahedral; so SF_4 is polar and SF_6 is not. SF_4 has the higher boiling point.

(b)

BF_3 bp = −99.9°C

ClF_3 bp = 11.3°C

VSEPR predicts that BF_3 is trigonal planar and nonpolar, whereas ClF_3 is T-shaped and polar. Therefore, ClF_3 has the higher boiling point.

(c)

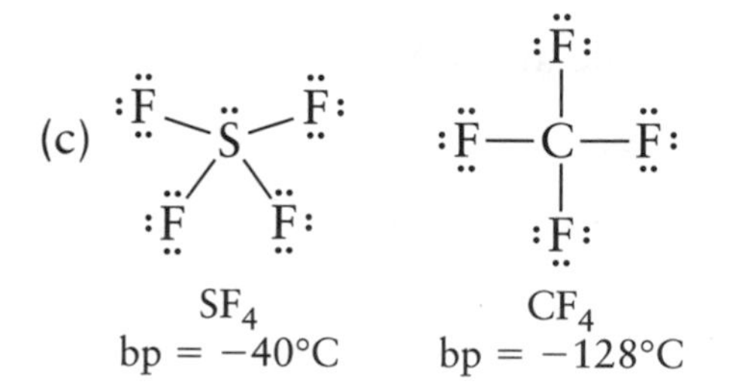

SF_4 is seesaw; CF_4 is tetrahedral. SF_4 has the higher boiling point; it has a small dipole moment; CF_4 has none.

(d)

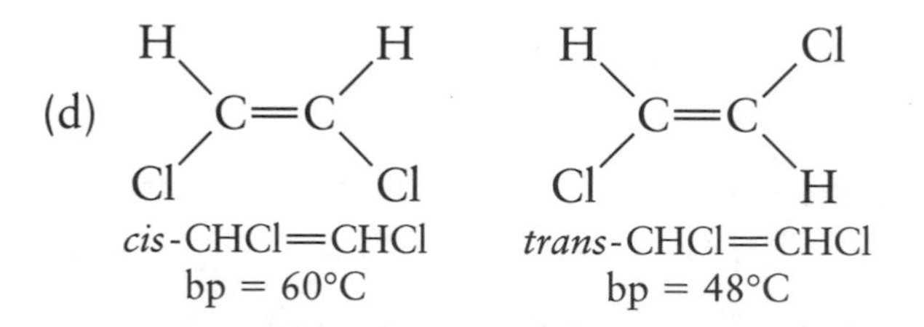

Both molecules have the same shape, but the atomic arrangement in the *cis* compound produces a dipole moment, and thus, a higher boiling point.

Liquid Structure

10.11 The intermolecular forces are stronger between water molecules in water than between water molecules and the hydrocarbon molecules in wax. Water molecules hydrogen bond to one another, but not to hydrocarbon molecules. These intermolecular forces in water are manifested in the physical property called surface tension.

10.13 (a) Ethanol has the greater viscosity, as a result of its stronger intermolecular forces due to hydrogen bonding, than has dimethyl ether, which cannot engage in hydrogen bonding.
(b) Propanone has the greater viscosity; it has stronger intermolecular forces, due to dipole-dipole interactions than has butane, which is nonpolar. Butane is, in fact, a gas at 0°C. It boils at −0.5°C.

Classification of Solids

10.15 (a) ionic (b) molecular (c) molecular (d) metallic

10.17 Use Table 10.3. A, ionic; B, metallic; C, molecular

Metallic Crystals

10.19 Refer to Fig. 10.17 and 10.20b, also Example 10.3.
(a) number of atoms per unit cell = 1 center atom × 1 atom per center = 1 atom
8 corner atoms × $\frac{1}{8}$ atom per corner = 1 atom
total = 2 atoms per unit cell
(b) Each atom is surrounded by 8 others; the coordination number is 8.

10.21 (a) See Example 10.4. The length of a side in an fcc structure is
side = $\sqrt{8}\,r = 2.828 \times 125$ pm = 354 pm = 3.54×10^{-8} cm
(b) volume of unit cell = side^3 = $(3.54 \times 10^{-8}\ \text{cm})^3 = 4.44 \times 10^{-23}\ \text{cm}^3$

$$\text{number of unit cells in } 1.00\ \text{cm}^3 = \frac{1.00\ \text{cm}^3}{4.44 \times 10^{-23}\ \text{cm}^3/\text{unit cell}} = 2.25 \times 10^{22}\ \text{unit cells}$$

10.23

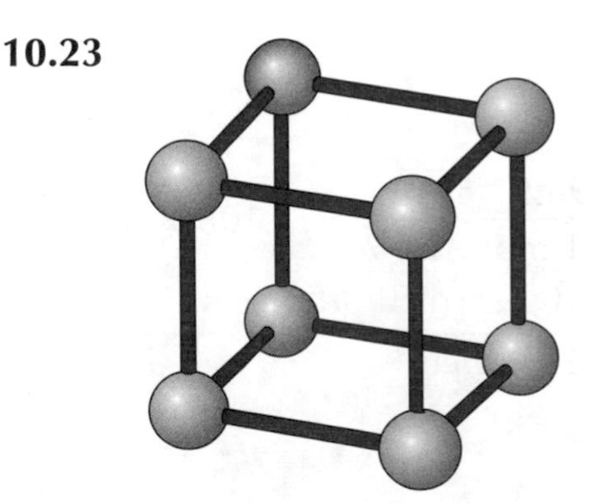

(a) There are 8 corners, so the number of atoms per unit cell = 8 corners × $\frac{1}{8}$ atom/corner = 1 atom.

(b) There are 6 nearest neighbors; so the coordination number is 6.

(c)

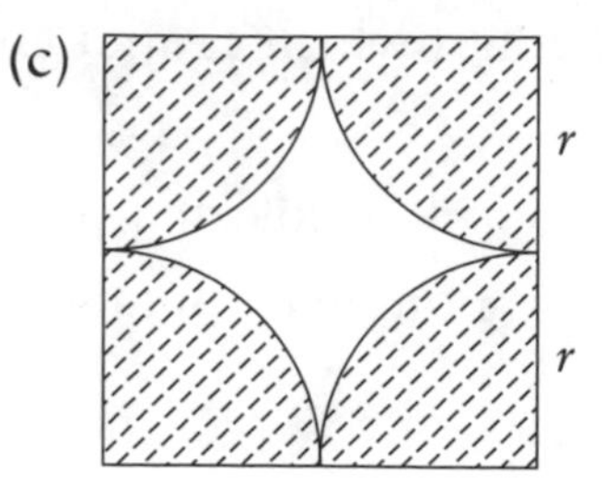

The length of side of unit cell is $2r$, and $r = 167$ pm.

Length $= 2 \times 167 \text{ pm} = 334 \text{ pm}$

10.25 (a) See Example 10.4 and the solutions to Exercises 10.25 and 10.26.

Let a = length of a side of the unit cell.

$a = \sqrt{8}r = 2.828 \times 125 \text{ pm} = 354 \text{ pm}$

volume of cell $= (354 \times 10^{-12} \text{ m})^3 = 4.44 \times 10^{-29} \text{ m}^3 = 4.44 \times 10^{-23} \text{ cm}^3$

$$\text{mass of one cell} = 4 \text{ Ni atoms} \times \frac{58.71 \text{ g/mol}}{6.022 \times 10^{23} \text{ Ni atoms/mol}}$$

$$= 3.900 \times 10^{-22} \text{ g}$$

$$\text{density} = \frac{\text{mass}}{\text{volume}} = \frac{3.900 \times 10^{-22} \text{ g}}{4.44 \times 10^{-23} \text{ cm}^3} = 8.78 \text{ g/cm}^3$$

(b) $a = \text{length of side} = \dfrac{4r}{\sqrt{3}} = \dfrac{4 \times 250. \text{ pm}}{\sqrt{3}} = 577 \text{ pm}$

volume of cell $= (577 \times 10^{-12} \text{ m})^3 = 1.92\overline{4} \times 10^{-28} \text{ m}^3 = 1.92\overline{4} \times 10^{-22} \text{ cm}^3$

$$\text{mass of one cell} = 2 \text{ Rb atoms} \times \frac{85.47 \text{ g/mol}}{6.022 \times 10^{23} \text{ Rb atoms/mol}}$$

$$= 2.838 \times 10^{-22} \text{ g}$$

$$\text{density} = \frac{\text{mass}}{\text{volume}} = \frac{2.838 \times 10^{-22} \text{ g}}{1.92\overline{4} \times 10^{-22} \text{ cm}^3} = 1.48 \text{ g/cm}^3$$

10.27 (a) $\text{mass of one unit cell} = (1 \text{ unit cell}) \times \left(\dfrac{4 \text{ atoms}}{1 \text{ unit cell}}\right)$

$$\times \left(\frac{1 \text{ mol Au}}{6.022 \times 10^{23} \text{ atoms Au}}\right) \times \left(\frac{196.97 \text{ g Au}}{1 \text{ mol Au}}\right) = 1.30\overline{8} \times 10^{-21} \text{ g}$$

$$V = \text{volume of unit cell} = \frac{\text{mass of unit cell}}{\text{density}} = \frac{1.30\overline{8} \times 10^{-21} \text{ g}}{19.3 \text{ g/cm}^3}$$

$$= 6.78 \times 10^{-23} \text{ cm}^3$$

$a = \text{length of side} = \sqrt[3]{V} = 4.08 \times 10^{-8} \text{ cm}$

$$\text{radius} = r = \frac{\sqrt{2}a}{4} = \frac{\sqrt{2}(4.08 \times 10^{-8} \text{ cm})}{4} = 1.44 \times 10^{-8} \text{ cm} = 144 \text{ pm}$$

(b) volume of unit cell $= \dfrac{\text{mass of unit cell}}{\text{density}}$

V = volume

$$= \frac{(1 \text{ unit cell}) \times \left(\dfrac{2 \text{ atoms}}{1 \text{ unit cell}}\right) \times \left(\dfrac{1 \text{ mol } V}{6.022 \times 10^{23} \text{ atoms } V}\right) \times \left(\dfrac{50.94 \text{ g } V}{1 \text{ mol } V}\right)}{6.11 \text{ g/cm}^3}$$

$$= 2.77 \times 10^{-23} \text{ cm}^3$$

$$a = \sqrt[3]{V} = 3.02 \times 10^{-8} \text{ cm}$$

$$\text{radius} = r = \frac{\sqrt{3}a}{4} = \frac{\sqrt{3} \times 3.02 \times 10^{-8} \text{ cm}}{4} = 1.31 \times 10^{-8} \text{ cm} = 131 \text{ pm}$$

Metals and Alloys

10.29 The fundamental difference is that the charge is carried by electrons in electronic conduction and by ions in ionic conduction. This difference is apparent from their names.

10.31 (a) Substitutional the atomic radii are similar.
(b) The melting point of the alloy will be lower than that of pure lead.

The Band Theory of Solids

10.33 (a) n-type, because P comes from a group with a higher number
(b) p-type, because In comes from a group with a lower number
(c) n-type, because Sb comes from a group with a higher number

10.35 In an electric conductor, the many molecular orbitals are very close together, forming a nearly continuous band (Fig. 10.26). If the molecular orbitals are not filled completely and are closely spaced, then electrons can be easily excited to empty molecular orbitals. An electric current then moves through the solid. In an electric insulator, a full band of orbitals, called a valence band, forms. There is a large energy gap between the full orbitals and the next available orbital band. Electrons are only excited to the higher energy orbitals if a large amount of energy is available. The electrons are not easily mobile, and we call the material an electric insulator.

10.37 Heating in a vacuum, where the partial pressure of oxygen is below its equilibrium value, causes partial loss of oxygen in the reaction

$ZnO(s) \longrightarrow ZnO_{1-x} + \frac{x}{2} O_2$

The zinc oxide is left nonstoichiometric, for example, $ZnO_{0.95}$. For every oxygen atom formed, two conducting electrons remain the zinc oxide lattice, and its conductivity increases. This trend is reversed when the ZnO is heated in oxygen.

Ionic Solids

10.39 (a)

Cl^-	Na^+
8 corners $\times \frac{1}{8}$ ion/corner = 1 ion	1 center $\times$ 1 ion/center = 1 ion
6 faces $\times \frac{1}{2}$ ion/face = 3 ions	12 edges $\times \frac{1}{4}$ ion/edge = 3 ions
total = 4 Cl^- ions	total = 4 Na^+ ions

Thus, there are 4 formula units of NaCl per unit cell.

(b)

Ca^{2+}	F^-
8 corners $\times \frac{1}{8}$ ion/corner = 1 ion	8 sites $\times$ 1 ion/site = 8 ions
6 faces $\times \frac{1}{2}$ ion/face = 3 ions	
total = 4 Ca^{2+} ions	total = 8 F^- ions

Thus, there are 4 formula units of CaF_2 per unit cell. The coordination numbers are 8 for Ca^{2+} and 4 for F^-.

10.41 Ti^{4+}: 8 corners $\times \frac{1}{8}$ ion per corner = 1 ion

Ca^{2+}: 1 center $\times$ 1 ion per center = 1 ion

O^{2-}: 12 edges $\times \frac{1}{4}$ ion per edge = 3 ions

Therefore, the formula is $CaTiO_3$.

10.43 FU = formula unit.

$$\text{number of unit cells} = (1\ \text{mm}^3) \times \left(\frac{0.1\ \text{cm}}{1\ \text{mm}}\right)^3 \times \left(\frac{2.17\ \text{g}}{1\ \text{cm}^3}\right) \times \left(\frac{1\ \text{mol}}{58.44\ \text{g}}\right) \times \left(\frac{6.022 \times 10^{23}\ \text{FU}}{1\ \text{mol}}\right) \times \left(\frac{1\ \text{unit cell}}{4\ \text{FU}}\right) = 5.59 \times 10^{18}\ \text{unit cells}$$

10.45 $\text{radius ratio} = \dfrac{\text{radius of cation}}{\text{radius of anion}}$

(a) radius ratio $= \frac{138 \text{ pm}}{196 \text{ pm}} = 0.704 > 0.7$

We predict the cesium chloride structure, but this is a close call. The cation is predicted to have a coordination number of 8 by the rules, but the actual structure is the rock-salt structure with a coordination number of 6.

(b) radius ratio $= \frac{58 \text{ pm}}{196 \text{ pm}} = 0.296 < 0.7$

We predict the rock-salt structure, with a coordination number of 6 for the cations.

(c) radius ratio $= \frac{136 \text{ pm}}{140 \text{ pm}} = 0.971 > 0.7$

Cesium chloride structure, coordination number 8.

Network Solids

10.47 B, C, Si, Ge, P, As

10.49 See Fig. 10.39. Each carbon atom is bonded to four other carbon atoms, but the average number of C—C bonds per carbon atom is obtained by dividing four by two, in order to avoid counting the same bond twice. Thus,

$$\left(\frac{713 \text{ kJ}}{1 \text{ mol C}}\right) \times \left(\frac{1 \text{ mol C}}{2 \text{ mol bonds}}\right) = 356 \text{ kJ/mol}$$

Molecular Solids

10.51 (a) London forces (b) dipole-dipole, London forces, and hydrogen bonding (c) dipole-dipole and London forces

10.53 $A_2B_2C_4$

Vapor Pressure

10.55 Dynamic equilibrium is a condition in which a forward process and its reverse are occurring simultaneously at equal rates. Thus, in a dynamic equilibrium, processes (at the molecular level) are occurring; it is only the net effect that is unchanging. In a static equilibrium, no process is occurring at any level. Static equilibrium is rare or nonexistent at the atomic or molecular level.

10.57 There are more lone pairs per molecule on the oxygen atoms in H_2O_2; its structure

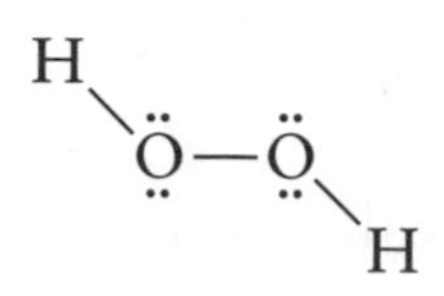

allows for stronger hydrogen bonding than in water. Its greater molecular weight allows for greater London forces. Both of these factors combine to produce lower vapor pressure and a higher boiling point.

10.59 $PV = nRT$

$$\text{amount (moles) of } H_2O = n = \frac{PV}{RT} = \frac{17.5 \text{ Torr} \times \left(\dfrac{1 \text{ atm}}{760. \text{ Torr}}\right) \times 1.0 \text{ L}}{0.08206 \text{ L·atm/K·mol} \times 293 \text{ K}}$$

$$= 9.5\overline{8} \times 10^{-4} \text{ mol}$$

$$\text{mass of } H_2O = 9.5\overline{8} \times 10^{-4} \text{ mol} \times 18.02 \text{ g/mol} = 0.017 \text{ g } H_2O$$

10.61 (a) 99.2°C (b) 99.7°C

Phase Changes

10.63 If we rearrange the Clausius-Clapeyron equation for $\frac{1}{T_1}$, we find that

$$\frac{R \ln\left(\dfrac{P_2}{P_1}\right)}{\Delta H_{vap}} + \frac{1}{T_2} = \frac{1}{T_1}$$

If T_1 = normal boiling temperature, then P_1 = 101.3 kPa

$T_2 = 34.9 + 273.15 = 308.0$ K and $P_2 = 13.3$ kPa

$\Delta H_{vap} = 43.5$ kJ/mol

$$\frac{(8.314 \text{ J/K·mol}) \ln\left(\dfrac{13.3 \text{ kPa}}{101.3 \text{ kPa}}\right)}{43.5 \times 10^3 \text{ J/mol}} + \frac{1}{308.0 \text{ K}} = \frac{1}{T_1}$$

$$-3.88\overline{0} \times 10^{-4}/\text{K} + 3.24\overline{7} \times 10^{-3}/\text{K} = \frac{1}{T_1}$$

$$T_1 = \frac{1}{2.85\overline{9} \times 10^{-3}/\text{K}} = 350. \text{ K or } 76.7°\text{C}$$

10.65 (a) vapor (b) liquid (c) vapor (d) vapor

10.67 (a) The liquid water existing at 0°C and 2 atm would freeze as the pressure drops.

(b) The liquid water existing at 50.°C and 2 atm would vaporize as the pressure drops.

10.69 (a) ~2.4 K (b) 10 atm (c) From figure, about 5 K; actual value is 4.2 K.
(d) No, there is no phase equilibrium line between solid and gas.

10.71 (a) At the lower pressure triple point, liquid helium I and II are in equilibrium with helium gas; at the higher pressure triple point, liquid helium I and II are in equilibrium with solid helium.
(b) The negative slope of the helium II/helium I phase boundary line suggests that—as in the case of the solid/liquid water phase boundary—helium I is the more dense, despite the fact that it is the higher temperature phase. Recall that solid water is less dense than liquid water. See the phase diagram of water, Fig. 10.51.

SUPPLEMENTARY EXERCISES

10.73 The intermolecular forces in isopropanol are not as strong as they are in water; so isopropanol has a lower boiling point (82.3°C) than water (100°C) and a higher vapor pressure, undoubtedly, than water at that temperature. Vaporization is an endothermic process, so heat is extracted from the skin; and because of its greater rate of evaporation than water, isopropanol has a more pronounced cooling effect.

10.75 (a) The attraction between atoms in both Xe and Ar is a result of London forces, but, because Xe has more electrons, it is bigger and therefore more polarizable; thus, it has stronger London forces and a higher melting point.
(b) HI and HCl have both London forces and dipole-dipole attractions, but the dominant attraction is the London force, which is greater in HI.
(c) There is strong hydrogen bonding in water, but none in $C_2H_5OC_2H_5$, so water has the lower vapor pressure at the same temperature.

10.77 (a) HCl has strong dipole-dipole interactions, but these forces are not nearly as strong as the ion-ion interactions in NaCl; thus, NaCl has the higher normal boiling point. It is a general rule that ionic compounds have higher melting and boiling points than molecular compounds.
(b) These compounds have the same structure, but SiH_4 has the greater molar mass; thus, we expect that it has the higher normal boiling point, and, in fact, it does: −112°C for SiH_4 versus −162°C for CH_4.
(c) HF, because of stronger hydrogen bonding
(d) H_2O, because of more opportunities for hydrogen bonding; there are two O—H bonds.

10.79 (a) H_2O; it has two O—H bonds, both of which are capable of strong hydrogen bonding. Each H_2O molecule is involved in about 4 hydrogen bonds.

(b) The surface tension of all these liquids is about the same. They are all capable of hydrogen bonding to about the same extent. However, London forces increase with increasing molar mass, so C_3H_7OH will have the greatest surface tension.

(c) CO_2, this is a nonpolar compound; SO_2 is polar and SiO_2 is a network solid.

(d) Very likely HCl; it has the weakest London forces, although it has the strongest dipole-dipole forces. The London forces predominate (see Example 10.2).

(e) H_2S; its hydrogen bonding forces are much less than those in H_2O, and its London forces are much less than those in H_2Te.

(f) H_2; it has the weakest London forces.

(g) NH_3, as a result of relatively strong hydrogen bonding forces.

(h) Na_2O; ion-ion forces are stronger than all other forces.

(i) All have the same shape, but the electronegativity difference is greatest in SF_2, so it probably has the strongest dipole-dipole forces.

(j) GeF_4, because it has the largest atoms with the most electrons.

10.81 (a) $\text{relative humidity} = \dfrac{25.0\text{ Torr}}{31.82\text{ Torr}} \times 100\% = 78.6\%$

(b) At 25°C, the vapor pressure of water is 23.76 Torr; therefore, some of the water vapor in the air would condense as dew or fog.

10.83 The spacing between the carbon layers increases, resulting in decreased perpendicular conductivity. The parallel conductivity increases, although the effect may not be dramatic, because the band is already half-filled.

10.85 r = atomic radius, a = length of side of unit cell, M = molar mass, d = density

for an fcc lattice; $a = \dfrac{4r}{\sqrt{2}} \quad V = a^3 = \left(\dfrac{4r}{\sqrt{2}}\right)^3$

$$\text{mass (g)} = 4\text{ atoms} \times \left(\frac{1\text{ mol atoms}}{6.022 \times 10^{23}\text{ atoms}}\right) \times \left(\frac{M\text{ g}}{1\text{ mol atoms}}\right)$$

$$d = \frac{\text{mass}}{V} = \frac{4M}{(6.022 \times 10^{23}) \times \left(\dfrac{4r}{\sqrt{2}}\right)^3} = \frac{2.94 \times 10^{-25} M}{r^3}$$

solving for r,

$$r = (6.65 \times 10^{-9})\sqrt[3]{\frac{M}{d}} \quad [M \text{ in g, } d \text{ in g/cm}^3]$$

$$r_{Ne} = (6.65 \times 10^{-9})\sqrt[3]{\frac{20.18}{1.21}} = 1.70 \times 10^{-8}\text{ cm} = 170\text{ pm}$$

$$r_{Ar} = (6.65 \times 10^{-9})\sqrt[3]{\frac{39.95}{1.66}} = 1.92 \times 10^{-8}\text{ cm} = 192\text{ pm}$$

$$r_{Kr} = (6.65 \times 10^{-9})\sqrt[3]{\frac{83.8}{2.82}} = 2.06 \times 10^{-8}\text{ cm} = 206\text{ pm}$$

$$r_{Xe} = (6.65 \times 10^{-9})\sqrt[3]{\frac{131.3}{3.56}} = 2.21 \times 10^{-8}\text{ cm} = 221\text{ pm}$$

$$r_{Rn} = (6.65 \times 10^{-9})\sqrt[3]{\frac{222}{4.4}} = 2.4\bar{6} \times 10^{-8}\text{ cm} = 24\bar{6}\text{ pm}$$

10.87 The relationship between the densities of a bcc structure and a ccp structure was derived in the solution to Exercise 10.86. It is

$d_{ccp} = 1.089 d_{bcc}$

Therefore, $d_{ccp}\,(\text{W}) = 1.089 d_{bcc}(\text{W}) = 1.089 \times 19.3\text{ g/cm}^3 = 21.0\text{ g/cm}^3$

Alternatively, we can rework the conversion as follows:

The mass of an atom of W is fixed; so

$$\text{mass per atom} = \frac{(19.3\text{ g/cm}^3)\left(\frac{4r}{\sqrt{3}}\right)\frac{\text{cm}^3}{\text{unit cell}}}{2\text{ atoms/unit cell}} = \frac{d_{ccp} \times \left(\frac{4r}{\sqrt{2}}\right)\frac{\text{cm}^3}{\text{unit cell}}}{4\text{ atoms/unit cell}}$$

where d_{ccp} = density in the ccp structure. Solving for d_{ccp}

$$d_{ccp} = (19.3\text{ g/cm}^3)(2)\left(\sqrt{\frac{2}{3}}\right)^3 = 21.0\text{ g/cm}^3$$

10.89 This material is a solid at room temperature and pressure, so it can't be nitrogen, oxygen, fluorine, or neon. The melting point is not high enough for carbon or boron. So the material must be beryllium or lithium. By consulting Appendix 2D, we find the phase diagram is for lithium.

10.91 $d = \lambda/2\sin\theta$, given in Investigating Matter 10.1

So $d = 152\text{ pm}/2\sin 12.1° = 362\text{ pm}$

APPLIED EXERCISES

10.93 (a) anisotropic (b) isotropic (c) anisotropic (d) anisotropic (e) anisotropic

10.95 Polar groups aid in the alignment of the molecules, the positive end of one molecule adhering to the negative end of another, and vice versa. That is, the dipoles align antiparallel to one another.

10.97 (a) Light of an appropriate wavelength may provide the energy needed for some electrons, or additional electrons, to jump the band gap. Electrical conductivity would increase.

(b) $E = h\nu$, and $\nu = \frac{c}{\lambda}$. Thus, $E = hc/\lambda$, or $\lambda = hc/E$

$\lambda = (6.626 \times 10^{-34}\ J{\cdot}s)(3.00 \times 10^{8}\ m/s)/2.9 \times 10^{-19}\ J$

$= 6.8 \times 10^{-7}$ m or $68\overline{5}$ nm

(c) Recall from Chapter 7 that infrared wavelengths are between about 1000 nm and 3×10^{6} nm, and the corresponding energy is between about 2×10^{-3} and 2×10^{-19} J. Infrared light does not contain enough energy to excite electrons across the band gap in selenium. So amorphous selenium can't be used in infrared burglar alarms.

INTEGRATED EXERCISES

10.99 (a) CH_3OH because it is a polar organic compound.

(b) KI; because the electronegativity difference is smaller between K and I; KI has more covalent character. I is also more polarized.

(c) LiBr; the electronegativity difference is smaller. Li and Br are also more polarizable, and therefore more covalent.

10.101 The net charge on iron must equal +4 overall to balance each oxygen's −2 charge. Let $x + y = 1.7$ and $3x + 2y = 4$, where x represents the fraction of Fe^{3+} and y the fraction of Fe^{2+}.

Then $x = 1.7 - y$ and $3(1.7 - y) + 2y = 4$

$5.1 - 3y + 2y = 4$ and $-y = -1.1$, or $y = 1.1$ Then $x = 0.6$

To check: $3(0.6) + 2(1.1) = 4$. The percentage of Fe^{2+} is $\frac{1.1}{1.7} \times 100 = 65\%$, and the percentage of Fe^{3+} must be 35%.

10.103 From Chapter 5, recall that a represents the effect of attractions. The more attractive forces, the larger a becomes. NH_3 molecules will hydrogen bond, so it has the largest value of a. CO_2 and O_2 have weak, intermolecular forces (London forces). CO_2 has somewhat stronger intermolecular forces because it has a larger molecular mass, and therefore a larger van der Waals a constant.

CHAPTER 11
CARBON-BASED MATERIALS

EXERCISES

Structures and Reactions of Aliphatic Compounds

11.1 The difference can be traced to the weaker London forces that exist between branched molecules. Atoms in neighboring branched molecules cannot lie as close together as they do in the unbranched isomers. As a result of the molecules' irregular shape, the atoms in neighboring branched molecules are more effectively shielded from one another than they are in neighboring unbranched molecules.

11.3 (a) four σ-type single bonds (b) two σ-type single bonds and one double bond, consisting of one σ bond and one π bond (c) one σ-type single bond, and one triple bond, consisting of one σ bond and two π bonds

11.5 (a) $CH_4 + Cl_2 \xrightarrow{\text{light}} CH_3Cl + HCl$, substitution
(b) $CH_2{=}CH_2 + Br_2 \longrightarrow CH_2Br{-}CH_2Br$, addition

11.7 (a) $CH_3CH_3 + 2\ Cl_2 \xrightarrow{\text{light}} CH_2ClCH_2Cl + 2\ HCl$, substitution
(b) $CH_2{=}CH_2 + Cl_2 \longrightarrow CH_2ClCH_2Cl$, addition
(c) $HC{\equiv}CH + 2\ Cl_2 \longrightarrow CHCl_2CHCl_2$, addition

Nomenclature of Aliphatic Compounds

11.9 (a) ethane (b) hexane (c) octane (d) heptane

11.11 (a) methyl (b) pentyl (c) propyl

11.13 (a) propane (b) ethane (c) pentane

11.15 (a) 4-methyl-2-pentene (b) 2,3-dimethyl-2-phenylpentane

11.17 (a) propene (no geometrical isomers) (b) *cis*-2-hexene, *trans*-2-hexene
(c) 1-butyne (no geometrical isomers) (d) 2-butyne (no geometrical isomers)

11.19 (a) $CH_2CHCH(CH_3)CH_2CH_3$ (b) $CH_3CH_2C(CH_3)_2CH(CH_2CH_3)(CH_2)_2CH_3$
(c) $CHC(CH_2)_2C(CH_3)_3$ (d) $CH_3CH(CH_3)CH(CH_2CH_3)CH(CH_3)_2$

11.21

```
        H   H   H   CH3   H   H   H   H   H
        |   |   |   |     |   |   |   |   |
(a) H—C—C—C—C——C—C—C—C—C—H
        |   |   |   |     |   |   |   |   |
        H   H   H   CH3   H   H   H   H   H
```

```
                          CH3
                          |
                  H   H   CH2   H   H   H   H   H
                  |   |   |     |   |   |   |   |
(b) H—C≡C—C—C——C——C—C—C—C—C—H
                  |   |   |     |   |   |   |   |
                  H   CH2 CH2   H   H   H   H   H
                      |   |
                      CH2 CH3
                      |
                      CH3
```

```
        H   CH3   H   CH3   H
        |   |     |   |     |
(c) H—C—C——C—C——C—H
        |   |     |   |     |
        H   CH3   H   H     H
```

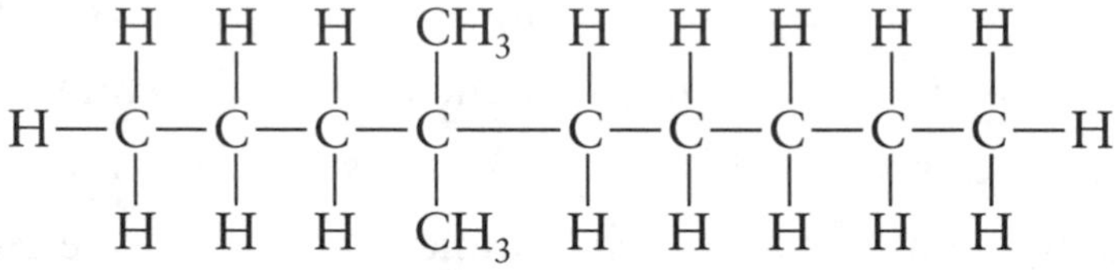

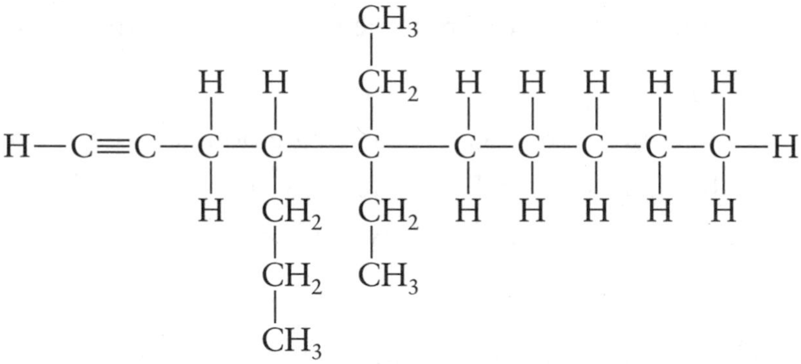

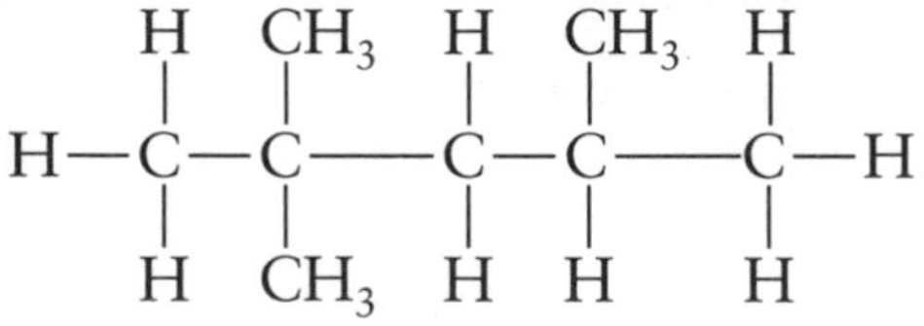

```
(d)  CH3CH2         H
           \       /
            C = C
           /       \
          H         CH2CH3
```

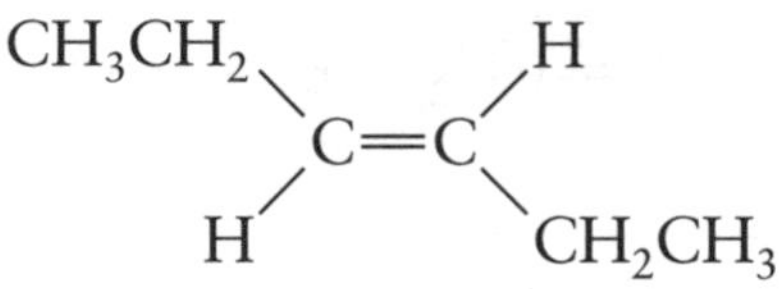

Aromatic Compounds

11.23 (a) 1-ethyl-3-methylbenzene (b) 1,2,3,4,5-pentamethylbenzene (or, because there is only one possible structure, pentamethylbenzene)

11.25 (a) 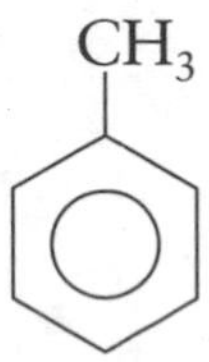 CH_3 (b)  CH_3, Cl (c) 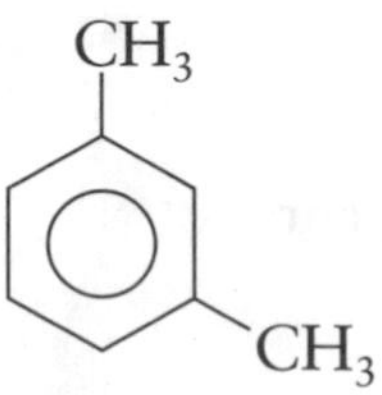 CH_3, CH_3 (d) 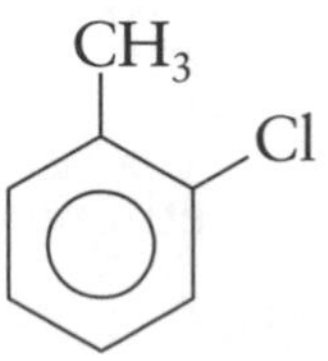 CH_3, Cl

11.27

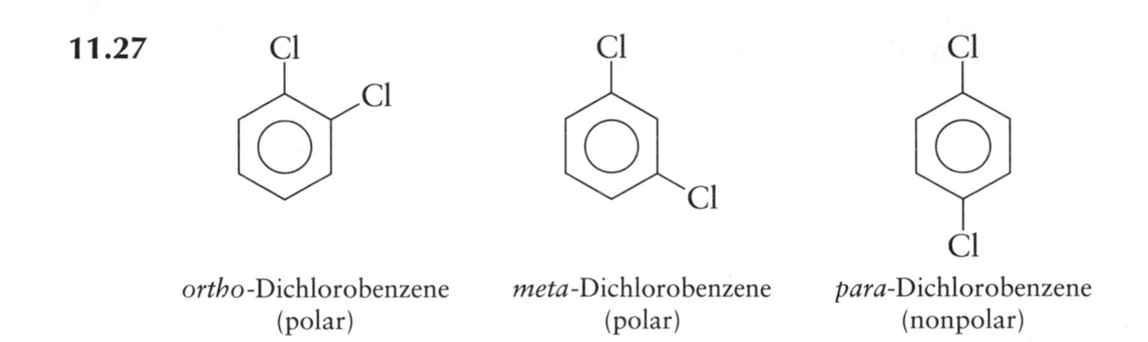

ortho-Dichlorobenzene (polar) *meta*-Dichlorobenzene (polar) *para*-Dichlorobenzene (nonpolar)

Identifying and Naming Functional Groups

11.29 (a) RNH_2, R_2NH, R_3N (b) ROH (c) $R{-}C({=}O){-}O{-}H$ or RCOOH

(d) $R{-}C({=}O){-}H$ or RCHO

11.31 (a) ether (b) ketone (c) primary amine (d) ester

11.33 (a) $CH_3CH_2CH_2CH_2OH$, primary alcohol (b) $CH_3CH_2CH(OH)CH_3$, secondary alcohol (c) $H_3C{-}C_6H_4{-}OH$, phenol

11.35 (a) diethyl ether (b) $CH_3OCH_2CH_3$

11.37 (a) aldehyde, ethanal
(b) ketone, propanone
(c) ketone, 3-pentanone

11.39 (a) ethanoic acid (b) butanoic acid (c) 2-aminoethanoic acid (glycine)

11.41 (a) methylamine (b) diethylamine (c) σ-methylaniline or 2-methylaniline or σ-methylphenylamine

11.43 (a) 2-propanol (b) dimethyl ether (c) methanal (d) 3-pentanone
(e) dimethylamine

11.45 (a) alcohol ($-OH$), ether ($-OCH_3$), aldehyde ($-CHO$)

(b) ketone $\left(\gt C{=}O \right)$

(c) tertiary amine $\left(\gt N{-}CH_3 \right)$, amide $\left(\gt N{-}C \lt^{O}_{} \right)$, ketone $\left(-C \lt^{O}_{} \right)$

Structures and Reactions of Functional Groups

11.47 (a) $H_2C{=}O$ (b) $(CH_3)_2C{=}O$ (c) $CH_3(CH_2)_3CH_2(CH_3)C{=}O$

11.49 (a) ethanol (b) 2-octanol (c) 5-methyl-1-octanol These reactions can be accomplished with an oxidizing agent such as acidified sodium dichromate, $Na_2Cr_2O_7$, though (a) and (c) may require a milder oxidizing agent, such as Ag and heat.

11.51 (a) $C_6H_5{-}C({=}O){-}OH$ (b) $CH_3{-}CH_2{-}CH(CH_3){-}C({=}O){-}OH$ (c) $CH_3{-}CH_2{-}C({=}O){-}OH$

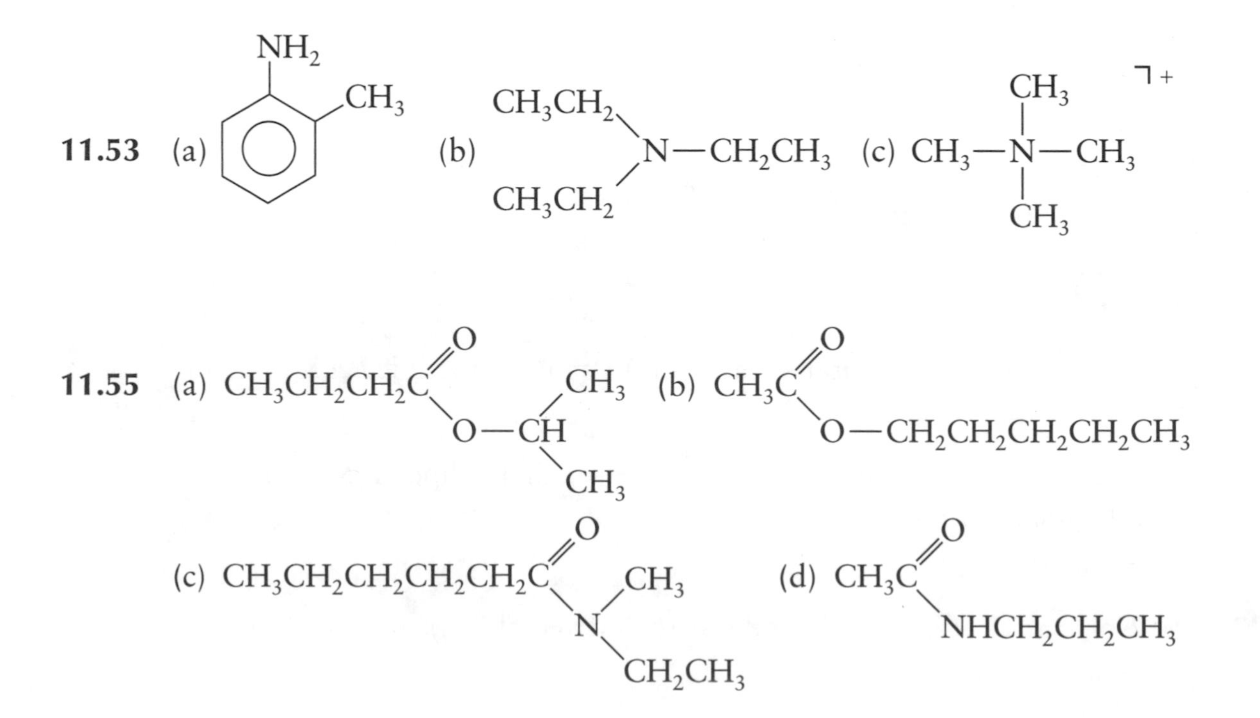

11.57 The following procedures may be used:

(1) Use an acid-base indicator and look for a color change.

(2) $CH_3CH_2CHO \xrightarrow{\text{Tollens reagent}} CH_3CH_2COOH + Ag(s)$

(3) $CH_3COCH_3 \xrightarrow{\text{Tollens reagent}}$ no reaction

Procedure (1) distinguishes ethanoic acid from propanal and 2-propanone. (2) and (3) distinguish propanal from 2-propanone.

Isomerism

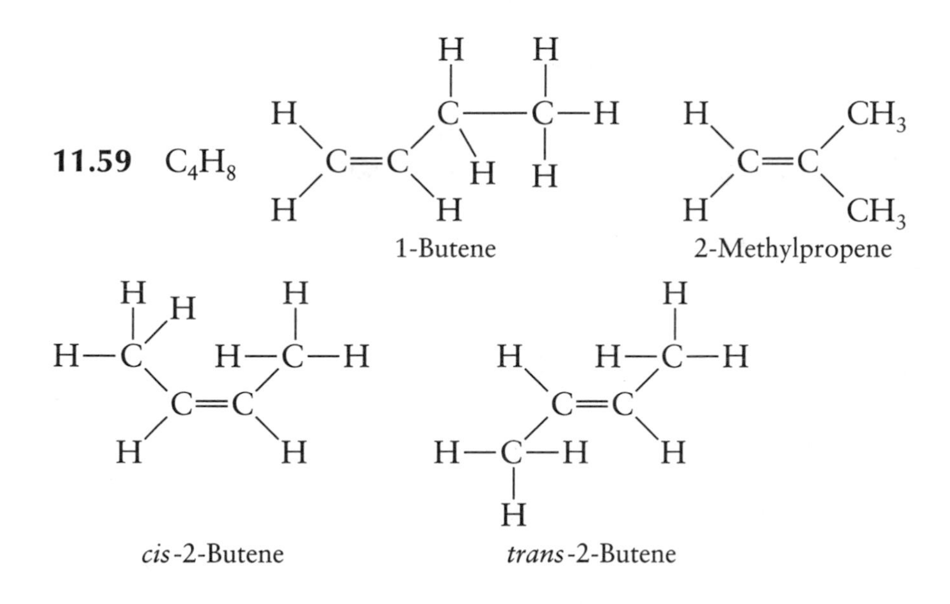

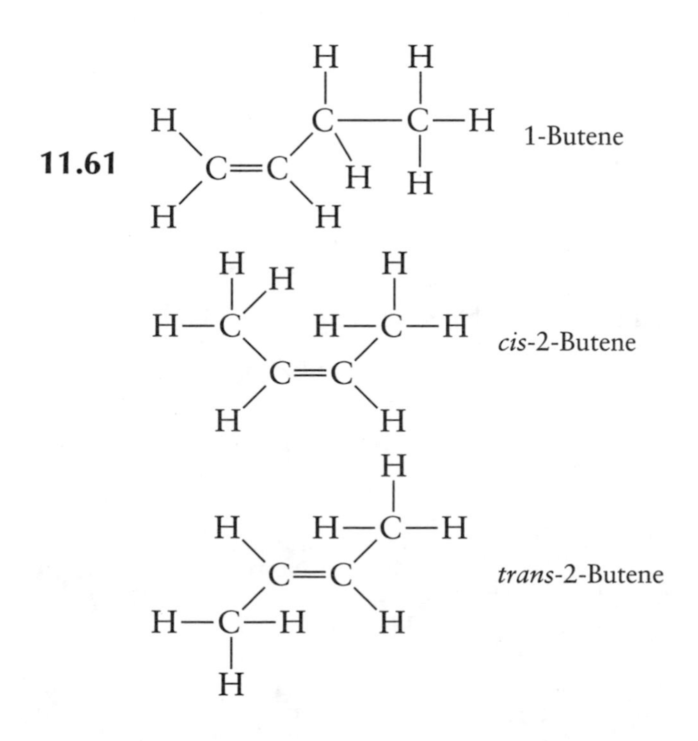

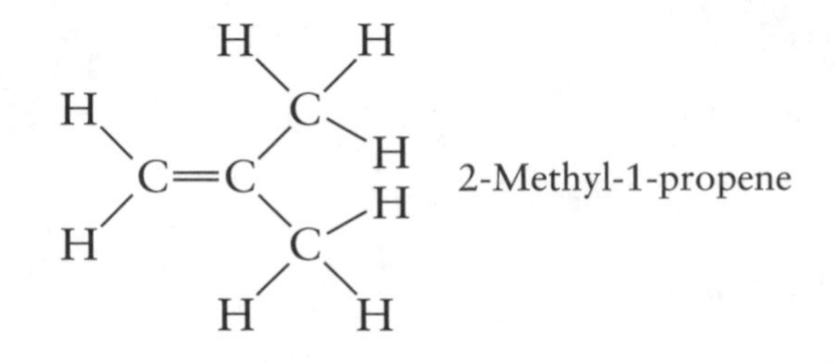

11.63 (a) Butane is C_4H_{10}, cyclobutane is C_4H_8. Because they have different formulas, they are not isomers.

(b) $CH_3CH_2CH_2CH_2CH_3$ $CH_3—C(CH_3)_2—CH_3$

Pentane (C_5H_{12})   2,2-Dimethylpropane (C_5H_{12})

Same formula, but different structures; therefore, they are structural isomers.

(c)

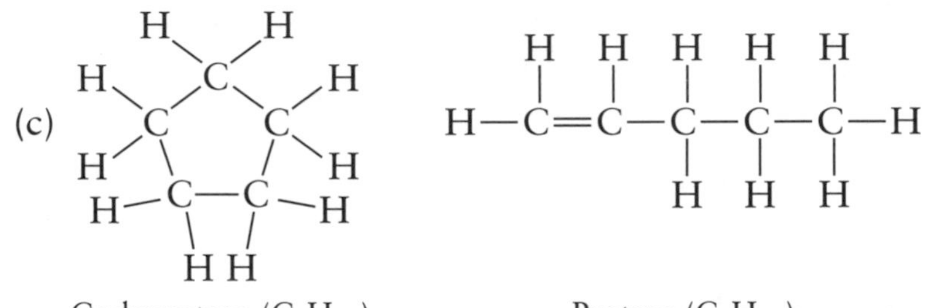

Cyclopentane (C_5H_{10})   Pentene (C_5H_{10})

Same formula, but different structures; therefore, they are structural isomers.

(d) Same formula (C_5H_{10}), same structure (bonding arrangement is the same), but different geometry; therefore, they are geometrical isomers.

(e) Not isomers, because only their positions in space are different and these positions can be interchanged. Same molecule.

11.65 If only two isomeric products are formed, and they are both branched, then the only possibilities are

(a) $CH_3—CH(CH_3)—CH_3$ (b) $CH_3—CCl(CH_3)—CH_3$ $Cl—CH_2—CH(CH_3)—CH_3$

Note: All methyl groups are equivalent.

11.67 (a) , (c), and (d) are optically active.

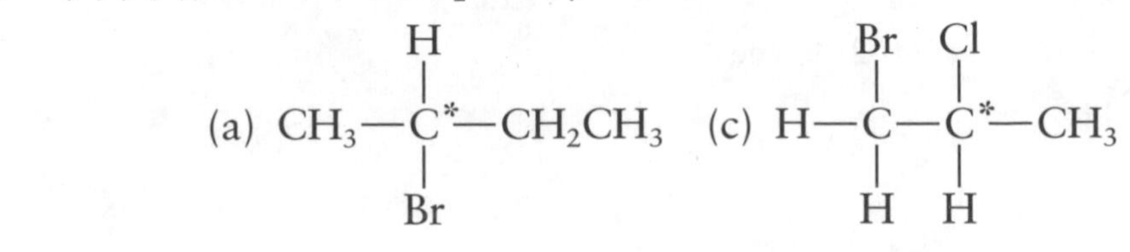

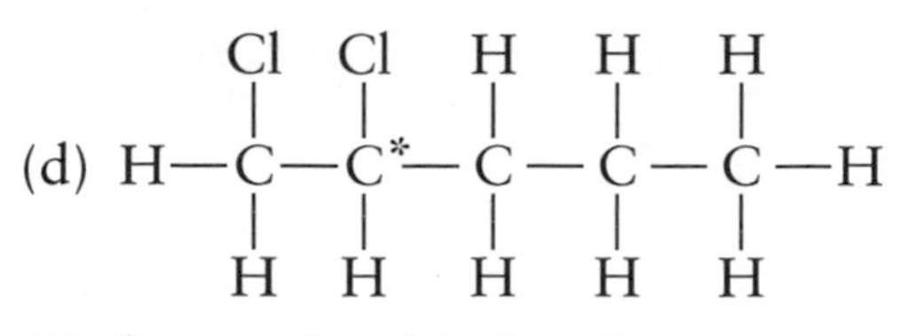

*Indicates the chiral carbon atoms.

11.69 (a)

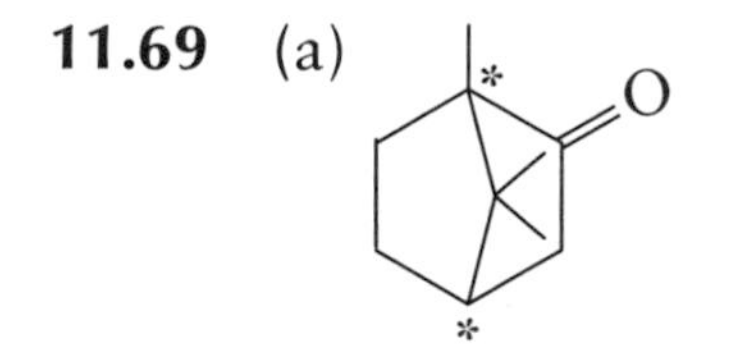

*Indicates the chiral carbon atoms.

Polymers

11.71

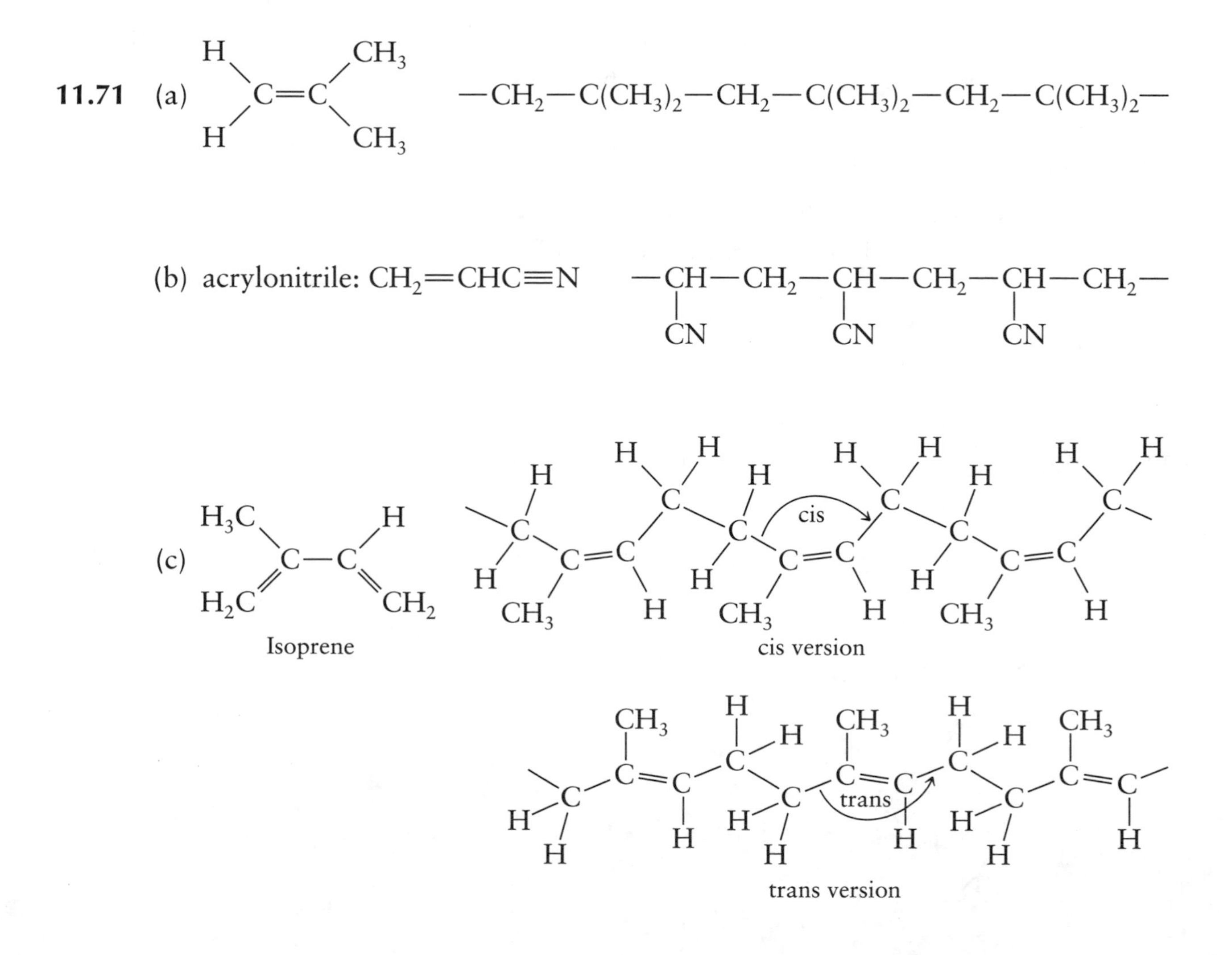

11.73 (a) $\underset{\displaystyle Cl}{\underset{|}{CH}}{=}CH_2$ (b) $\overset{\displaystyle F}{\overset{|}{\underset{\displaystyle Cl}{\underset{|}{C}}}}{=}CF_2$

11.75 (a) $-OC-CO-NH-(CH_2)_4-NHCO-CO-NH-(CH_2)_4-NH-$
(b) $-OC-\underset{\displaystyle CH_3}{\underset{|}{CH}}-NH-OC-\underset{\displaystyle CH_3}{\underset{|}{CH}}-NH-$

11.77 An isotactic polymer is a polymer in which the substituents are all on the same side of the chain.
A syndiotactic polymer is a polymer in which the substituent groups alternate, one side of the chain to the other.
An atactic polymer is a polymer in which the groups are randomly attached, one side or the other, along the chain.

11.79 Longer chain length allows for greater intertwining of the chains, making them more difficult to pull apart. This twining results in (a) higher softening points, (b) greater viscosity, and (c) greater mechanical strength.

11.81 Highly linear, unbranched chains allow for maximum interaction between chains. The greater the intermolecular contact between chains, the stronger the forces between them, and the greater the strength of the material.

Biopolymers

11.83 (a) $-C(=O)-NH-$ (b) amide (c) condensation

11.85 $H_2N-CH(COOH)-CH_2-CH_2-CH_2-CH_2-NH_2$

11.87 Side groups that contain hydroxyl, carbonyl, and amino groups are all potentially capable of participating in hydrogen bonding that could contribute to the tertiary structure of the protein. Serine, threonine, tyrosine, aspartic acid, glutamic acid, lysine, arginine, histidine, asparagine, and glutamine satisfy the criteria. Proline and tryptophan generally do not contribute through hydrogen bonding because they are typically found in hydrophobic regions of proteins.

11.89

```
        NH2   O
        |    //
     H—C—C
        |    \
       CH2    NH—CH2—COOH
        |
      (benzene ring)
        |
       OH
```

11.91

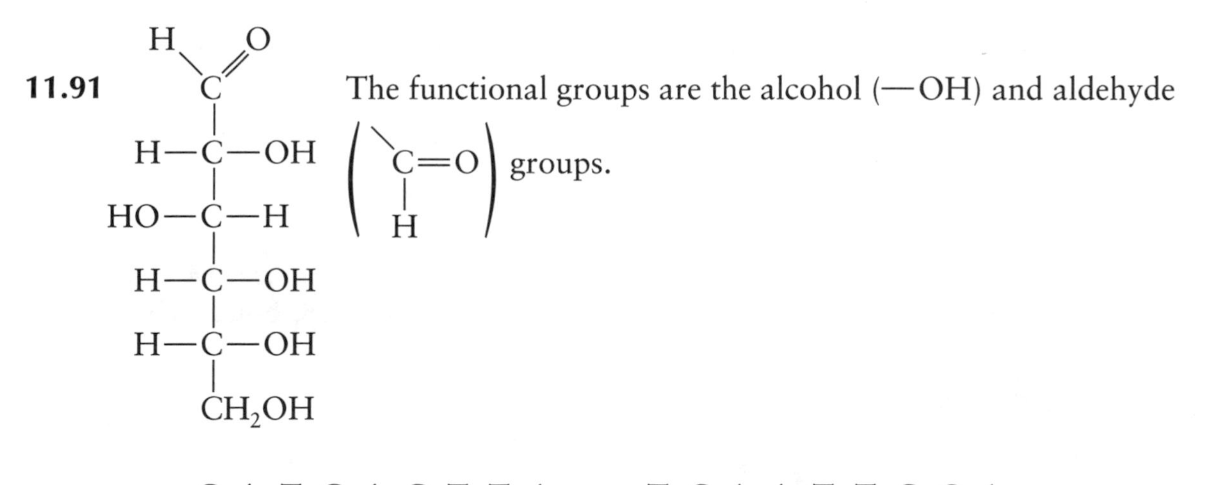

The functional groups are the alcohol (—OH) and aldehyde (—CH=O) groups.

11.93 (a)

```
C A T G A G T T A
| | | | | | | | |
G T A C T C A A T
```

(b)

```
T G A A T T G C A
| | | | | | | | |
A C T T A A C G T
```

SUPPLEMENTARY EXERCISES

11.95 There are a large number and a wide variety of organic compounds because of the ability of carbon to form four bonds with as many as four different atoms or groups of atoms, the ability of carbon to bond directly to other carbon atoms to form long chains of carbon-carbon bonds (the basis for polymerization), and the ability of carbon-containing compounds to form a variety of isomers: compounds with the same molecular formula but different structural formulas.

11.97 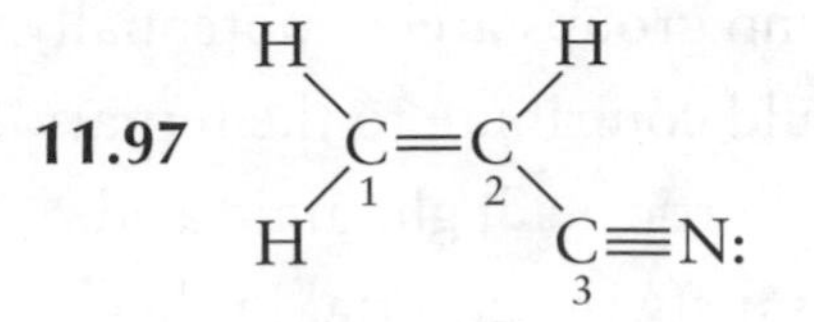

Carbon 1 has 3 σ bonds and 1 π bond; Carbon 2 has 3 σ bonds and one π bond; Carbon 3 has 2 σ bonds and 2 π bond. The CCC bond angle is approximately 120°. Carbon 1 and Carbon 2 have sp^2 hybridization. Carbon 3 has sp hybridization.

11.99 $H_3C{-}C{\equiv}C{-}CH_3 + 2\ HBr \longrightarrow H_3C{-}CHBr{-}CHBr{-}CH_3$

11.101 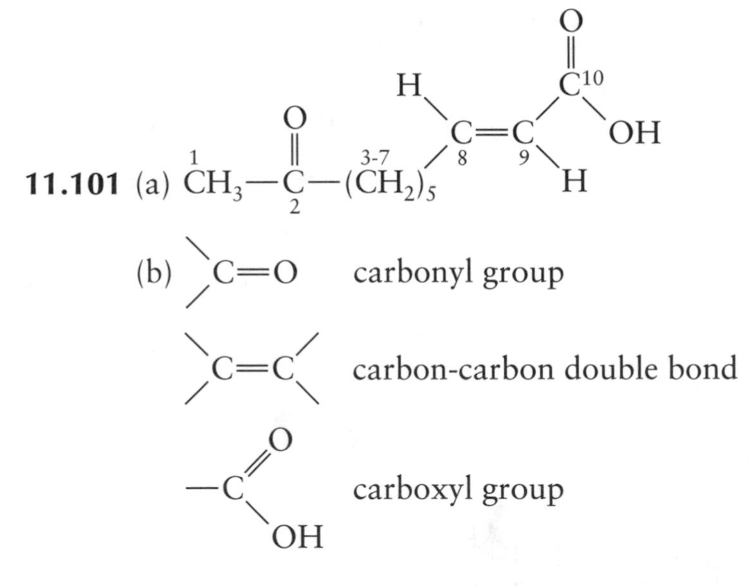

(c) 28 σ bonds, 3π bonds

(d) Carbons 1 and 3–7 are sp^3 hybridized; Carbons 2 and 8–10 are sp^2 hybridized. Both carbonyl oxygens are also sp^2 hybridized.

11.103 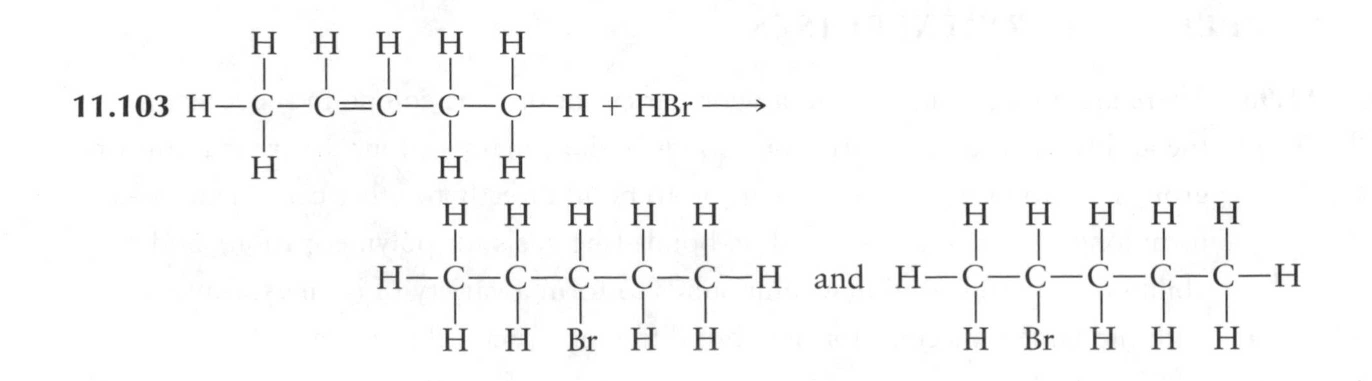

11.105

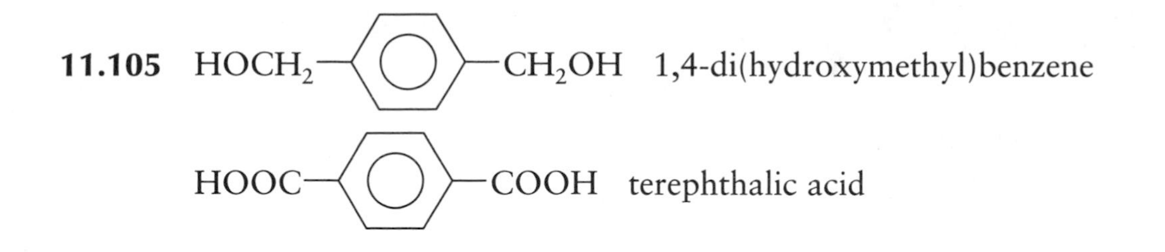

11.107 $CH_3(CH_2)_2CH(OH)CH_3 \xrightarrow[120°C]{H_2SO_4} H_2O(g) + CH_3(CH_2)_2CH{=}CH_2$

APPLIED EXERCISES

11.109 (a) The Lewis structure of ethyne is H—C≡C—H. See the figure below for its valence bond structure. The top portion of the figure shows the σ-bond framework and the bottom portion shows the overlapping of atomic p-orbitals that result in the two π-bonds.

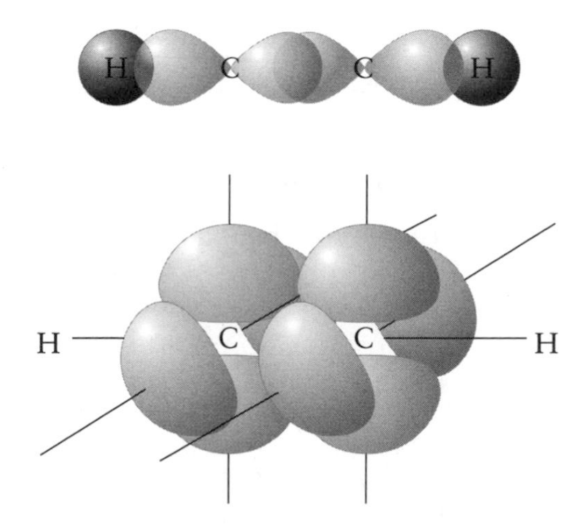

The carbon atoms are *sp* hybrids, as shown in the top part of the figure.

(b) —C=C—C= ⟷ =C—C=C— (each carbon bonded to an H below: H H H, H H H)

Note: The second structure has the same number of C—H units, but the single and double bonds have exchanged places.

(c) Each carbon is sp^2 hybridized.

(d) Because of resonance, every carbon-carbon bond in polyacetylene has considerable double-bond character. Free rotation cannot occur about double bonds; even though these bonds are only partially double, we expect polyacetylenes to be fairly rigid.

11.111 (a) NH_2 (benzene ring with NH_2 substituent)

(b) sp^3 hybridized

(c) sp^2 hybridized

(d) Each nitrogen has a lone pair.

(e) The double and single bonds on the nitrogen atoms are linked to the alternating double bonds on the carbon atoms. Electrons are delocalized along this long chain, so yes, the nitrogens do help carry the current.

INGEGRATED EXERCISES

11.113
$$\text{number of moles of H} = \left(\frac{3.32 \text{ g } H_2O}{18.02 \text{ g } H_2O/\text{mol } H_2O}\right) \times \left(\frac{2 \text{ mol H}}{1 \text{ mol } H_2O}\right)$$
$$= 0.368 \text{ mol H}$$
$$\text{number of moles of C} = \left(\frac{6.48 \text{ g } CO_2}{44.01 \text{ g } CO_2/\text{mol } CO_2}\right) \times \left(\frac{1 \text{ mol C}}{1 \text{ mol } CO_2}\right)$$
$$= 0.147 \text{ mol C}$$
$$\frac{0.368 \text{ mol H}}{0.147 \text{ mol C}} = 2.50\left(\frac{\text{mol H}}{\text{mol C}}\right) = \frac{5 \text{ mol H}}{2 \text{ mol C}}$$

Therefore, the empirical formula is C_2H_5. The molecular formula might be C_4H_{10}, which matches the general formula for alkanes, C_nH_{2n+2}. The compound cannot be an alkene or alkyne, because they all have mol H/mol C ratios less than 2.5.

11.115 (a)

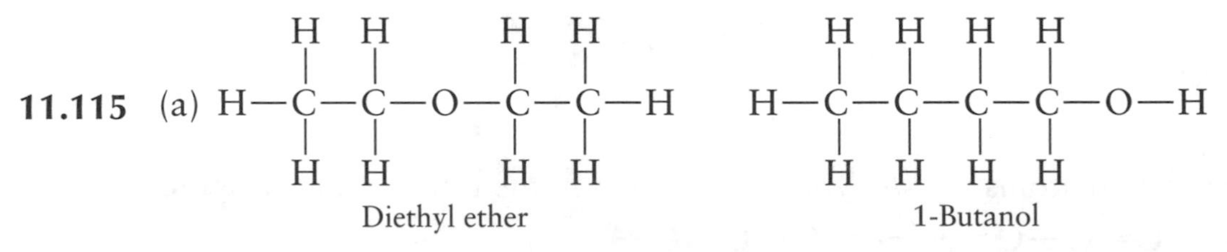

Diethyl ether 1-Butanol

(b) 1-Butanol is capable of hydrogen bonding, but diethyl ether is not. The stronger interactions between 1-butanol molecules leads to its higher boiling point.

11.117 (a) Cyclohexene has one double bond, so one mole of H_2 is needed to hydrogenate 1.00 mol of cyclohexene.

$V = nRT/P$ so $V = (1.000 \text{ mol } H_2)(0.082\ 06 \text{ L}\cdot\text{atm/K}\cdot\text{mol})(298 \text{ K})/1.00 \text{ atm}$
$= 24.4 \text{ L } H_2$

(b) Benzene has three double bonds, so three moles of H_2 are needed to hydrogenate 1.00 mol of benzene.

$V = (3.00\ \text{mol}\ H_2)(0.082\ 06\ \text{L}\cdot\text{atm/K}\cdot\text{mol})(298\ \text{K})/1.00\ \text{atm} = 73.4\ \text{L}\ H_2$

(c) For cyclohexene,

$$\Delta H° = [\Delta H_b(\text{C}{=}\text{C}) + \Delta H_b(\text{H}{-}\text{H})] - [2\Delta H_b(\text{C}{-}\text{H}) + \Delta H_b(\text{C}{-}\text{C})]$$
$$= (612 + 436)\ \text{kJ/mol} - (2 \times 412 + 348)\ \text{kJ/mol} = -124\ \text{kJ/mol}$$

For benzene,

$$\Delta H° = [3\Delta H_b(\text{C}{\overset{\cdots}{-}}\text{C}) + 3\Delta H_b(\text{H}{-}\text{H})] - [6\Delta H_b(\text{C}{-}\text{H}) + 3\Delta H_b(\text{C}{-}\text{C})]$$
$$= [6(518) + 3(436)]\ \text{kJ/mol} - [6(412) + 6(348)]\ \text{kJ/mol} = -144\ \text{kJ/mol}$$

The calculations do not support this statement. The enthalpy of benzene is lower than its Lewis structures suggest because of resonance. Without resonance, the enthalpy of hydrogenation for benzene would be $3 \times (-124\ \text{kJ/mol}) = -376$ kJ/mol.

11.119 (a) $H_2C = CHCH_3(g) + Br_2(l) \longrightarrow CH_2BrCHBrCH_3$

(b) $n = PV/RT \quad n = (10.0\ \text{atm})(0.400\ \text{L})/(0.082\ 06\ \text{L}\cdot\text{atm/K}\cdot\text{mol})(298\ \text{K})$
$= 0.164$ mol propene

$$0.500\ \text{L}\ Br_2 \left(\frac{0.220\ \text{mol}\ Br_2}{\text{L}}\right) = 0.110\ \text{mol}\ Br_2$$

Because the stoichiometry is 1 : 1, Br_2 limits.

(c) $$0.110\ \text{mol}\ Br_2 \left(\frac{1\ \text{mol}\ CH_2BrCHBrCH_3}{1\ \text{mol}\ Br_2}\right)\left(\frac{201.90\ \text{g}\ CH_2BrCHBrCH_3}{1\ \text{mol}\ CH_2BrCHBrCH_3}\right)$$
$$= 22.2\ \text{g}\ CH_2BrCHBrCH_3$$

CHAPTER 12
THE PROPERTIES OF SOLUTIONS

EXERCISES

Solubility

12.1 Both water and methanol are capable of hydrogen bonding, and they readily hydrogen bond to each other. Thus, they intermingle at the molecular level. Toluene, on the other hand, cannot hydrogen bond to methanol. The much weaker London forces between methanol and toluene result in only limited miscibility.

12.3 (a) water, because of strong ion-dipole interactions (b) benzene, because of stronger London forces (c) water, because of the hydrophilic head group ($—CO_2^-$) of CH_3COOH in water

12.5 (a) SF_4 will be more soluble in water; it is a nonpolar molecule and dipole-dipole interactions are stronger.
(b) AsF_5 is nonpolar; dipole-dipole interactions are responsible for solubility in water, so AsF_5.

12.7 (a) hydrophilic, because of hydrogen bonding (b) hydrophobic, because it is nonpolar (c) hydrophilic, because of the polar —COOH group

12.9 (a) Grease is composed principally of nonpolar hydrocarbons, which are expected to dissolve in gasoline (also a nonpolar hydrocarbon mixture) but not in the very polar water.
(b) Soaps are long-chain molecules with both polar and nonpolar ends. The polar end of the soap dissolves in water, and the nonpolar end dissolves in the nonpolar grease. As such, the soap serves as a "bridge" to "connect" the grease to the water so that it can be washed away. See Fig. 12.8 and the similar figure here.

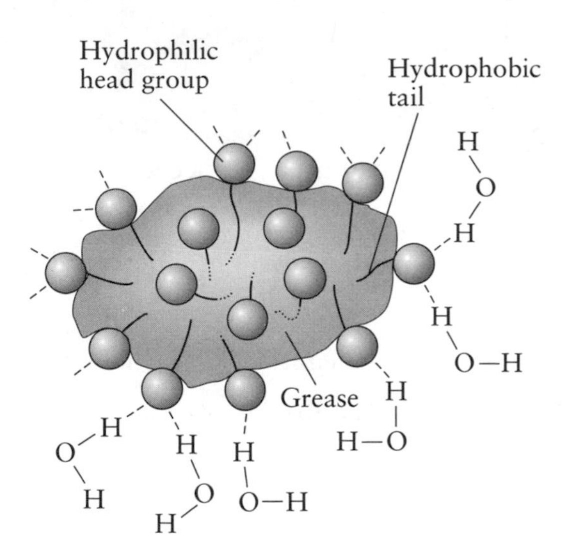

12.11 (a) a sol and an emulsion (b) foam

Gas Solubility

12.13 In each case, solubility = $S = P \times k_H$

(a) $S = (50.\text{ kPa O}_2) \times \left(\dfrac{1\text{ atm}}{101\text{ kPa}}\right) \times \left(\dfrac{1.3 \times 10^{-3}\text{ mol}}{(\text{L}\cdot\text{atm})}\right) = 6.4 \times 10^{-4}\text{ mol/L}$

(b) $S = (500.\text{ Torr CO}_2) \times \left(\dfrac{1\text{ atm}}{760\text{ Torr}}\right) \times \left(\dfrac{2.3 \times 10^{-2}\text{ mol}}{(\text{L}\cdot\text{atm})}\right)$

$= 1.5 \times 10^{-2}\text{ mol/L}$

(c) $S = (0.10\text{ atm CO}_2) \times \left(\dfrac{2.3 \times 10^{-2}\text{ mol}}{(\text{L}\cdot\text{atm})}\right) = 2.3 \times 10^{-3}\text{ mol/L}$

12.15 amount (moles) of CO_2 released $= 0.355\text{ L} \times 2.3 \times 10^{-2}\text{ mol/L}\cdot\text{atm}$
$\times\ 3.00\text{ atm} = 2.4\bar{4} \times 10^{-2}\text{ mol}$

$$V = \frac{nRT}{P} = \frac{2.4\bar{4} \times 10^{-2}\text{ mol} \times 0.082\,06\text{ L}\cdot\text{atm/K}\cdot\text{mol} \times 293\text{ K}}{1.00\text{ atm}} = 0.59\text{ L}$$

12.17 We represent the equilibrium as $CO_2(aq) \rightleftharpoons CO_2(g)$.

(a) If the partial pressure of CO_2 in the air above the solution is doubled, the equilibrium (above) will shift to the left and the concentration of CO_2 in solution will double.

(b) If the total pressure of the gas is increased by the addition of nitrogen, no change in the equilibrium (above) will occur; the partial pressure of CO_2 is unchanged and the concentration is unchanged.

Enthalpy of Solution

12.19 Ion hydration enthalpies for ions of the same charge within a group of elements become progressively less negative (or more positive) as we proceed down a group. This trend parallels increasing ion size. Because hydration energy is an ion-dipole interaction, it should be highest for small, high charge density ions.

12.21 The enthalpy of hydration is larger, this quantity is negative, as is the enthalpy of solution. The lattice enthalpy, if positive, must be the smaller value.

12.23 Table 12.5 applies to very dilute solutions. We assume here that 10.0 g of compound in 100.0 g of water is sufficiently dilute.

(a) $NaCl \longrightarrow Na^{+}(aq) + Cl^{-}(aq) \qquad \Delta H_{sol} = +3.9\text{ kJ/mol}$

$$\Delta H = (10.0\text{ g NaCl}) \times \left(\frac{3.9\text{ kJ}}{1\text{ mol NaCl}}\right) \times \left(\frac{1\text{ mol NaCl}}{58.44\text{ g NaCl}}\right) \times \left(\frac{10^3\text{ J}}{1\text{ kJ}}\right)$$

$$= 6.7 \times 10^2\text{ J} = +6.7 \times 10^2\text{ J for NaCl} = -6.7 \times 10^2\text{ J for water}$$

(b) $NaBr \longrightarrow Na^{+}(aq) + Br^{-}(aq) \qquad \Delta H_{sol} = -0.6\text{ kJ/mol}$

$$\Delta H = (10.0\text{ g NaBr}) \times \left(\frac{-0.6\text{ kJ}}{1\text{ mol NaBr}}\right) \times \left(\frac{1\text{ mol NaBr}}{102.9\text{ g NaBr}}\right) \times \left(\frac{10^3\text{ J}}{1\text{ kJ}}\right)$$

$$= -60\text{ J (1 sf) for NaBr} = +60\text{ J (1 sf) for water}$$

(c) $AlCl_3 \longrightarrow Al^{3+}(aq) + 3\ Cl^{-}(aq) \qquad \Delta H_{sol} = -329\text{ kJ/mol}$

$$\Delta H = (10.0\text{ g AlCl}_3) \times \left(\frac{-329\text{ kJ}}{1\text{ mol AlCl}_3}\right) \times \left(\frac{1\text{ mol AlCl}_3}{133.3\text{ g AlCl}_3}\right) \times \left(\frac{10^3\text{ J}}{1\text{ kJ}}\right)$$

$$= -2.47 \times 10^4\text{ J} = -2.47 \times 10^4\text{ J for AlCl}_3 = +2.47 \times 10^4\text{ J for water}$$

(d) $NH_4NO_3 \longrightarrow NH_4^{+}(aq) + NO_3^{-}(aq) \qquad \Delta H_{sol} = 25.7\text{ kJ/mol}$

$$\Delta H = (10.0\text{ g NH}_4\text{NO}_3) \times \left(\frac{25.7\text{ kJ}}{1\text{ mol NH}_4\text{NO}_3}\right)$$

$$\times \left(\frac{1\text{ mol NH}_4\text{NO}_3}{80.05\text{ g NH}_4\text{NO}_3}\right) \times \left(\frac{10^3\text{ J}}{1\text{ kJ}}\right)$$

$$= 3.21 \times 10^3\text{ J} = +3.21 \times 10^3\text{ J for NH}_4\text{NO}_3 = -3.21 \times 10^3\text{ J for water}$$

12.25 The enthalpy of solution of $SrCl_2$ is the sum of the lattice enthalpy and the enthalpy of hydration.

$SrCl_2(s) \longrightarrow Sr^{2+}(g) + 2\ Cl^{-}(g)$	ΔH_L	$= +2153$ kJ
$Sr^{+}(g) + 2\ Cl^{-}(g) \longrightarrow Sr^{+}(aq) + 2\ Cl^{-}(aq)$	ΔH_{hyd}	$= -2204$ kJ
$SrCl_2(s) \longrightarrow Sr^{+}(aq) + 2\ Cl^{-}(aq)$	ΔH_{sol}	$= -51$ kJ

12.27 LiCl has a higher hydration enthalpy and will likely have a higher enthalpy of solution and therefore, a more disordered solution.

12.29 (a) The disorder in the system increases. This is true of any solution process, be it exothermic or endothermic. In dissolving exothermically, the energy (enthalpy) of the system decreases.
(b) The disorder in the surroundings increases as energy is dispersed into it from the system. Therefore, both the energy and disorder increase when a solute dissolves exothermically.

Measures of Concentration

12.31 general formula: $\text{mass \% of A} = \dfrac{\text{mass of A in solution}}{\text{total mass of solution}} \times 100\%$

(a) $\text{mass \% NaCl} = \dfrac{4.0 \text{ g NaCl}}{100. \text{ g solution}} \times 100\% = 4.0\% \text{ NaCl}$

(b) $\text{mass \% NaCl} = \dfrac{4.0 \text{ g NaCl}}{4.0 \text{ g NaCl} + 100. \text{ g } H_2O} \times 100\% = 3.8\% \text{ NaCl}$

(c) $\text{mass \% } C_{12}H_{22}O_{11} = \dfrac{1.66 \text{ g } C_{12}H_{22}O_{11}}{1.66 \text{ g } C_{12}H_{22}O_{11} + 200. \text{ g } H_2O} \times 100\%$
$= 0.823\% \; C_{12}H_{22}O_{11}$

12.33 general formula: $x_{\text{solute}} = \dfrac{n_{\text{solute}}}{n_{\text{solvent}} + n_{\text{solute}}}$

(a) $n_{H_2O} = (25.0 \text{ g } H_2O) \times \left(\dfrac{1 \text{ mol } H_2O}{18.02 \text{ g } H_2O}\right) = 1.39 \text{ mol } H_2O \quad \text{(solute)}$

$n_{C_2H_5OH} = (50.0 \text{ g } C_2H_5OH) \times \left(\dfrac{1 \text{ mol } C_2H_5OH}{46.07 \text{ g } C_2H_5OH}\right)$
$= 1.08 \text{ mol } C_2H_5OH \quad \text{(solvent)}$

Then, $x_{H_2O} = \dfrac{1.39 \text{ mol } H_2O}{1.39 \text{ mol } H_2O + 1.08 \text{ mol } C_2H_5OH} = 0.563$

Therefore, $x_{C_2H_5OH} = 0.437$

(b) $n_{H_2O} = (25.0 \text{ g } H_2O) \times \left(\dfrac{1 \text{ mol } H_2O}{18.02 \text{ g } H_2O}\right) = 1.39 \text{ mol } H_2O$

$n_{CH_3OH} = (50.0 \text{ g } CH_3OH) \times \left(\dfrac{1 \text{ mol } CH_3OH}{32.04 \text{ g } CH_3OH}\right) = 1.56 \text{ mol } CH_3OH$

Then, $x_{H_2O} = \dfrac{1.39 \text{ mol } H_2O}{1.39 \text{ mol } H_2O + 1.56 \text{ mol } CH_3OH} = 0.471$

Therefore, $x_{CH_3OH} = 0.529$

12.35 general formula: $\text{molality} = \dfrac{\text{number of moles of solute}}{\text{mass of solvent (kg)}} = \dfrac{n_{\text{solute}}}{\text{mass}_{\text{solvent}}}$

This formula can be used to solve for n_{solute}.

(a) $n_{\text{NaCl}} = (10.0\text{ g NaCl}) \times \left(\dfrac{1\text{ mol NaCl}}{58.44\text{ g NaCl}}\right) = 1.71 \times 10^{-1}\text{ mol NaCl}$

$$\text{molality} = \frac{1.71 \times 10^{-1}\text{ mol NaCl}}{(250.\text{ g } H_2O) \times \left(\dfrac{1\text{ kg}}{10^3\text{ g}}\right)} = 6.84 \times 10^{-1}\ m$$

(b) $\text{molality} = \dfrac{0.48\text{ mol KOH}}{(50.0\text{ g } H_2O) \times \left(\dfrac{1\text{ kg}}{10^3\text{ g}}\right)} = 9.6\ m$

(c) $n_{\text{urea}} = 1.94\text{ g } CO(NH_2)_2 \times \left(\dfrac{1\text{ mol } CO(NH_2)_2}{60.06\text{ g } CO(NH_2)_2}\right)$

$= 3.23 \times 10^{-2}\text{ mol } CO(NH_2)_2$

$$\text{molality} = \frac{3.23 \times 10^{-2}\text{ mol } CO(NH_2)_2}{(200.\text{ g } H_2O) \times \left(\dfrac{1\text{ kg}}{10^3\text{ g}}\right)} = 0.162\ m$$

12.37 general formula: $\text{molality} = \dfrac{\text{number of moles of solute}}{\text{mass of solvent (kg)}} = \dfrac{n_{\text{solute}}}{\text{mass}_{\text{solvent}}\text{ (kg)}}$

n_{solute} can be solved from this formula.

(a) $3.0\text{ mass \%} = \dfrac{\text{mass } KClO_3}{\text{mass } KClO_3 + 20.0\text{ g } H_2O} \times 100\%$

mass $KClO_3 = 0.62$ g (2 sf)

(b) $3.0\ m\text{ solute} = \dfrac{n_{KClO_3}}{(20.0\text{ g } H_2O) \times \left(\dfrac{1\text{ kg}}{10^3\text{ g}}\right)}$

$n_{KClO_3} = 0.060$ and $(0.060\text{ mol } KClO_3) \times \left(\dfrac{122.6\text{ g } KClO_3}{1\text{ mol } KClO_3}\right) = 7.4\text{ g } KClO_3$

12.39 (a) $x_{\text{solute}} = \dfrac{n_{\text{solute}}}{n_{\text{solvent}} + n_{\text{solute}}}$

$n_{C_2H_5OH} = (25.0\text{ g } C_2H_5OH) \times \left(\dfrac{1\text{ mol } C_2H_5OH}{46.07\text{ g } C_2H_5OH}\right) = 0.543\text{ mol } C_2H_5OH$

$n_{H_2O} = (150.\text{ g } H_2O) \times \left(\dfrac{1\text{ mol } H_2O}{18.02\text{ g } H_2O}\right) = 8.32\text{ mol } H_2O$

$$x_{C_2H_5OH} = \frac{0.543\text{ mol } C_2H_5OH}{0.543\text{ mol } C_2H_5OH + 8.32\text{ mol } H_2O} = 0.0613$$

$x_{H_2O} = 1.000 - 0.0613 = 0.938$

(b) $\text{molality} = \dfrac{n_{\text{solute}}}{\text{mass solvent (kg)}}$

$$\text{molality of ethanol} = \left(\frac{0.543 \text{ mol } C_2H_5OH}{150. \text{ g } H_2O}\right) \times \left(\frac{10^3 \text{ g } H_2O}{1 \text{ kg } H_2O}\right)$$
$$= 3.62\ m\ C_2H_5OH$$

12.41 general formula: $x_A = \dfrac{n_A}{n_A + n_B + n_C + \ . \ . \ .}$

(a) $0.10\ m$ NaCl(aq) $\equiv$ 0.10 mol Na^+ + 0.10 mol Cl^-

+ 10^3 g H_2O (or 55.5 mol H_2O)

$$x_{Na+} = x_{Cl-} = \frac{0.10 \text{ mol}}{0.10 \text{ mol } Na^+ + 0.10 \text{ mol } Cl^- + 55.5 \text{ mol } H_2O} = 1.8 \times 10^{-3}$$

$$x_{H_2O} = \frac{55.5 \text{ mol}}{0.10 \text{ mol } Na^+ + 0.10 \text{ mol } Cl^- + 55.5 \text{ mol } H_2O} = 0.996 \text{ or } \sim 1.0 \text{ (2 sf)}$$

(b) $0.20\ m$ $Na_2CO_3 \equiv$ 0.40 mol Na^+ + 0.20 mol CO_3^{2-}

+ 10^3 g H_2O (or 55.5 mol)

$$x_{Na^+} = \frac{0.40 \text{ mol } Na^+}{0.40 \text{ mol } Na^+ + 0.20 \text{ mol } CO_3^{2-} + 55.5 \text{ mol } H_2O} = 7.1 \times 10^{-3}$$

$$x_{CO_3^{2-}} = \frac{7.1 \times 10^{-3}}{2} = 3.6 \times 10^{-3}$$

$$x_{H_2O} = \frac{55.5 \text{ mol } H_2O}{0.40 \text{ mol } Na^+ + 0.20 \text{ mol } CO_3^{2-} + 55.5 \text{ mol } H_2O} = 0.99$$

12.43 (a) Assume 1 L of solution. The mass of the solute, $(NH_4)_2SO_4$, is then

$$\text{mass of solute} = (0.35 \text{ mol } (NH_4)_2SO_4) \times \left(\frac{132.1 \text{ g } (NH_4)_2SO_4}{1 \text{ mol } (NH_4)_2SO_4}\right) \times \left(\frac{1 \text{ kg}}{10^3 \text{ g}}\right)$$
$$= 0.046\bar{2} \text{ kg } (NH_4)_2SO_4$$

$$\text{mass of solution} = (1 \text{ L}) \times \left(\frac{1 \text{ mL}}{10^{-3} \text{ L}}\right) \times \left(\frac{1.027 \text{ g}}{1.0 \text{ mL}}\right) \times \left(\frac{1 \text{ kg}}{10^3 \text{ g}}\right)$$
$$= 1.027 \text{ kg solution}$$

mass of solvent = 1.027 kg soln − $0.046\bar{2}$ kg solute = 0.981 kg solvent

$$\text{molality} = \frac{n_{solute}}{\text{mass of solvent (kg)}} = \frac{0.35 \text{ mol } (NH_4)_2SO_4}{0.981 \text{ kg solvent}} = 0.36 \text{ mol/kg}$$

(b) The number of moles of solvent (water) is

$$0.981 \text{ kg} \times \left(\frac{10^3 \text{ g}}{1 \text{ kg}}\right) \times \left(\frac{1 \text{ mol } H_2O}{18.02 \text{ g } H_2O}\right) = 54.4 \text{ mol } H_2O$$

$$x_{(NH_4)_2SO_4} = \frac{0.35 \text{ mol}}{0.35 \text{ mol} + 54.4 \text{ mol}} = 0.0064$$

Vapor-Pressure Lowering

12.45 Let $P^* = P_{pure}$, sucrose = solute, H_2O = solvent

(a) $x_{solute} = 0.100, \quad x_{solvent} = 0.900$

$P = (x_{solvent})(P^*)$

$P^*_{H_2O} = 760$ Torr at 100°C

$P = (0.900) \times (760 \text{ Torr}) = 684 \text{ Torr}$

(b) $\text{molality} = \dfrac{n_{sucrose}}{\text{mass } H_2O \text{ (kg)}} = 0.100\ m = \dfrac{0.100 \text{ mol sucrose}}{1 \text{ kg } H_2O}$

$$n_{H_2O} = (1 \text{ kg } H_2O) \times \left(\frac{10^3 \text{ g}}{1 \text{ kg}}\right) \times \left(\frac{1 \text{ mol } H_2O}{18.02 \text{ g } H_2O}\right) = 55.5 \text{ mol } H_2O$$

$$x_{solute} = \frac{0.100 \text{ mol sucrose}}{0.100 \text{ mol sucrose} + 55.5 \text{ mol } H_2O} = 1.8 \times 10^{-3}$$

Therefore, $x_{solvent} = 1 - 1.8 \times 10^{-3} = 0.998$

$P = (x_{solvent})(P^*)$

$P^*_{H_2O} = 760$ Torr at 100°C

$P = (0.998) \times (760 \text{ Torr}) = 759 \text{ Torr}$

12.47 Each calculation requires the determination of the mole fraction of the solvent, $x_{solvent}$, to be used in Raoult's law:

$P = x_{solvent} \times P_{pure}$ (let $P_{pure} = P^*$ below) and $x_{solvent} = \dfrac{n_{solvent}}{n_{solute} + n_{solvent}}$

(a) 1.0% ethylene glycol, $C_2H_4(OH)_2$, in water is 1.0 g ethylene glycol and 99 g H_2O, in 100. g of solution

$x_{solvent} =$

$$\frac{(99 \text{ g } H_2O) \times \left(\dfrac{1 \text{ mol } H_2O}{18.02 \text{ g } H_2O}\right)}{(1.0 \text{ g } C_2H_4(OH)_2) \times \left(\dfrac{1 \text{ mol } C_2H_4(OH)_2}{62.07 \text{ g } C_2H_4(OH)_2}\right) + (99 \text{ g } H_2O) \times \left(\dfrac{1 \text{ mol } H_2O}{18.02 \text{ g } H_2O}\right)}$$

$= 0.99\overline{7}$

$P = (x_{solvent}) \times (P^*)$

$P^*_{H_2O} = 4.58$ Torr at 0°C

$P = (0.99\overline{7}) \times (4.58 \text{ Torr}) = 4.5\overline{7} \text{ Torr} \approx 4.6 \text{ Torr}$

(b) $0.10\ m$ NaOH = 0.10 mol NaOH in 1 kg H_2O;

2 mol particles per 1 mol NaOH ($i = 2$); 1000 g H_2O = 55.5 mol H_2O

$$x_{solvent} = \frac{55.5 \text{ mol } H_2O}{(0.10 \text{ mol NaOH}) \times (2) + 55.5 \text{ mol } H_2O} = 0.99\overline{6}$$

$P = (x_{solvent}) \times (P^*)$

$P^*_{H_2O} = 355$ Torr at 80°C

$P = (0.99\overline{6}) \times (355 \text{ Torr}) = 3.5 \times 10^2$ Torr

(c) $x_{solvent} =$

$$\frac{(100.\text{ g H}_2\text{O}) \times \left(\dfrac{1 \text{ mol H}_2\text{O}}{18.02 \text{ g H}_2\text{O}}\right)}{6.6 \text{ g CO(NH}_2)_2 \times \left(\dfrac{1 \text{ mol CO(NH}_2)_2}{60.06 \text{ g CO(NH}_2)_2}\right) + 100.\text{ g H}_2\text{O} \times \left(\dfrac{1 \text{ mol H}_2\text{O}}{18.02 \text{ g H}_2\text{O}}\right)}$$

$= 0.98$

$P = (x_{solvent}) \times (P^*)$

$P^*_{H_2O} = 9.21$ Torr at 10°C

$P = (0.98) \times (9.21 \text{ Torr}) = 9.03$ Torr

Therefore, $\Delta P = 9.21 \text{ Torr} - 9.03 \text{ Torr} = 0.18$ Torr

12.49 NaCl(aq)

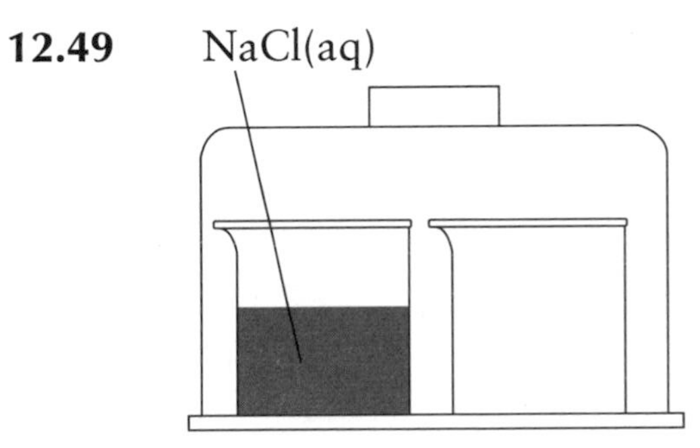

The vapor pressure of the 0.010 *m* NaCl solution is lower than that of pure water. More water will evaporate from the pure water beaker as the system approaches equilibrium. The solute molecules in the 0.010 *m* NaCl solution prevent as much water from evaporating because the vapor pressure of the pure water is always higher than that of the water in the solution. At equilibrium, it has all evaporated.

12.51 Let $P^* = P_{\text{pure solvent}}$, x = unknown molar mass of X (do not confuse with mole fraction)

(a) $P = (x_{solvent}) \times (P^*)$

$P = 94.8$ Torr $\quad P^* = 100.$ Torr

$x_{solvent} = \dfrac{P}{P^*} = \dfrac{94.8}{100.} = 0.948$ Therefore, $x_X = 1 - 0.948 = 0.052$

(b) $x_{solvent} = \dfrac{\text{no. of mol solvent}}{\text{no. of mol benzene} + \text{no. of mol X}}$

$$= \frac{(100.\text{ g}) \times \left(\dfrac{1 \text{ mol benzene}}{78.11 \text{ g benzene}}\right)}{(100.\text{ g}) \times \left(\dfrac{1 \text{ mol benzene}}{78.11 \text{ g benzene}}\right) + (8.05 \text{ g}) \times \left(\dfrac{1 \text{ mol X}}{x \text{ grams}}\right)}$$

$$0.948 = \frac{1.28 \text{ mol}}{1.28 \text{ mol} + \left(\frac{8.05}{x}\right) \text{mol}}$$

Solve for x (take reciprocals of both sides).

After taking reciprocals,

$$1.05\bar{5} = 1 + \frac{8.05 \text{ g}}{1.28\, x \text{ g}}$$

$$0.055 = \frac{8.05 \text{ g}}{1.28\, x \text{ g}}$$

$$x = 11\bar{4} \text{ g}$$

$$\text{molar mass of X} = \frac{x \text{ g}}{\text{mol}} = 11\bar{4} \text{ g/mol} = 1.1 \times 10^2 \text{ g/mol}$$

Boiling-Point Elevation and Freezing-Point Depression

12.53 ΔT = boiling-point elevation, m = molality

(a) $\Delta T = k_b m = 0.51 \text{ K·kg/mol} \times 0.10 \text{ mol/kg} = 0.051 \text{ K} = 0.051°\text{C}$

new b.p. = 100.0°C + 0.051°C = 100.051°C

(b) $\Delta T = ik_b m, \quad i = 2$

$\Delta T = 2 \times 0.51 \text{ K·kg/mol} \times 0.22 \text{ mol/kg} = 0.22 \text{ K} = 0.22°\text{C}$

new b.p. = 100.0°C + 0.22°C = 100.22°C

(c) $\text{Solubility} = \frac{230. \text{ mg}}{100. \text{ g}} = \frac{2.30 \text{ g}}{1.00 \text{ kg}}$

$$\text{molality} = \left(\frac{2.30 \text{ g}}{1.000 \text{ kg}}\right) \times \left(\frac{1 \text{ mol LiF}}{25.94 \text{ g LiF}}\right) = 0.0887 \text{ mol/kg}$$

$\Delta T = ik_b m, \quad i = 2$

$\Delta T = 2 \times 0.51 \text{ K·kg/mol} \times 0.0887 \text{ mol/kg} = 0.090 \text{ K} = 0.090°\text{C}$

new b.p. = 100.0°C + 0.090°C = 100.090°C

12.55 $\Delta T_b = \text{boiling-point elevation} = k_b m, \quad k_b\ (CCl_4) = 4.95 \text{ K·kg/mol}$

$$m = \frac{\Delta T_b}{k_b} = \frac{(334.66 - 334.35) \text{ K}}{4.95 \text{ K·kg/mol}} = 0.062\bar{6} \text{ mol/kg}$$

Let X = solute and n_X = number of moles of X in 100.0 g CCl_4 (0.100 kg CCl_4)

$$n_X = 0.1000 \text{ kg } CCl_4 \times \left(\frac{0.0626 \text{ mol}}{1 \text{ kg } CCl_4}\right) = 6.26 \times 10^{-3} \text{ mol X}$$

$$\text{molar mass of X} = \frac{1.05 \text{ g X}}{6.26 \times 10^{-3} \text{ mol X}} = 168 \text{ g/mol}$$

12.57 (a) ΔT = freezing-point depression

$\Delta T = k_f m = 1.86\ \text{K}\cdot\text{kg/mol} \times 0.10\ \text{mol/kg} = 0.18\overline{6}\ \text{K} = 0.19°\text{C}$

new f.p. = $0°\text{C} - 0.19°\text{C} = -0.19°\text{C}$

(b) $\Delta T = ik_f m, \quad i = 2$

$\Delta T = 2 \times 1.86\ \text{K}\cdot\text{kg/mol} \times 0.22\ \text{mol/kg} = 0.82\ \text{K} = 0.82°\text{C}$

new f.p. = $0°\text{C} - 0.82°\text{C} = -0.82°\text{C}$

(c) 120. mg LiF per 100.0 g H_2O 1.20 g LiF per 1.000 kg H_2O

$$m = (1.20\ \text{g LiF}) \times \left(\frac{1\ \text{mol LiF}}{25.94\ \text{g LiF}}\right) = 0.0463\ \text{mol LiF in 1 kg H}_2\text{O}$$

$$= 0.0463\ \text{mol/kg}$$

$\Delta T = ik_f m = 2 \times 1.86\ \text{K}\cdot\text{kg/mol} \times 0.0463\ \text{mol/kg} = 0.172\ \text{K} = 0.172°\text{C}$

new f.p. = $0°\text{C} - 0.172°\text{C} = -0.172°\text{C}$

12.59 Let X = molecular substance

$$\left(\frac{1.14\ \text{g X}}{100.0\ \text{g camphor}}\right) \times \left(\frac{10^3\ \text{g camphor}}{1\ \text{kg camphor}}\right) = \frac{11.4\ \text{g X}}{1\ \text{kg camphor}}$$

$\Delta T = k_f m = 179.8°\text{C} - 177.3°\text{C} = 2.5°\text{C} = 2.5\ \text{K}$

$$2.5\ \text{K} = 39.7\ \text{K}\cdot\text{kg/mol} \times \frac{11.4\ \text{g X}}{\left(\frac{\text{molar mass X}}{1\ \text{kg camphor}}\right)}$$

After rearranging, $0.0630\ \text{mol X} = \dfrac{11.4\ \text{g X}}{\text{molar mass X}}$

$$\text{molar mass X} = \frac{11.4\ \text{g X}}{0.0630\ \text{mol X}} = 181\ \text{g/mol}$$

12.61 (a) $\Delta T_f = k_f m \quad \Delta T_b = k_b m$

$$\frac{\Delta T_b}{k_b} = m = \frac{\Delta T_f}{k_f} \quad \text{and} \quad \frac{\Delta T_b}{k_b} = \frac{\Delta T_f}{k_f}$$

$$\Delta T_f = \frac{k_f \Delta T_b}{k_b} = \frac{(1.90) \times (5.12°\text{C})}{2.53} = 3.8°\text{C} \quad [\text{units of } k_f \text{ and } k_b \text{ cancel}]$$

Therefore, f.p. = $5.5°\text{C} - 3.8°\text{C} = 1.7°\text{C}$

(b) $\Delta T_f = k_f m$. Assume that the solute does not dissociate. Then,

$$m = \frac{\Delta T_f}{k_f} = \frac{3.04\ \text{K}}{1.86\ \text{K}\cdot\text{kg/mol}} = 1.63\ \text{mol/kg}$$

12.63 Assume 100. g of solution.

1.00% NaCl ≡ 1.00 g NaCl per 100. g solution; therefore,

mass H_2O = mass solution − mass NaCl

= 100. g − 1.00 g = 99.0 g H_2O

$$\text{molality} = m = \left(\frac{1.00\text{ g NaCl}}{99.0\text{ g H}_2\text{O}}\right) \times \left(\frac{1\text{ mol NaCl}}{58.44\text{ g NaCl}}\right) \times \left(\frac{10^3\text{ g}}{1\text{ kg}}\right) = 0.173\text{ mol/kg}$$

(a) $\Delta T = ik_f m$, $\Delta T = 0.593°\text{C} = 0.593\text{ K}$

$$i = \frac{\Delta T}{k_f m} = \frac{0.593\text{ K}}{1.86\text{ K}\cdot\text{kg/mol} \times 0.173\text{ mol/kg}} = 1.84$$

(b) $i = 1.84$ corresponds to 84% dissociation. This can be seen by considering that 84% dissociation of one mole of NaCl leads to 0.84 mol of Na^+ ions and 0.84 mol of Cl^- ions, and 0.16 mol of undissociated NaCl (0.84 + 0.84 + 0.16 = 1.84).

12.65 $i = 1.075$ for 7.5% dissociation. This is seen by considering that 7.5% dissociation leads to 0.075 for fraction of positive ion, 0.075 for fraction of negative ion, and 0.925 for fraction of undissociated electrolyte (0.075 + 0.075 + 0.925 = 1.075).

$\Delta T = ik_f m = 1.075 \times 1.86\text{ K}\cdot\text{kg/mol} \times 0.10\text{ mol/kg} = 0.20\text{ K} = 0.20°\text{C}$

Therefore, f.p. = 0.00°C − 0.20°C = −0.20°C

Osmosis and Osmometry

12.67 In each case $\Pi = i \times RT \times \text{molarity}$. At 20°C, $T = 293\text{ K}$, $RT = 0.082\,06\text{ L}\cdot\text{atm/K}\cdot\text{mol} \times 293\text{ K} = 24.04\text{ L}\cdot\text{atm/mol}$

(a) $\Pi = 1 \times 24.04\text{ L}\cdot\text{atm/mol} \times 0.010\text{ mol/L} = 0.24\text{ atm}$

(b) $\Pi = 2 \times 24.04\text{ L}\cdot\text{atm/mol} \times 1.0\text{ mol/L} = 48\text{ atm}$

(c) $\Pi = 3 \times 24.04\text{ L}\cdot\text{atm/mol} \times 0.010\text{ mol/L} = 0.72\text{ atm}$

12.69 $\Pi = i \times RT \times M$, M = molarity

$$M = \frac{\Pi}{iRT} = \frac{(3.74\text{ Torr}) \times \left(\frac{1\text{ atm}}{760\text{ Torr}}\right)}{(1) \times \left(\frac{0.082\,06\text{ L}\cdot\text{atm}}{\text{K}\cdot\text{mol}}\right) \times (273 + 27)\text{ K}} = 2.0 \times 10^{-4}\text{ mol/L}$$

$$\text{and} \quad \left(\frac{0.40\text{ g polypeptide}}{1.0\text{ L}}\right) \times \left(\frac{1\text{ L}}{2.0 \times 10^{-4}\text{ mol}}\right) = 2.0 \times 10^3\text{ g/mol}$$

12.71 $\Pi = i \times RT \times M$, M = molarity

$$M = \frac{\Pi}{iRT} = \frac{(5.4\text{ Torr}) \times \left(\frac{1\text{ atm}}{760\text{ Torr}}\right)}{(1) \times (0.082\,06\text{ L}\cdot\text{atm/K}\cdot\text{mol}) \times (273 + 20)\text{ K}}$$

$$= 3.0 \times 10^{-4}\text{ mol/L}$$

Then, $(3.0 \times 10^{-4}\ \text{mol/L}) \times \left(\frac{10^{-3}\ \text{L}}{1\ \text{mL}}\right) \times (100.\ \text{mL}) = 3.0 = 10^{-5}$ mol polymer

and molar mass $= \frac{0.10\ \text{g}}{3.0 \times 10^{-5}\ \text{mol}} = 3.3 \times 10^{3}$ g/mol

12.73 (a) $\Pi = i\,RTM = 1 \times 0.082\ 06\ \text{L·atm/K·mol} \times 293\ \text{K} \times 0.050\ \text{mol/L}$
$= 1.2$ atm

(b) $\Pi = 2 \times 0.082\ 06\ \text{L·atm/K·mol} \times 293\ \text{K} \times 0.0010\ \text{mol/L}$
$= 0.048$ atm

(c) molarity of solution $= \left(\frac{2.3 \times 10^{-5}\ \text{g}}{100.0\ \text{g H}_2\text{O}}\right) \times \left(\frac{1\ \text{mol AgCN}}{133.9\ \text{g AgCN}}\right) \times \left(\frac{0.998\ \text{g}}{1\ \text{mL}}\right)$
$\times \left(\frac{1\ \text{mL}}{10^{-3}\ \text{L}}\right) = 1.7 \times 10^{-6}$ mol/L

$\Pi = i \times RT \times M = (2) \times (0.082\ 06\ \text{L·atm/K·mol}) \times (273 + 20.)\ \text{K}$
$\times (1.7 \times 10^{-6}\ \text{mol/L}) = 8.2 \times 10^{-5}$ atm

SUPPLEMENTARY EXERCISES

12.75 $CuSO_4$ (s) $\rightleftharpoons$ $CuSO_4$ (aq)

The equilibrium between solid and aqueous $CuSO_4$ is a dynamic process. Cu^{2+} and SO_4^{2-} ions continually leave the solid and are replaced with ions returning to the solid from solution. Because the surface area of the smaller crystals is larger, per gram, than that of the bigger crystals (see below), the ions leaving the solid surface preferentially come from the former (smaller) crystals. However, those returning to the solid surface have a greater chance of landing on a large crystal (per crystal) than a small one. So the large crystals grow at the expense of the small crystals without a change in the concentration of the solution.

The ratio of surface area to mass of a crystal, for a cubic crystal, can be determined as follows:

$$\frac{\text{surface area}}{\text{mass}} = \frac{\text{surface area}}{\text{density} \times \text{volume}} = \frac{6a^2}{d \times a^3} \propto \frac{1}{a}$$

a is the length of a side of the cube; $6/d$ is a constant. Thus, smaller crystals (smaller a) have a greater surface area to mass ratio.

12.77 Assume 100. g of solution.

(a) 10.0% H_2SO_4 ≡ 10.0 g H_2SO_4 per 100. g solution

and $\left(\frac{10.0\ \text{g H}_2\text{SO}_4}{100.\ \text{g soln}}\right) \times \left(\frac{1.07\ \text{g soln}}{1\ \text{mL}}\right) = \frac{10.7\ \text{g H}_2\text{SO}_4}{100.\ \text{mL soln}}$

Therefore, $(6.32 \text{ g } H_2SO_4) \times \left(\frac{100. \text{ mL soln}}{10.7 \text{ g } H_2SO_4}\right) = 59.1 \text{ mL soln}$

(b) $100. \text{ g soln} - 10.0 \text{ g } H_2SO_4 = 90.0 \text{ g } H_2O = 0.0900 \text{ kg } H_2O$

$$n_{H_2SO_4} = \frac{10.0 \text{ g } H_2SO_4}{98.08 \text{ g } H_2SO_4/\text{mol } H_2SO_4} = 0.102 \text{ mol } H_2SO_4$$

$$\text{molality} = \frac{n_{H_2SO_4}}{\text{mass of soln (kg)}} = \frac{0.102 \text{ mol}}{0.0900 \text{ kg}} = 1.13\ m$$

(c) From part (a), $\left(\frac{10.7 \text{ g } H_2SO_4}{100. \text{ mL soln}}\right) \times (300.0 \text{ mL soln}) = 32.1 \text{ g } H_2SO_4$

12.79 $2.0\ m\ HNO_3 \equiv \frac{2 \text{ mol } HNO_3}{1 \text{ kg } H_2O}$ (assume density of the $2.0\ m\ HNO_3 = 1.0$ g/mL)

$$\left(\frac{2.0 \text{ mol } HNO_3}{1 \text{ kg } H_2O}\right) \times \left(\frac{1 \text{ kg}}{10^3 \text{ g}}\right) \times \left(\frac{1 \text{ g}}{1 \text{ mL}}\right) \times (250. \text{ mL})$$

$$= 0.50 \text{ mol } HNO_3 \text{ required}$$

Then, $70.\%\ HNO_3 = \left(\frac{70. \text{ g } HNO_3}{100. \text{ g soln}}\right) \times \left(\frac{1 \text{ mol } HNO_3}{63.0 \text{ g } HNO_3}\right)$

$$= 0.011 \text{ mol } HNO_3 \text{ per g soln}$$

and $(0.50 \text{ mol } HNO_3) \times \left(\frac{1 \text{ g soln}}{0.011 \text{ mol } HNO_3}\right) = 45 \text{ g of } 70.\%\ HNO_3$

12.81 $\Delta T = k_f m,\ m = \frac{n_{solute}}{\text{mass}_{solvent\ (kg)}},\ n_{solute} = \frac{\text{mass}_{solute}}{M_{solute}}$

or

$$\Delta T = k_f \frac{\text{mass}_{solute}}{M_{solute} \times \text{mass}_{solvent\ (kg)}}$$

Solving for M_{solute},

$$M_{solute} = \frac{k_f \times \text{mass}_{solute}}{\Delta T \times \text{mass}_{solvent\ (kg)}}$$

(a) If mass_{solute} appears greater, M_{solute} appears greater than actual molar mass, as mass_{solute} occurs in the numerator above.

(b) If $d_{solvent} < 1.00 \text{ g/cm}^3$, true $\text{mass}_{solvent} = d \times V <$ assumed mass, or assumed mass > true mass, and M_{solute} appears less than actual M_{solute}, as $\text{mass}_{solvent}$ occurs in the denominator in the above expression.

(c) If true freezing point is higher than the recorded freezing point, true $\Delta T <$ assumed ΔT, or assumed $\Delta T >$ true ΔT, and M_{solute} appears less than actual M_{solute}, as ΔT occurs in the denominator.

(d) If not all solute dissolved, the true $\text{mass}_{solute} <$ assumed mass_{solute}, or assumed $\text{mass}_{solute} >$ true mass_{solute}, and M_{solute} appears greater than the actual M_{solute}, as mass_{solute} occurs in the numerator.

12.83 $\Pi = i \times RT \times M$

$$\text{molarity} = M = \frac{\Pi}{iRT} = \frac{(1.2\ \text{Torr}) \times \left(\dfrac{1\ \text{atm}}{760\ \text{Torr}}\right)}{(1) \times \left(\dfrac{0.082\,06\ \text{L}\cdot\text{atm}}{\text{K}\cdot\text{mol}}\right) \times (273 + 20.)\ \text{K}}$$

$$= 6.6 \times 10^{-5}\ \text{mol/L}$$

$$\text{and} \left(\frac{0.166\ \text{g catalase}}{10.0\ \text{mL H}_2\text{O}}\right) \times \left(\frac{1\ \text{mL}}{10^{-3}\ \text{L}}\right) \times \left(\frac{1\ \text{L}}{6.6 \times 10^{-5}\ \text{mol}}\right)$$

$$= 2.5 \times 10^{5}\ \text{g/mol (2 sf)}$$

12.85 The process of osmosis tries to equalize the concentration of water in the solution, both in the cells of the fish and the surrounding water; water from the aquarium passes through the cell membranes, causing them to expand and rupture.

12.87 (a) $\Pi = ghd$; recall from Ch. 5 (Table 5.2) that $1\ \text{Pa} = \text{kg/m}\cdot\text{s}^2$

$$\Pi = \left(\frac{9.807\ \text{m}}{\text{s}^2}\right) \times (5.22\ \text{cm}) \times \left(\frac{1\ \text{m}}{10^2\ \text{cm}}\right) \times \left(\frac{0.998\ \text{g}}{\text{cm}^3}\right) \times \left(\frac{1\ \text{kg}}{10^3\ \text{g}}\right)$$

$$\times \left(\frac{10^2\ \text{cm}}{1\ \text{m}}\right)^3 = 5.11 \times 10^2\ \text{Pa}$$

$$5.11 \times 10^2\ \text{Pa} \times \left(\frac{1\ \text{atm}}{1.013 \times 10^5\ \text{Pa}}\right) = 5.04 \times 10^{-3}\ \text{atm}$$

$$M = \frac{\Pi}{iRT} = \frac{5.04 \times 10^{-3}\ \text{atm}}{(1) \times (0.082\,06\ \text{L}\cdot\text{atm/K}\cdot\text{mol}) \times (273 + 20.)\text{K}}$$

$$= 2.10 \times 10^{-4}\ \text{mol/L}$$

$$M = \left(\frac{0.010\ \text{mol protein}}{10.0\ \text{mL H}_2\text{O}}\right) \times \left(\frac{1\ \text{mL}}{10^{-3}\ \text{L}}\right) \times \left(\frac{1\ \text{L}}{2.10 \times 10^{-4}\ \text{mol}}\right)$$

$$= 4.76 \times 10^3\ \text{g/mol}$$

(b) $$\text{molality} = m = \left(\frac{2.10 \times 10^{-4}\ \text{mol}}{1\ \text{L}}\right) \times \left(\frac{10^{-3}\ \text{L}}{1\ \text{mL}}\right) \times \left(\frac{1\ \text{mL}}{0.998\ \text{g}}\right) \times \left(\frac{10^3\ \text{g}}{1\ \text{kg}}\right)$$

$$= 2.10 \times 10^{-4}\ \text{mol/kg}$$

$$\Delta T = ik_f m = 1 \times 1.86\ \text{K}\cdot\text{kg/mol} \times 2.10 \times 10^{-4}\ \text{mol/kg}$$

$$= 3.91 \times 10^{-4}\ \text{K}$$

$$\text{fp} = 0.00°\text{C} - 3.91 \times 10^{-4}°\text{C} = -3.91 \times 10^{-4}°\text{C}$$

(c) The small ΔT of part (b) cannot be measured accurately; the molar mass determined from it cannot be accurate. Therefore, the osmotic pressure, which can be measured accurately (3 sf), is the superior method for the determination of the molar mass.

APPLIED EXERCISES

12.89 (a) panel (a) shows carboxylate and quaternary ammonium ions; panel (c) shows carboxylic acids and amides; panel (d) shows amides.

(b) The interactions are strongest in panel (a).

(c) The interactions are weakest in panel (b), since the compounds are hydrocarbons, which interact only by London forces.

12.91 volume of plasma = $0.45 \times 6.00\text{ L} = 2.7\text{ L}$

$S = k_H \times P = 5.8 \times 10^{-4}\text{ mol/L·atm} \times 10.0\text{ atm} = 5.8 \times 10^{-3}\text{ mol/L}$

amount (moles) of N_2 = $5.8 \times 10^{-3}\text{ mol/L} \times 2.7\text{ L} = 1.5\overline{7} \times 10^{-2}\text{ mol}$

$$V = \frac{nRT}{P} = \frac{1.5\overline{7} \times 10^{-2}\text{ mol} \times 0.082\ 06\text{ L·atm/K·mol} \times 310\text{ K}}{1.00\text{ atm}}$$

$= 0.40\text{ L (2 sf)}$

12.93 Amino acid (a) is more polar, because it has two carboxylic acid groups and an amine group. It should be absorbed by the stationary phase more strongly than (b). Amino acid (b) will thus have traveled more rapidly and further than amino acid (a).

INTEGRATED EXERCISES

12.95 (a) $\Pi = iRTM$

$$\text{molarity} = M = \frac{\Pi}{iRT} = \frac{0.0112\text{ atm}}{(1) \times (0.082\ 06\text{ L·atm/K·mol}) \times (273 + 25)\text{K}}$$

$= 4.58 \times 10^{-4}\text{ mol/L}$

Assuming 1 L of solution, $\dfrac{3.16\text{ g sample}}{4.58 \times 10^{-4}\text{ mol}} = 6.90 \times 10^{3}\text{ g/mol}$

(b) Propene is $CH_2{=}CH{-}CH_3$ and has a molecular mass of 42.08 g/mol. The number of monomer units is $\dfrac{6.90 \times 10^{3}\text{ g/mol}}{42.08\text{ g/mol}} = 164.$

(c) The polymer looks like $\left(\!-CH_2-\underset{\displaystyle CH_3}{\underset{|}{C}}-\!\right)_n$; each monomer has three C—C bonds, each with an average length of 154 pm (Table 9.4). The average chain length is

$(154\text{ pm}) \times (3) \times (164\text{ units}) = 7.58 \times 10^{4}\text{ pm or 75.8 nm}$

12.97

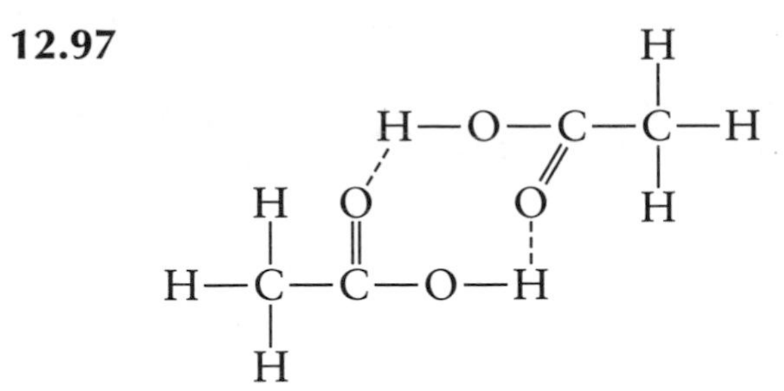

Molecules of acetic acid can attract one another. The carboxyl groups attract the hydroxyl groups, reducing the apparent number of solute species. In acetone, the acetic acid molecules interact more strongly with each other than they do with acetone molecules; thus, dimerization is more likely in acetone.

12.99 Let p = dichlorobenzene = cmpd

(a) $\Delta T = 5.48°\text{C} - 1.20°\text{C} = 4.28°\text{C} = 4.28\text{K}$ and $\Delta T = k_f m$

$$m = \frac{\Delta T}{k_f} = \frac{4.28\text{ K}}{5.12\text{ K}\cdot\text{kg/mol}} = 0.836\text{ (mol cmpd)/kg}$$

$$0.836\text{ (mol cmpd)/kg} \times \left(\frac{1\text{ kg}}{10^3\text{ g}}\right) = 8.36 \times 10^{-4}\text{ (mol cmpd)/(1 g benzene)}$$

$$\text{Then, molar mass cmpd} = \left(\frac{10.0\text{ g cmpd}}{80.0\text{ g benzene}}\right) \times \left(\frac{1\text{ g benzene}}{8.36 \times 10^{-4}\text{ mol cmpd}}\right)$$

$$= 150.\text{ g/mol}$$

(b) $\dfrac{\text{molar mass cmpd}}{\text{molar mass } C_3H_2Cl} = \dfrac{150.\text{ g/mol}}{73.5\text{ g/mol}} \approx 2$

Therefore, the formula is $C_6H_4Cl_2$.

(c) For $C_6H_4Cl_2$, molar mass = 146.99 g/mol

CHAPTER 13
THE RATES OF REACTIONS

EXERCISES

Reaction Rates

13.1 (a) $\text{rate}(N_2) = \text{rate}(H_2) \times \left(\dfrac{1 \text{ mol } N_2}{3 \text{ mol } H_2}\right) = \dfrac{1}{3} \times \text{rate}(H_2)$

(b) $\text{rate }(NH_3) = \text{rate}(H_2) \times \left(\dfrac{2 \text{ mol } NH_3}{3 \text{ mol } H_2}\right) = \dfrac{2}{3} \times \text{rate}(H_2)$

(c) $\text{rate}(NH_3) = \text{rate}(N_2) \times \left(\dfrac{2 \text{ mol } NH_3}{1 \text{ mol } N_2}\right) = 2 \times \text{rate}(N_2)$

13.3 (a) rate of decomposition of ozone $= (1.5 \text{ mmol } O_2/\text{L}\cdot\text{s}) \times \left(\dfrac{2 \text{ mol } O_3}{3 \text{ mol } O_2}\right)$

$$= 1.0 \text{ mmol } O_3/\text{L}\cdot\text{s}$$

(b) rate of formation of dichromate ions =

$$\left(0.14 \, \frac{\text{mol } Cr_2O_7^{\,2-}}{\text{L}\cdot\text{s}}\right) \times \left(\frac{2 \text{ mol } CrO_4^{\,2-}}{1 \text{ mol } Cr_2O_7^{\,2-}}\right) = 0.28 \text{ (mol } Cr_2O_4^{\,2-})/\text{L}\cdot\text{s}$$

Rate Laws

13.5 The concentrations of NO_2 and O_2 at time t, $[NO_2]_t$ and $[O_2]_t$, are related to the amount of N_2O_5 decomposed by the stoichiometry of the reaction; that is,

$$[NO_2]_t \left(\frac{\text{mol } NO_2}{\text{L}}\right) =$$

$$([N_2O_5]_{t=0} - [N_2O_5]_t) \times \left(\frac{\text{mol } N_2O_5}{\text{L}}\right) \times \left(\frac{4 \text{ mol } NO_2}{2 \text{ mol } N_2O_5}\right) \text{ and}$$

$$[O_2]_t \left(\frac{\text{mol } O_2}{\text{L}}\right) = ([N_2O_5]_{t=0} - [N_2O_5]_t) \times \left(\frac{\text{mol } N_2O_5}{\text{L}}\right) \times \left(\frac{1 \text{ mol } O_2}{2 \text{ mol } N_2O_5}\right)$$

For example, after 1.11 h,

$$[NO_2]\left(\frac{\text{mol } NO_2}{\text{L}}\right) = (2.15 - 1.88) \times 10^{-3} \times 2\left(\frac{\text{mol } NO_2}{\text{L}}\right)$$

$$= 5.4 \times 10^{-4} \text{ (mol } NO_2)/\text{L}$$

All other values in the table below are obtained in a similar manner. The concentrations are plotted as a function of time in the following figure. The rates at a time t can be found by determining the slope of the concentration versus time curve at time t. Methods of determining the slope are discussed in Section 13.2 and in more detail in Appendix 1E. Here, we adopt the two-point method described in the appendix, because it is likely to be more accurate than a strictly graphical method based on drawing tangents to the curve. The tangent method would work well if a very accurate curve with many data points could be constructed.

(b)

Time, h	$[N_2O_5]^a$	$[NO_2]^a$	$[O_2]^a$	Rate, $\frac{\Delta[N_2O_5]}{\Delta t}^b$
0	2.15	0.00	0.00	2.4
1.11	1.88	0.54	0.14	2.3
2.22	1.64	1.02	0.26	2.0
3.33	1.43	1.44	0.36	1.8
4.44	1.25	1.80	0.45	1.6

a. Units = 10^{-3} mol/L b. Units = 10^{-4} mol/L·h

(a, c)

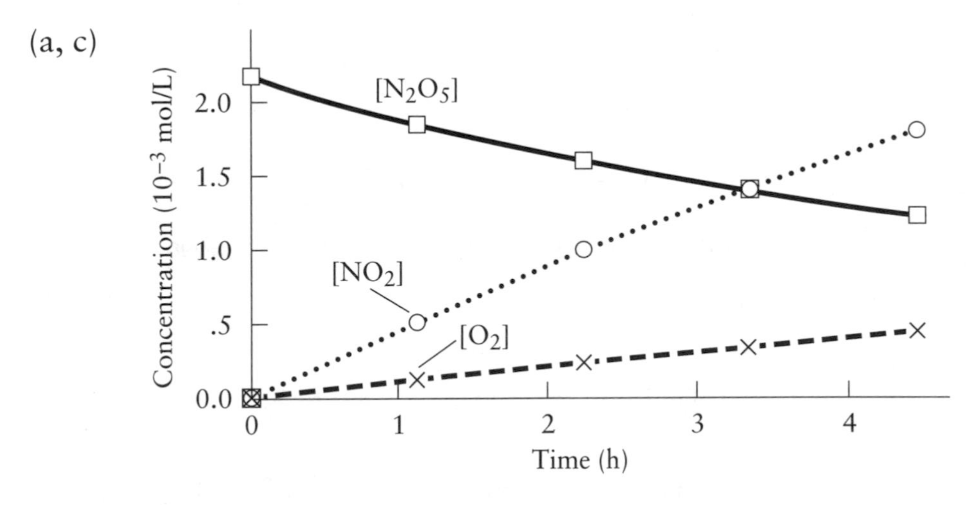

The rate at time t is found from the two closest values of $[N_2O_5]$ at time t_2 and t_1, such that t lies between t_2 and t_1; that is,

$$\text{Rate at } t = \frac{\Delta[N_2O_5]_t}{\Delta t} = \frac{[N_2O_5]_{t_1} - [N_2O_5]_{t_2}}{t_2 - t_1}$$

$$\text{For example, the rate at 1.11 h} = \frac{(2.15 - 1.64) \times 10^{-3} \text{ mol/L}}{(2.22 - 0) \text{ h}} = 2.3 \times 10^{-4} \text{ mol/L·h}$$

In the table above, the rates at times 1.11 h, 2.22 h, and 3.33 h are determined in this manner. The rates at times 0 h and 4.44 h are estimated by extrapolation.

13.7 (a) rate $= k[X]^2[Y]^{1/2}$

(b) rate $= k[A][B][C]^{-1/2}$

sum of orders $= 1 + 1 + x = \frac{3}{2}$, thus $x = -\frac{1}{2}$. The rate law is then

rate $= k[A][B][C]^{-1/2}$

13.9 For A $\longrightarrow$ products, rate = (mol A)/L·s

(a) rate (mol A/L·s) $= k_0[A]^0 = k_0$, so units of k_0 are (mol A)/L·s (same as the units for the rate in this case)

(b) rate (mol A/L·s) $= k_1[A]$, so units of k_1 are $\dfrac{\text{(mol A)/L·s}}{\text{(mol A)/L}} = \text{/s}$

(c) rate (mol A/L·s) $= k_2[A]^2$, so units of k_2 are $\dfrac{\text{(mol A)/L·s}}{[\text{(mol A)/L}]^2} = \text{L/(mol A)·s}$

13.11 In general, we can write

rate [mol/L·(unit of time)] $= k\ (\text{mol/L})^n$, or

$$k = \frac{[\text{mol/L·(unit of time)}]}{(\text{mol/L})^n} = (\text{L}^{n-1}/\text{mol}^{n-1}\cdot\text{(unit of time)}$$

where n = order. In this case, $n - 1 = 2$, so $n = 3$.

13.13 From the units of the rate constant, k, and the solution of Exercise 13.11, it follows that the reaction is first order, so rate $= k[N_2O_5]$.

$$[N_2O_5] = \left(\frac{2.0\text{ g } N_2O_5}{1.0\text{ L}}\right) \times \left(\frac{1\text{ mol } N_2O_5}{108.02\text{ g } N_2O_5}\right) = 0.018\overline{5}\text{ mol/L}$$

rate $= 5.2 \times 10^{-3}\text{/s} \times 0.018\overline{5}\text{ mol/L} = 9.6 \times 10^{-5}\ (\text{mol } N_2O_5)\text{/L·s}$

13.15 (a) From the units of the rate constant and the solution to Exercise 13.11, it follows that the reaction is second order, thus

rate $= k[H_2][I_2] =$

$$0.063\text{ L/mol·s} \times \left(\frac{0.15\text{ g } H_2}{0.500\text{ L}}\right) \times \left(\frac{1\text{ mol } H_2}{2.016\text{ g } H_2}\right) \times \left(\frac{0.32\text{ g } I_2}{0.500\text{ L}}\right) \times \left(\frac{1\text{ mol } I_2}{253.8\text{ g } I_2}\right)$$
$$= 2.4 \times 10^{-5}\text{ mol/L·s}$$

(b) rate(new) $= k \times 2 \times [H_2]_{\text{initial}}[I_2] = 2 \times$ rate(initial)

Reaction rate increases by a factor of 2.

13.17 (a) first order in H_2; first order in I_2; second order overall

(b) first order in SO_2; zero order in O_2; negative one-half order in SO_3; one-half order overall

(c) second order in A; zero order in B; negative two order in C; zero order overall

13.19 Because the rate increased in direct proportion to the concentrations of both reactants, the rate is first order in both reactants. rate $= k[CH_3Br][OH^-]$

13.21 (a) rate $= k[A]^a[B]^b$, where the orders a and b are to be determined. When the concentration of A was decreased by a factor of 3, the rate decreased by a factor of 9; thus $(\frac{1}{3})^a = (\frac{1}{9})$, giving $a = 2$, and the reaction is second order in A. When the concentration of B was decreased by a factor of 3, the rate decreased by a factor of 3; thus $(\frac{1}{3})^b = (\frac{1}{3})$, giving $b = 1$, and the reaction is first order in B. The overall order $= 2 + 1 = 3$.

(b) rate $= k[A]^2[B]$

(c) $$k = \frac{\text{rate}}{[A]^2[B]} = \left(\frac{12.6\ \text{mol}}{\text{L}\cdot\text{s}}\right) \times \left(\frac{\text{L}}{0.60\ \text{mol}}\right)^2 \times \left(\frac{\text{L}}{0.30\ \text{mol}}\right)$$
$$= 1.1\overline{7} \times 10^2\ \text{L}^2/\text{mol}^2\cdot\text{s} = 1.2 \times 10^2\ \text{L}^2/\text{mol}^2\cdot\text{s}$$

The data from any of the experiments 1, 2, or 3 could have been used.

(d) $$\text{rate} = \left(\frac{1.1\overline{7} \times 10^2\ \text{L}^2}{\text{mol}^2\cdot\text{s}}\right) \times \left(\frac{0.17\ \text{mol}}{\text{L}}\right)^2 \times \left(\frac{0.25\ \text{mol}}{\text{L}}\right)$$
$$= 0.84\ \text{mol/L}\cdot\text{s}$$

13.23 When the concentration of ICl was doubled, the rate doubled (experiments 1 and 2). Thus, the reaction is first order in ICl. When the concentration of H_2 was tripled, the rate tripled (experiments 2 and 3). Thus, the reaction is first order in H_2.

(a) rate $= k[ICl][H_2]$

(b) $$k = \left(\frac{22 \times 10^{-3}\ \text{mol}}{\text{L}\cdot\text{s}}\right) \times \left(\frac{\text{L}}{3.0 \times 10^{-3}\ \text{mol}}\right) \times \left(\frac{\text{L}}{4.5 \times 10^{-3}\ \text{mol}}\right)$$
$$= 1.6 \times 10^2\ \text{L/mol}\cdot\text{s}$$

(c) $$\text{rate} = \left(\frac{1.6 \times 10^2\ \text{L}}{\text{mol}\cdot\text{s}}\right) \times \left(\frac{4.7 \times 10^{-3}\ \text{mol}}{\text{L}}\right) \times \left(\frac{2.7 \times 10^{-3}\ \text{mol}}{\text{L}}\right)$$
$$= 2.1\ \text{mmol/L}\cdot\text{s}$$

13.25 (a) Doubling the concentration of A (experiments 1 and 2) doubled the rate; so the reaction is first order in A. Increasing the concentration of B by the ratio 3.02/1.25 (experiments 2 and 3) increased the rate by $(3.02/1.25)^2$; so the reaction is second order in B. Tripling the concentration of C (experiment 3 and 4) increased the rate by $3^2 = 9$; so the reaction is second order in C. Therefore, rate $= k[A][B]^2[C]^2$

(b) overall order $= 5$

(c) $$k = \frac{\text{rate}}{[A][B]^2[C]^2}$$

Using the data from experiment 4, we get

$$k = \left(\frac{0.457 \text{ mol}}{\text{L}\cdot\text{s}}\right) \times \left(\frac{\text{L}}{1.25 \times 10^{-3} \text{ mol}}\right) \times \left(\frac{\text{L}}{3.02 \times 10^{-3} \text{ mol}}\right)^2 \times \left(\frac{\text{L}}{3.75 \times 10^{-3} \text{ mol}}\right)^2 = 2.85 \times 10^{12} \text{ L}^4/\text{mol}^4\cdot\text{s}$$

From experiment 3, we get

$$k = \left(\frac{5.08 \times 10^{-2} \text{ mol}}{\text{L}\cdot\text{s}}\right) \times \left(\frac{\text{L}}{1.25 \times 10^{-3} \text{ mol}}\right) \times \left(\frac{\text{L}}{3.02 \times 10^{-3} \text{ mol}}\right)^2 \times \left(\frac{\text{L}}{1.25 \times 10^{-3} \text{ mol}}\right)^2 = 2.85 \times 10^{12} \text{ L}^4/\text{mol}^4\cdot\text{s} \quad \text{(Checks!)}$$

(d) $\text{rate} = \left(\frac{2.85 \times 10^{12} \text{ L}^4}{\text{mol}^4\cdot\text{s}}\right) \times \left(\frac{3.01 \times 10^{-3} \text{ mol}}{\text{L}}\right) \times \left(\frac{1.00 \times 10^{-3} \text{ mol}}{\text{L}}\right)^2 \times \left(\frac{1.15 \times 10^{-3} \text{ mol}}{\text{L}}\right)^2 = 1.13 \times 10^{-2} \text{ mol/L}\cdot\text{s}$

Integrated Rate Laws

13.27 (a) $k = \frac{0.693}{t_{1/2}} = \frac{0.693}{1000 \text{ s}} = 6.93 \times 10^{-4} \text{ /s}$

(b) We use $\ln\left(\frac{[\text{A}]_0}{[\text{A}]_t}\right) = kt$ and solve for k.

$$k = \frac{\ln([\text{A}]_0/[\text{A}]_t)}{t} = \frac{\ln\left(\frac{0.33 \text{ mol/L}}{0.14 \text{ mol/L}}\right)}{47 \text{ s}} = 1.8 \times 10^{-2} \text{ /s}$$

(c) $[\text{A}]_t = \left(\frac{0.050 \text{ mol A}}{\text{L}}\right) - \left[\left(\frac{2 \text{ mol A}}{1 \text{ mol B}}\right) \times \left(\frac{0.015 \text{ mol B}}{\text{L}}\right)\right]$

$= 0.020 \text{ (mol A)/L}$

$$k = \frac{\ln\left(\frac{0.050 \text{ mol/L}}{0.020 \text{ mol/L}}\right)}{120. \text{ s}} = 7.6 \times 10^{-3} \text{ /s}$$

13.29 (a) This is the half-life itself, 200. s.

(b) $\frac{[\text{A}]}{[\text{A}]_0} = \frac{1}{16} = \left(\frac{1}{2}\right)^4$; so the time elapsed is 4 half-lives.

$t = 4 \times 200. \text{ s} = 800. \text{ s}$

(c) Because $\frac{1}{9}$ is not a multiple of $\frac{1}{2}$, we cannot work directly from the half-life. But $k = 0.693 / t_{1/2}$

so $k = \frac{0.693}{200. \text{ s}} = 3.46\bar{5} \times 10^{-3} \text{ /s}$

Then

$$t = \frac{\ln\left(\frac{[A]_0}{[A]_t}\right)}{k} = \frac{\ln\left(\frac{9}{1}\right)}{3.46\overline{5} \times 10^{-3}\ /\text{s}} = 634\ \text{s}$$

13.31 (a) $t_{1/2} = \frac{0.693}{k} = \left(\frac{0.693\ \text{s}}{3.7 \times 10^{-5}}\right) \times \left(\frac{1\ \text{min}}{60\ \text{s}}\right) \times \left(\frac{1\ \text{h}}{60\ \text{min}}\right) = 5.2\ \text{h}$

(b) $[A]_t = [A]_0\, e^{-kt}$

$t = 2.0\ \text{h} = 2.0\ \text{h} \times 3600\ \text{s/h} = 7.2 \times 10^3\ \text{s}$

$[N_2O_5] = 2.33 \times 10^{-2}\ \text{mol/L} \times e^{-(3.7\times10^{-5}\ /\text{s}\times7.2\times10^{3}\ \text{s})} = 1.78 \times 10^{-2}\ \text{mol/L}$

(c) Solve for t from $\ln\left(\frac{[A]_0}{[A]_t}\right) = kt$, which gives

$$t = \frac{\ln\left(\frac{[A]_0}{[A]_t}\right)}{k} = \frac{\ln\left(\frac{[N_2O_5]_0}{[N_2O_5]_t}\right)}{k} = \frac{\ln\left(\frac{2.33 \times 10^{-2}}{1.76 \times 10^{-2}}\right)}{3.7 \times 10^{-5}\ /\text{s}} = 7.5\overline{8} \times 10^3\ \text{s}$$

$$= 7.5\overline{8} \times 10^3\ \text{s} \times \left(\frac{1\ \text{min}}{60\ \text{s}}\right) = 1.3 \times 10^2\ \text{min}$$

13.33 (a) $t_{1/2} = \frac{0.693}{k} = \frac{0.693}{2.81 \times 10^{-3}\ /\text{min}} = 247\ \text{min}$

(b) See the solutions to Exercises 10.29(c) and 10.31(c).

$$t = \frac{\ln\left(\frac{[SO_2Cl_2]_0}{[SO_2Cl_2]_t}\right)}{k} = \frac{\ln 10.}{2.81 \times 10^{-3}\ /\text{min}} = 819\ \text{min}$$

(c) $[A]_t = [A]_0\, e^{-kt}$

Because the vessel is sealed, masses and concentrations are proportional, and we write

$(\text{mass left})_t = (\text{mass})_0\, e^{-kt}$

$= 14.0\ \text{g} \times e^{-(2.81\times10^{-3}\ /\text{min}\times60\ \text{min/h}\times1.5\ \text{h})}$

$= 11\ \text{g}$

Note: Knowledge of the volume of the vessel is not required. However, one could have converted mass to concentration, solved for the new concentration at 1.5 h, and finally converted back to the new (remaining) mass. But this is not necessary.

13.35 (a) $\ln\left(\frac{[A]_0}{[A]_t}\right) = kt$

$\ln\left(\frac{100.\%}{20.\%}\right) = k \times 120\ \text{s}$

$$k = \frac{\ln 5.0}{120.\ \text{s}} = 0.0134\ /\text{s}$$

(b) Solving for t as we did in Exercise 13.31(c), we get

$$t = \frac{\ln\left(\frac{[\text{A}]_0}{[\text{A}]_t}\right)}{k} = \frac{\ln\left(\frac{100.\%}{10.\%}\right)}{0.013\overline{4}\ /\text{s}} = 1.7 \times 10^2\ \text{s}$$

13.37 For a first-order reaction, $\ln[\text{A}]_t = \ln[\text{A}]_0 - kt$ [Eq. 2]
The description of the kinetic data fits this equation, which is the equation of a straight line with a negative slope, $-k$. Therefore, the reaction is first order.

13.39 We draw up the following table:

Time, s	$[N_2O_5]$, 10^{-3} mol/L	$[\text{A}]_0/[\text{A}]$	$\ln([\text{A}]_0/[\text{A}])$
0	2.15	1.00	0.000
4000.	1.88	1.14	0.131
8000.	1.64	1.31	0.270
12 000.	1.43	1.50	0.405
16 000.	1.25	1.72	0.542

(a) The integrated rate equation for a first-order reaction is

$$\ln[\text{A}]_t = \ln[\text{A}]_0 - kt \quad \text{or} \quad \ln\left(\frac{[\text{A}]_0}{[\text{A}]_t}\right) = kt$$

We then plot $\ln([\text{A}]_0/[\text{A}]_t)$ versus t and see whether a straight line through the origin results.

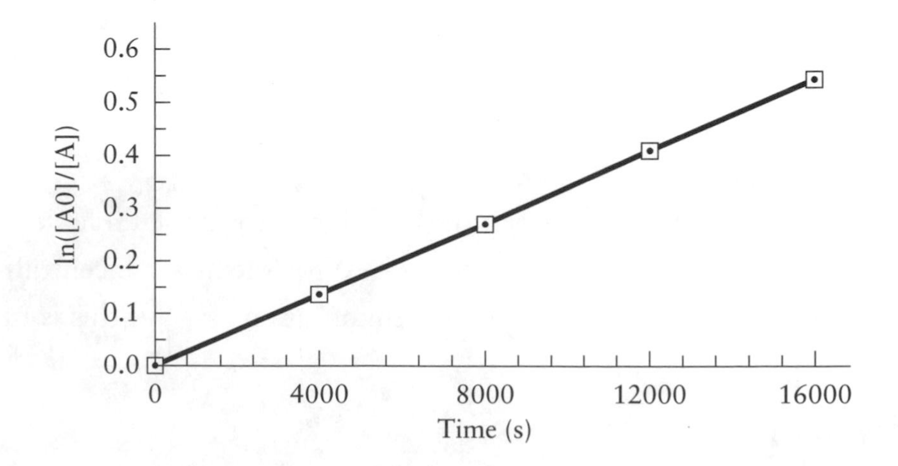

The data fit the first-order equation well.

(b) The rate constant is the slope of this line. It is easily determined with plotting software, with a graphing calculator, or by the methods of Appendix 1E and Example 13.5. Using first and last points in the graph yields

$$\text{slope} = \frac{0.542 - 0.000}{(16\ 000 - 0)\ \text{s}} = 3.39 \times 10^{-5}\ /\text{s} = k$$

13.41 (a) Draw up the following table and plot 1/[HI] against time.

Time, s	[HI], 10^{-3} mol/L	1/[HI], 10^{3} L/mol
0	1000.	0.00100
1000.	112	0.00893
2000.	61	$0.016\overline{4}$
3000.	41	$0.024\overline{4}$
4000.	31	$0.032\overline{2}$

Equation 4 in the text can be rearranged as

$$\frac{1}{[\text{A}]_t} = \frac{1 + [\text{A}]_0\, kt}{[\text{A}]_0} = \frac{1}{[\text{A}]_0} + kt$$

Thus, if the reaction is second order, a plot of 1/[HI] against time should give a straight line of slope k.

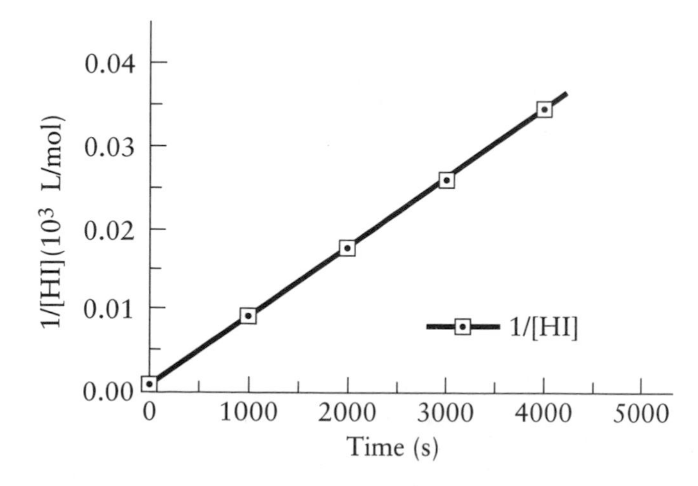

As can be seen from the graph, the data fit the equation for a second-order reaction quite well. The slope can be determined from any pair of data points, but we choose the points at 4000. and 1000. s.

(b) $$\text{slope} = \frac{(0.032\overline{2} - 0.008\ 33) \times 10^{3}\ \text{L/mol}}{(4000. - 1000.)\ \text{s}}$$

$$= 8.0 \times 10^{-3}\ \text{L/mol}\cdot\text{s} = k$$

Collision Theory and Arrhenius Behavior

13.43 As temperature increases, the average speed of the reacting molecules also increases; therefore, the number of collisions between molecules increases as well. The rate of reaction is proportional to the number of collisions between molecules; thus, the rate increases in proportion to this factor. This collision frequency factor is contained within the pre-exponential factor, A, in the Arrhenius equation. Not all molecules that collide will react; only those that have sufficient collision energy to surmount the energy barrier, E_a, between reactants and products will react. The fraction of molecules having the necessary energy to react increases with temperature in a manner governed by the Maxwell distribution of speeds. This fraction is proportional to $e^{-E_a/RT}$. Therefore, we can write
$k \propto \text{collision frequency} \times e^{-E_a/RT}$ or $k = Ae^{-E_aRT}$.

13.45

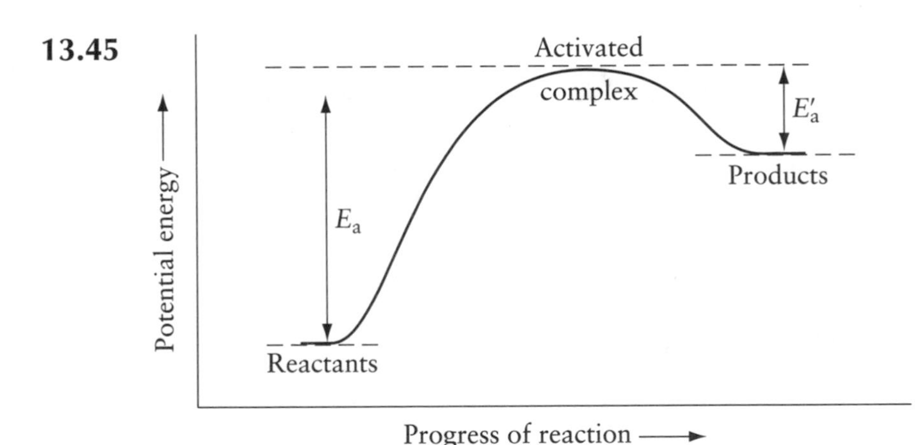

E_a' is related to E_a as follows:
$E_a = E_a' + \Delta H$, or $E_a' = E_a - \Delta H$

13.47 (a) Make the following table and graph.

T, K	$1/T$, /K	k, /s	ln k
750.	0.001 33	1.8×10^{-4}	−8.62
800.	0.001 25	2.7×10^{-3}	−5.91
850.	0.001 18	3.0×10^{-2}	−3.51
900.	0.001 11	2.6×10^{-1}	−1.35

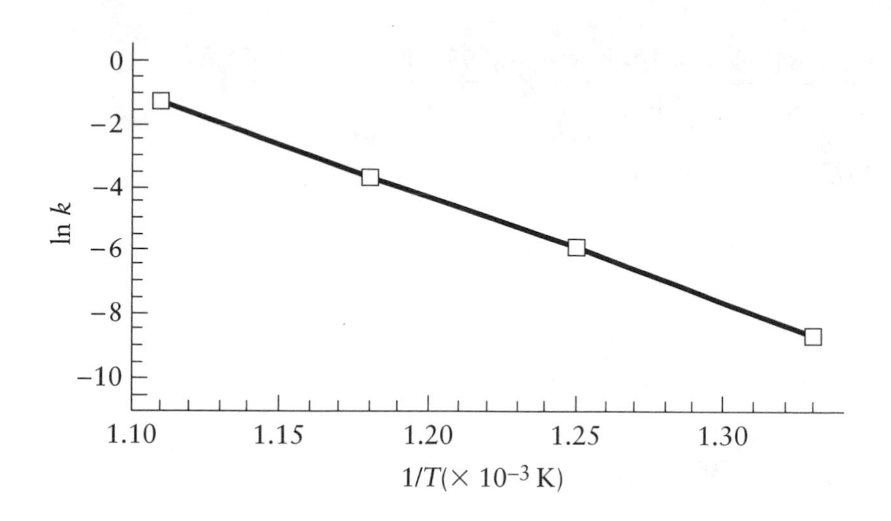

$\ln k = \ln(A - E_a/RT)$

$$\text{slope of plot} = -E_a/R = \frac{[-1.35 - (-8.62)]\ \text{K}}{0.001\ 11 - 0.001\ 33} = 3.3 \times 10^4\ \text{K}$$

$E_a = (3.3 \times 10^4\ \text{K}) \times (8.31 \times 10^{-3}\ \text{kJ/mol·K}) = 2.7 \times 10^2\ \text{kJ/mol}$

(b) $\ln\left(\frac{k'}{k}\right) = \frac{E_a}{R}\left(\frac{1}{T} - \frac{1}{T'}\right)$

$$= \left(\frac{2.7 \times 10^2\ \text{kJ/mol}}{8.31 \times 10^{-3}\ \text{kJ/K·mol}}\right) \times \left(\frac{1}{750.\ \text{K}} - \frac{1}{873\ \text{K}}\right) = 6.1$$

$\frac{k'}{k} = 44\bar{8}$

$k' = 44\bar{8} \times (1.8 \times 10^{-4}\ \text{/s}) = 8.0 \times 10^{-2}\ \text{/s}$

An approximate value of k' can also be obtained from the plot itself and yields the same value.

13.49 We use $\ln\left(\frac{k'}{k}\right) = \frac{E_a}{R}\left(\frac{1}{T} - \frac{1}{T'}\right) = \frac{E_a}{R}\left(\frac{T' - T}{T'T}\right)$

$$\ln\left(\frac{k'}{k}\right) = \ln\left(\frac{0.87\ \text{/s}}{0.38\ \text{/s}}\right) = \left(\frac{E_a}{8.31 \times 10^{-3}\ \text{kJ/K·mol}}\right) \times \left(\frac{1030.\ \text{K} - 1000.\ \text{K}}{1030.\ \text{K} \times 1000.\ \text{K}}\right)$$

$$E_a = \frac{(8.31 \times 10^{-3}\ \text{kJ/K·mol})(1000.\ \text{K})(1030.\ \text{K})(0.83)}{30.\ \text{K}} = 2.4 \times 10^2\ \text{kJ/mol}$$

13.51 We use $\ln\left(\frac{k'}{k}\right) = \frac{E_a}{R}\left(\frac{1}{T} - \frac{1}{T'}\right) = \frac{E_a}{R}\left(\frac{T' - T}{TT'}\right)$

k' = rate constant at 700.°C, $T' = (700. + 273)\ \text{K} = 973\ \text{K}$

$$\ln\left(\frac{k'}{k}\right) = \left(\frac{315 \text{ kJ/mol}}{8.31 \times 10^{-3} \text{ kJ/K}\cdot\text{mol}}\right) \times \left(\frac{973 \text{ K} - 1073 \text{ K}}{973 \text{ K} \times 1073 \text{ K}}\right)$$

$$= -3.63; \quad \frac{k'}{k} = 0.026\bar{5}$$

$$k' = 0.026\bar{5} \times 9.7 \times 10^{10} \text{ L/mol}\cdot\text{s} = 2.6 \times 10^{9} \text{ L/mol}\cdot\text{s}$$

13.53 $$\ln\left(\frac{k'}{k}\right) = \frac{E_a}{R}\left(\frac{1}{T} - \frac{1}{T'}\right) = \frac{E_a}{R}\left(\frac{T' - T}{TT'}\right)$$

$$= \left(\frac{103 \text{ kJ/mol}}{8.31 \times 10^{-3} \text{ kJ/K}\cdot\text{mol}}\right) \times \left(\frac{323 \text{ K} - 318 \text{ K}}{318 \text{ K} \times 323 \text{ K}}\right) = 0.603$$

$$\frac{k'}{k} = 1.82\bar{8}$$

$$k' = 1.82\bar{8} \times 5.1 \times 10^{-4} \text{ /s} = 9.3 \times 10^{-4} \text{ /s}$$

Catalysis

13.55 cat = catalyzed, uncat = uncatalyzed $\quad E_{a,cat} = \frac{1}{2} E_a = \frac{1}{2} E_{a,uncat}$

$$\frac{\text{rate(cat)}}{\text{rate(uncat)}} = \frac{k_{cat}}{k_{uncat}} = \frac{Ae^{-E_{a,cat}/RT}}{Ae^{-E_a/RT}} = \frac{e^{-(1/2)E_a/RT}}{e^{-E_a/RT}} = e^{(1/2)E_a/RT}$$

$$e^{[1/2(100. \text{ kJ/mol})/(8.31\times10^{-3} \text{ kJ/K}\cdot\text{mol}\times400. \text{ K})]} = e^{15.0} = 3.4 \times 10^{6}$$

13.57 cat = catalyzed, uncat = uncatalyzed

$$\frac{\text{rate(cat)}}{\text{rate(uncat)}} = \frac{k_{cat}}{k_{uncat}} = 1000. = \frac{Ae^{-E_{a,cat}/RT}}{Ae^{-E_a/RT}} = \frac{e^{-E_{a,cat}/RT}}{e^{-E_a/RT}}$$

$$\ln 1000. = \frac{-E_{a,cat}}{RT} + \frac{E_a}{RT}$$

$$E_{a,cat} = E_a - RT \ln 1000.$$

$$= 98 \text{ kJ/mol} - (8.31 \times 10^{-3} \text{ kJ/K}\cdot\text{mol})(298 \text{ K})(\ln 1000.) = 81 \text{ kJ/mol}$$

Reaction Mechanisms

13.59 (a) rate = $k[NO]^2$, bimolecular
(b) rate = $k[Cl_2]$, unimolecular
(c) rate = $k[NO_2]^2$, bimolecular
(d) (b) and (c), because Cl and NO are radicals

13.61 (a) $O_3 + O \longrightarrow 2\ O_2$
(b) Step 1: rate = $k[O_3][NO]$, bimolecular
Step 2: rate = $k[NO_2][O]$, bimolecular

(c) NO is the catalyst.

(d) NO_2, because it appears only in the course of the reaction and is neither a reactant nor a product.

13.63 $2\ HA + B^{2-} \longrightarrow H_2B + 2\ A^-$; HB^- is a reaction intermediate.

13.65 The first elementary reaction is the rate-controlling step, because it is the slow step. The second elementary reaction is fast and does not affect the overall reaction order, which is second order as a result of the fact that the rate-controlling step is bimolecular.

rate $= k[NO][Cl_2]$

13.67 The overall rate is determined by the slow step; rate $= k_3[COCl][Cl_2]$. But COCl is an intermediate and its concentration has to be eliminated.

$k_1[Cl_2] = k_1'[Cl]^2$, giving $[Cl] = \sqrt{\frac{k_1}{k_1'}[Cl_2]}$

and $k_2 = [Cl][CO] = k_2'[COCl]$; substituting for [Cl] gives

$$[COCl] = \left(\frac{k_2}{k_2'}\right)\sqrt{\frac{k_1}{k_1'}}\ [Cl_2]^{1/2}[CO], \text{ thus}$$

rate $= k_3(k_2/k_2')(k_1/k_1')^{1/2}[CO][Cl_2]^{3/2} = k[CO][Cl_2]^{3/2}$

13.69 Mechanism (a) gives a rate law of rate $= k\ [O_3]^2$. This does not agree with the expression given. Mechanism (b) gives a rate law of rate $= k\ [O_3]$, because the slow step governs the overall rate law. This mechanism does not agree with the expression given. Mechanism (c) remains; its rate law is

rate $= k_2[O_3][O]$

and $k_1[O_3] = k_{-1}[O_2][O]$, solve for [O]

$$[O] = \frac{k_1}{k_{-1}}\frac{[O_3]}{[O_2]} \text{ and}$$

$$\text{rate} = k_2\frac{k_1}{k_{-1}}[O_3]\frac{[O_3]}{[O_2]} \text{ or rate} = k[O_3]^2[O_2]^{-1}$$

This agrees with the experimental rate law given.

SUPPLEMENTARY EXERCISES

13.71 (a) rate $= k[A][B]^{1/2}$

(b) $$k = \frac{\text{units of rate}}{\text{units of }([A][B]^{1/2})} = \frac{\text{mol/L}\cdot\text{min}}{(\text{mol/L})\cdot(\text{mol/L})^{1/2}}$$

$= L^{1/2}/\text{mol}^{1/2}\cdot\text{min}$

13.73 (a) At t_0, there are 12 molecules of X. At t_5, there are 6 molecules of X; therefore, $t_{1/2} = 5.00$ s.

(b) For first-order processes, use the equation $\ln[A]_t = \ln[A]_o - kt$ to determine $[A]_t$. Note that $k = \ln 2 / t_{1/2}$.

$k = 0.693/5s = 0.139$ /s and

$\ln[A]_t = \ln(12) - (0.139\ /s)(8s) = 1.37$

$[A]_t = e^{1.37} = 3.9 \simeq 4$ molecules. So

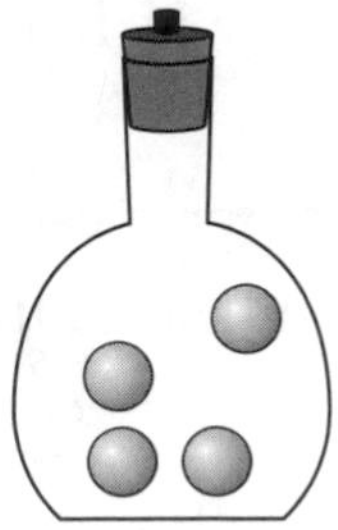

13.75 (a) exothermic

(b) the reverse reaction; the greater the activation energy, the more rapidly k increases with temperature

13.77 x = amount of original sample = 1.0 μg

n = number of half-lives

$$\left(\frac{1}{2}\right)^n \times x = \text{amount remaining}$$

$$\frac{5.2}{12} = 0.43 \text{ half-lives}$$

$$\left(\frac{1}{2}\right)^{0.43} \times 1.0\ \mu g = 0.74\ \mu g$$ (Use the y^x key on your calculator.)

13.79 Not every collision occurs with the correct geometrical arrangement or orientation of the atoms required for the products to form. Those collisions with incorrect placement of the atoms will not produce products.

13.81 (a) not linear; it is ln[A] against time that is linear

(b) linear; see Fig. 13.7(a)

(c) linear; see Fig. 13.14(a)

(d) linear, because $\dfrac{1}{[A]} = \dfrac{1}{[A]_0} + kt$; see Fig. 13.14(b)

(e) not linear; it is $\ln k$ against $1/T$ that is linear; see Fig. 13.14(a)

(f) linear, because $(\text{rate})_0 = k[A]$

13.83 (a) The objective is to reproduce the observed rate law. If step 2 is the slow step, if step 1 is a rapid equilibrium, and if step 3 is fast also, then our proposed rate law will be rate = $k_2[N_2O_2][H_2]$. Consider the equilibrium of step 1:

$$k_1[NO]^2 = k_1'[N_2O_2] \quad [N_2O_2] = \frac{k}{k_1'}[NO]^2$$

Substituting in our proposed rate law, we have

$$\text{rate} = k_2\left(\frac{k_1}{k_1'}\right)[NO]^2[H_2] = k[NO]^2[H_2], \text{ where } k = k_2\left(\frac{k_1}{k_1'}\right)$$

The assumptions made above reproduce the observed rate law; therefore, step 2 is the slow step.

(b)

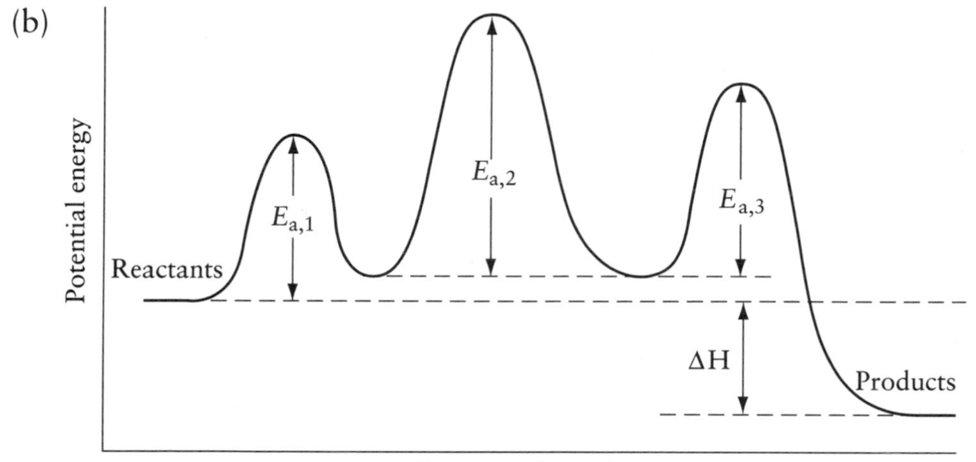

13.85 cat = catalyzed, uncat = uncatalyzed

The rates are proportional to the rate constants.

$$\frac{\text{rate (cat)}}{\text{rate (uncat)}} = \frac{k_{cat}}{k_{uncat}} = \frac{Ae^{-Ea(cat)/RT}}{Ae^{-Ea(uncat)/RT}} = e^{-[E_{a(cat)} - E_{a(uncat)}]/RT}$$

$$\frac{E_{a(cat)} - E_{a(uncat)}}{RT} = \frac{(162 - 350)\text{ kJ/mol}}{8.314 \times 10^{-3}\text{ kJ/K}\cdot\text{mol} \times 973\text{ K}} = -23.2\overline{4}$$

$$\frac{\text{rate(cat)}}{\text{rate(uncat)}} = e^{-(-23.2\overline{4})} = 1.2 \times 10^{10}$$

= factor by which reaction rate is increased

13.87 (a) Make the following table and graph.

T, K	$1/T$, /K	$(k \times 10^3)$ L/mol·s	$\ln(k \times 10^3)$
297	0.003 37	1.3	$0.26\overline{2}$
301	0.003 32	2.0	$0.69\overline{3}$
305	0.003 28	3.0	$1.1\overline{0}$
309	0.003 24	4.4	$1.4\overline{8}$
313	0.003 19	6.4	$1.8\overline{6}$

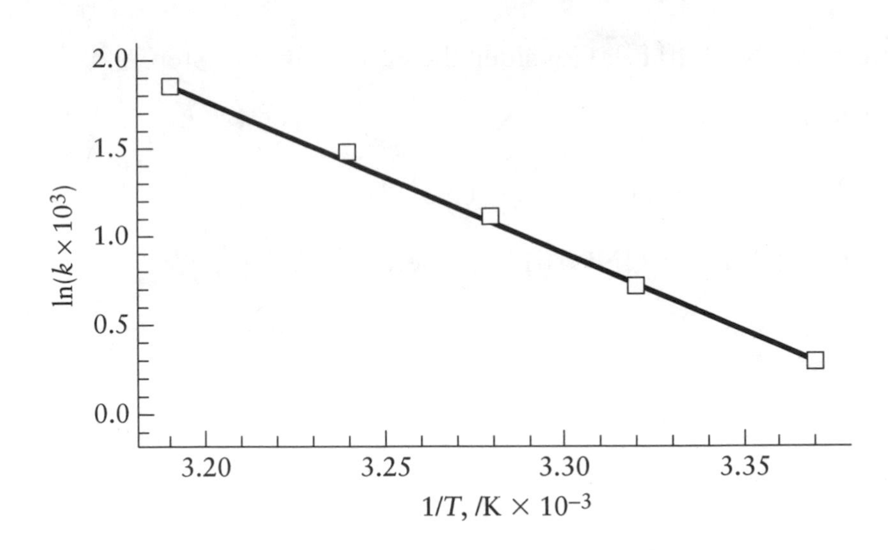

$$\text{slope of straight line} = -\left(\frac{E_a}{R}\right) = \frac{(1.8\overline{6} - 0.26)\ \text{K}}{0.003\,19 - 0.003\,37} = -8.9 \times 10^3\ \text{K}$$

$E_a = (8.9 \times 10^3\ \text{K})(8.31 \times 10^{-3}\ \text{kJ/K}\cdot\text{mol}) = 74\ \text{kJ/mol}$

(b) $T = 298\ \text{K}, \quad 1/T = 3.36 \times 10^{-3}\ /\text{K}$

From the plot, $\ln(k \times 10^3) \approx 0.37, \quad k = 1.4 \times 10^{-3}\ \text{L/mol}\cdot\text{s}$

13.89 It is convenient to obtain an expression for the half-life of a second-order reaction. We work with Eq. 4.

$$[\text{A}]_t = \frac{[\text{A}]_0}{1 + [\text{A}]_0\, kt}$$

$$\frac{[\text{A}]_{t_{1/2}}}{[\text{A}]_0} = \frac{1}{1 + [\text{A}]_0\, kt_{1/2}} = \frac{1}{2}$$

Therefore, $1 + [\text{A}]_0\, kt_{1/2} = 2$, or $[\text{A}]_0\, kt_{1/2} = 1$, or $t_{1/2} = \dfrac{1}{k[\text{A}]_0}$

It is also convenient to solve Eq. 4 for k.

$$\frac{1}{[\text{A}]_t} = \frac{1}{[\text{A}]_0} + kt$$

$$k = \frac{\dfrac{1}{[\text{A}]_t} - \dfrac{1}{[\text{A}]_0}}{t}$$

(a) $k = \dfrac{1}{t_{1/2}\,[\text{A}]_0} = \dfrac{1}{(100.\ \text{s}) \times (2.5 \times 10^{-3}\ \text{mol/L})} = 4.0\ \text{L/mol}\cdot\text{s}$

(b) $k = \dfrac{\dfrac{1}{[\text{A}]} - \dfrac{1}{[\text{A}]_0}}{t}$

$$[A] = \frac{0.30 \text{ mol A}}{\text{L}} - \left[\left(\frac{0.010 \text{ mol C}}{\text{L}}\right) \times \left(\frac{3 \text{ mol A}}{1 \text{ mol C}}\right)\right] = 0.27 \text{ mol A/L}$$

$$k = \frac{\dfrac{1 \text{ L}}{0.27 \text{ mol}} - \dfrac{1 \text{ L}}{0.30 \text{ mol}}}{200. \text{ s}} = 1.9 \times 10^{-3} \text{ L/mol·s}$$

13.91 $[A]_t = [A_0]e^{-kt}$ or $t = -\dfrac{\ln([A]_t/[A]_0)}{k}$, so $k = \dfrac{0.693}{50.5} = 0.0137/\text{s}$

(a) $\dfrac{-\ln[(0.84/16)/0.84]}{0.0137/\text{s}} = 202 \text{ s}$

(b) $\dfrac{-\ln[(0.84/4)/0.84]}{0.0137/\text{s}} = 101 \text{ s}$

(c) $\dfrac{-\ln[(0.84/5)/0.84]}{0.0137/\text{s}} = 117 \text{ s}$

13.93 (a) A reasonable approach is to assume that the reaction is either first or second order. Then, attempt to fit the data to the first-order case. If the fit is good, the reaction is first order. If not, try a second-order plot and see if the fit is good. If neither a first order nor a second order fit is good, other orders would have to be tried. In this case, the fit to the second-order integrated rate law is good, as can be seen in the following graph.

Make up the following table and graph.

Time, s	[A], 10^{-3} mol/L	$\dfrac{1}{[A]}$, L/mol
0	100.	10.
5	14.1	70.9
10.	7.8	128
15	5.3	189
20.	4.0	250

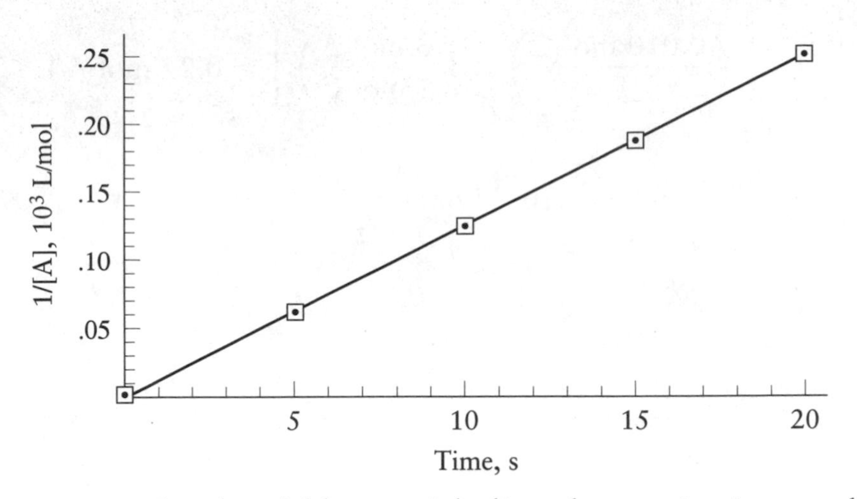

Because the plot yields a straight line, the reaction is second order. Using values of $\frac{1}{[A]}$ in L/mol, the slope is $\frac{(250 - 10.)\ \text{L/mol}}{20.\ \text{s}} = 12\ \text{L/mol}\cdot\text{s}$.

(b) The slope, 12 L/mol·s = k; the rate constant.

13.95 For a second-order reaction,

$$t_{1/2} \propto \frac{1}{[A_0]} \quad \text{or} \quad t_{1/2} = \frac{\text{constant}}{[A_0]}$$

(a) The time necessary for the concentration to fall to one-half of the initial concentration is one half-life:

$$\text{first half-life} = t_1 = t_{1/2} = \frac{\text{constant}}{[A_0]}$$

(b) This time, $t_{1/4}$, is two half-lives, but because of different starting concentrations, the half-lives are not the same:

$$\text{second half-life} = t_2 = \frac{\text{constant}}{(\frac{1}{2}[A_0])} = \frac{2(\text{constant})}{[A_0]} = 2t_1$$

$$\text{total time} = t_1 + t_2 = t_1 + 2t_1 = 3t_{1/2}$$

(c) This time, $t_{1/16}$, is four half-lives, again the half-lives are not the same:

$$\text{third half-life} = t_3 = \frac{\text{constant}}{(\frac{1}{4}[A_0])} = \frac{4(\text{constant})}{[A_0]} = 14t_1$$

$$\text{fourth half-life} = t_4 = \frac{\text{constant}}{(\frac{1}{8}[A_0])} = \frac{8(\text{constant})}{[A_0]} = 8t_1$$

$$\text{total time} = t_1 + t_2 + t_3 + t_4 = t_1 + 2t_1 + 4t_1 + 8t_1 = 15t_{1/2}$$

If t_1 is known, the times $t_{1/4}$ and $t_{1/16}$ can be calculated easily.

13.97 See the solution to Exercise 13.91 for the derivation of the formulas needed here.

(a) $$t = \frac{\dfrac{1}{[A]} - \dfrac{1}{[A]_0}}{k} = \frac{\dfrac{1\text{ L}}{0.080\text{ mol}} - \dfrac{1\text{ L}}{0.10\text{ mol}}}{0.010\text{ L/mol}\cdot\text{min}} = 2.5 \times 10^2\text{ min}$$

(b) $$[A] = \frac{0.45\text{ mol A}}{\text{L}} - \left[\left(\frac{0.45\text{ mol B}}{\text{L}}\right) \times \left(\frac{1\text{ mol A}}{2\text{ mol B}}\right)\right]$$

$$= 0.22\bar{5}(\text{mol A})/\text{L} = \tfrac{1}{2}[A]_0$$

$$t = t_{1/2} = \frac{1}{k[A]_0} = \frac{1}{(0.0045\text{ L/mol}\cdot\text{min}) \times (0.45\text{ mol/L})}$$

$$= 4.9 \times 10^2\text{ min}$$

APPLIED EXERCISES

13.99 At high concentrations of substrate, the plot of rate against substrate concentration levels off to a constant rate. Therefore, the expression

$$\text{rate} \propto [S]^a = \text{constant}$$

implies that $a = 0$ and $[S]^0 = 1 = \text{constant}$. Therefore, the reaction is zero order in substrate at high substrate concentrations.

13.101 There are many types of inhibitors. Suppose, for example, that the inhibitor functions by competing with the substrate to form an EI complex; then the slope of the curve will be reduced, and the rate of product formation for a given substrate concentration will also be less.

13.103 (a) Cl is the catalyst.

(b) ClO is the reaction intermediate.

(c) Cl, ClO, O

(d) Step 1 is initiating; step 2 is propagating.

(e) $Cl + Cl \longrightarrow Cl_2$

INTEGRATED EXERCISES

13.105 (a) rate $= k[H_2SeO_3][I^-]^3[H^+]^2 \quad k = 5.0 \times 10^5\text{ L}^5/\text{mol}^5\cdot\text{s}$

$$\text{rate}_0 = \left(\frac{5.0 \times 10^5\text{ L}^5}{\text{mol}^5\cdot\text{s}}\right) \times \left(\frac{0.020\text{ mol}}{\text{L}}\right) \times \left(\frac{0.020\text{ mol}}{\text{L}}\right)^3 \times \left(\frac{0.010\text{ mol}}{\text{L}}\right)^2$$

$$= 8.0 \times 10^{-6}\text{ mol/L}\cdot\text{s}$$

(b) $[H_2SeO_3]$: 0.020 mol/L

$$[I^-]: \frac{0.020 \text{ mol}}{\text{L}} \times \left(\frac{1 \text{ mol } H_2SeO_3}{6 \text{ mol } I^-}\right) = 3.3 \times 10^{-3} \text{ mol/L}$$

$$[H^+]: \frac{0.010 \text{ mol}}{\text{L}} \times \left(\frac{1 \text{ mol } H_2SeO_3}{4 \text{ mol } H^+}\right) = 2.5 \times 10^{-3} \text{ mol/L}$$

H^+ is the limiting reactant.

(c) $$\left(\frac{0.010 \text{ mol } H^+}{\text{L}}\right) \times (1.00 \text{ L}) \times \left(\frac{1 \text{ mol Se}}{4 \text{ mol } H^+}\right) \times \left(\frac{78.96 \text{ g Se}}{1 \text{ mol Se}}\right) = 0.20 \text{ g Se}$$

13.107 (a) $$k = \frac{0.693}{t_{1/2}} = \frac{0.693}{1.02 \text{ s}} = 0.679 \text{ /s}$$

Because volume is a constant, $P_{CH_3N{=}NCH_3} \propto \text{mass} \propto [CH_3N{=}NCH_3]$; therefore,

$$\ln\left(\frac{P_0}{P_t}\right) = \ln\left(\frac{[CH_3N{=}NCH_3]_0}{[CH_3N{=}NCH_3]_t}\right) = \ln\left(\frac{\text{mass}_0}{\text{mass}_t}\right) = kt$$

$$\ln\left(\frac{\text{mass}_0}{\text{mass}_t}\right) = 0.679 \text{ /s} \times 10. \text{ s} = 6.79$$

$$\frac{\text{mass}_0}{\text{mass}_t} = 889, \quad \text{mass} = \frac{45.0 \text{ mg}}{889} = 0.0506 \text{ mg}$$

(b) $$\ln\left(\frac{\text{mass}_0}{\text{mass}_t}\right) = 0.679 \text{ /s} \times 3.0 \text{ s} = 2.04$$

$$\frac{\text{mass}_0}{\text{mass}_t} = 7.67, \quad \text{mass} = \frac{45.0 \text{ mg}}{7.67} = 5.87 \text{ mg } CH_3N{=}NCH_3$$

$$n_{N_2} = \text{amount of } N_2(g) = [(45.0 - 5.9) \times 10^{-3} \text{ g } CH_3N{=}NCH_3]$$
$$\times \left(\frac{1 \text{ mol } CH_3N{=}NCH_3}{58.09 \text{ g } CH_3N{=}NCH_3}\right) \times \left(\frac{1 \text{ mol } N_2}{1 \text{ mol } CH_3N{=}NCH_3}\right)$$
$$= 6.73 \times 10^{-4} \text{ mol } N_2(g)$$

$$P_{N_2} = \frac{n_{N_2}RT}{V} = \frac{(6.73 \times 10^{-4} \text{ mol})(0.0821 \text{ L}\cdot\text{atm/K}\cdot\text{mol})(573 \text{ K})}{0.300 \text{ L}}$$
$$= 0.106 \text{ atm}$$

CHAPTER 14
CHEMICAL EQUILIBRIUM

EXERCISES

Equilibrium and Equilibrium Constants

14.1 (a) At the molecular level, chemical equilibrium is a dynamic process and interconversion of reactants and products continues to occur. However, at the macroscopic level, no change in concentrations of reactants and products is observable at equilibrium, because the rates of product and reactant formation are equal.
(b) At a fixed temperature, the equilibrium constant is truly a constant, independent of concentrations. However, if the concentration of a reactant is increased, the concentrations of other chemical species will adjust because the equilibrium constant remains the same.

14.3 The concentrations of reactants and products stop changing when equilibrium has been attained. Equilibrium occurs when the forward and reverse rates of reaction are equal, a condition that will always be reached eventually, given enough time. At equilibrium, the concentrations of reactants and products have reached the values dictated by the equilibrium constant. We can think of the difference of concentrations of reactants and products from their equilibrium values as a kind of driving force for the reaction to proceed. When that difference is 0, as it is at equilibrium, this driving force is also 0, and the concentrations no longer continue to change.

14.5 The equilibrium composition in each system will be that which results in the same value of K. Thus, (d) and its inverse are identical in the two containers, but (a), (b), and (c) will be different, because the second container contains five times as much material.
(e) The time required to reach equilibrium might be different in the two containers. There is no fixed simple relation between the time required to reach equilibrium and the concentration of reactant.
A more detailed analysis of (a) through (d) follows. The equilibrium constant for this reaction in this exercise is

$$K_c = \frac{[O_2(g)]^3}{[O_3(g)]^2}$$

In terms of the change in concentration, x, that has occurred at equilibrium, we can write

$$K_c = \frac{(\frac{3}{2}x)^3}{(0.10 - x)^2} \quad \text{in the first container}$$

$$K_c = \frac{(\frac{3}{2}x)^3}{(0.50 - x)^2} \quad \text{in the second container}$$

The change, x, is different in the two cases.

(a) Because the volume is the same, the amount (moles) of O_2, $n_{O_2} = [O_2] \times$ volume, is larger in the second experiment.

(b) Because K_c is a constant and the denominator is larger in the second case, the numerator must also be larger. Therefore, the concentration of O^2 ($= \frac{3}{2}x$) is larger in the second case.

(c) Although $[O_2]^3/[O_3]^2$ is the same, $[O_2]/[O_3]$ will be different. This can be seen by solving the two K_c expressions above for x and then calculating the ratio. The ratio depends on K_c, which is needed to obtain its numerical value.

(d) Because K_c is a constant, $[O_2]^3/[O_3]^2$ is the same.

14.7 (a) $K_c = \frac{[COCl][Cl]}{[CO][Cl_2]}$ (b) $K_c = \frac{[HBr]^2}{[H_2][Br_2]}$ (c) $K_c = \frac{[SO_2]^2[H_2O]^2}{[O_2]^3[H_2S]^2}$

14.9 $N_2(g) + 3\,H_2(g) \rightleftharpoons 2\,NH_3(g)$ $\quad K_p = \frac{P_{NH_3}{}^2}{P_{N_2}P_{H_2}{}^3} = 41$

(a) $K_{p(a)} = \frac{P_{N_2}P_{H_2}{}^3}{P_{NH_3}{}^2}$ Therefore, $K_{p(a)} = 1/K_p = 1/41 = 0.024$

(b) $K_{p(b)} = \frac{P_{NH_3}}{P_{N_2}{}^{1/2}P_{H_2}{}^{3/2}}$ Therefore, $K_{p(b)} = \sqrt{K_p} = 6.4$

(c) $K_{p(c)} = \frac{P_{NH_3}{}^4}{P_{N_2}{}^2P_{H_2}{}^6}$ Therefore, $K_{p(c)} = K_p{}^2 = 41^2 = 1.7 \times 10^3$

Rate Constants and Equilibrium

14.11 (a) At equilibrium, the rates of the forward and reverse reactions are the same. The rate constants, k, and k_{-1}, do not change (unless temperature changes).

(b) If K_c is large for an elementary reaction, then the rate constant for the forward reaction is much larger than the rate constant of the reverse reaction. $K_c = k_1/k_{-1}$.

Equilibrium Constants from Equilibrium Amounts

14.13 $K_c = \dfrac{[HI]^2}{[H_2][I_2]}$

first case: $K_c = \dfrac{(1.37 \times 10^{-2})^2}{(6.47 \times 10^{-3})(5.94 \times 10^{-4})} = 48.8$

second case: $K_c = \dfrac{(1.69 \times 10^{-2})^2}{(3.84 \times 10^{-3})(1.52 \times 10^{-3})} = 48.9$

third case: $K_c = \dfrac{(1.00 \times 10^{-2})^2}{(1.43 \times 10^{-3})(1.43 \times 10^{-3})} = 48.9$

14.15 (a) $2\ NOCl(g) \rightleftharpoons 2\ NO(g) + Cl_2(g) \qquad \Delta n = 1$

$$K_c = \frac{K_p}{(RT)^{\Delta n}} = \frac{K_p}{[0.082\ 06\ (T/K)]^{\Delta n}} = \frac{1.8 \times 10^{-2}}{(0.082\ 06 \times 500.)} = 4.4 \times 10^{-4}$$

(b) $CaCO_3(s) \rightleftharpoons CaO(s) + CO_2(g) \qquad \Delta n = 1$

$$K_c = \frac{167}{(0.082\ 06 \times 1073)} = 1.90$$

Heterogeneous Equilibria

14.17 All the equilibria are heterogeneous, because in each case more than one phase is involved in the equilibrium.

(a) $K_p = P_{H_2S}P_{NH_3}$; (b) $K_p = P_{NH_3}{}^2P_{CO_2}$; (c) $K_p = P_{O_2}$

14.19 (a) $K_c = 1/[Cl_2]$; (b) $K_c = [N_2O][H_2O]^2$; (c) $K_c = [CO_2]$

The Extent and Direction of Reactions

14.21 $H_2(g) + I_2(g) \rightleftharpoons 2\ HI(g) \qquad (K_c = 160 \text{ at } 500\text{ K}) \quad K_c = \dfrac{[HI]^2}{[H_2][I_2]} = 160$ and

$$[H_2] = \frac{[HI]^2}{K_c[I_2]} = \frac{(2.21 \times 10^{-3})^2}{(160)(1.46 \times 10^{-3})} = 2.09 \times 10^{-5} \text{ mol/L}$$

14.23 $PCl_5(g) \rightleftharpoons PCl_3(g) + Cl_2(g)$

$$K_p = \frac{P_{PCl_3}P_{Cl_2}}{P_{PCl_5}} = 25 \quad \text{Therefore, } P_{PCl_3} = \frac{K_pP_{PCl_5}}{P_{Cl_2}} = \frac{(25)(0.15)}{0.20} = 19 \text{ atm}$$

14.25 (a) $K_c = 160.$ at 500.°C; the reaction is $H_2(g) + I_2(g) \rightleftharpoons 2\ HI(g)$

$$Q_c = \frac{[HI]^2}{[H_2][I_2]} = \frac{(2.4 \times 10^{-3})^2}{(4.8 \times 10^{-3}) \times (2.4 \times 10^{-3})} = 0.50$$

(b) no, $Q_c \neq K_c$

(c) $Q_c < K_c$; tendency to form products

14.27 For the reaction $2\ SO_2(g) + O_2(g) \rightleftharpoons 2\ SO_3(g)$

(a) $$Q_c = \frac{[SO_3]^2}{[SO_2]^2[O_2]} = \frac{(1.00 \times 10^{-4}/0.500)^2}{(1.20 \times 10^{-3}/0.500)^2 \times (5.0 \times 10^{-4}/0.500)} = 6.9$$

(b) $Q_c < K_c$; therefore, more $SO_3(g)$ will form.

Equilibrium Constants from Initial Amounts

14.29 $2\ HI(g) \rightleftharpoons H_2(g) + I_2(g) \qquad \left(K_c = \frac{[H_2][I_2]}{[HI]^2}\right)$

$$[HI] = \left(\frac{1.90\text{ g HI}}{2.00\text{ L}}\right) \times \left(\frac{1\text{ mol HI}}{127.91\text{ g HI}}\right) = 7.43 \times 10^{-3}\text{ mol/L}$$

$[I_2] = [H_2] = (\frac{1}{2}) \times$ (mol HI reacted)

$$= \frac{1}{2} \times \left(\frac{0.0172\text{ mol}}{2.00\text{ L}} - 7.43 \times 10^{-3}\text{ mol/L}\right) = 5.85 \times 10^{-4}\text{ mol/L}$$

Then, $$K_c = \frac{5.85 \times 10^{-4} \times 5.85 \times 10^{-4}}{(7.43 \times 10^{-3})^2} = 6.20 \times 10^{-3}$$

14.31 $NH_2CO_2NH_4(s) \rightleftharpoons 2\ NH_3(g) + CO_2(g) \qquad (K_c = [NH_3]^2[CO_2])$

$$[CO_2] = \left(\frac{17.4\text{ mg }CO_2}{250\text{ mL}}\right) \times \left(\frac{10^{-3}\text{ g}}{1\text{ mg}}\right) \times \left(\frac{1\text{ mol }CO_2}{44.01\text{ g }CO_2}\right) \times \left(\frac{1\text{ mL}}{10^{-3}\text{ L}}\right)$$

$$= 1.58 \times 10^{-3}\ (\text{mol }CO_2)/\text{L}$$

$$[NH_3] = 1.58 \times 10^{-3}\ (\text{mol }CO_2)/\text{L} \times \left(\frac{2\text{ mol }NH_3}{1\text{ mol }CO_2}\right) = 3.16 \times 10^{-3}\ (\text{mol }NH_3)/\text{L}$$

Substituting in the expression for K_c,

$K_c = (3.16 \times 10^{-3})^2 \times (1.58 \times 10^{-3}) = 1.58 \times 10^{-8}$.

14.33 $CH_3COOH(l) + C_2H_5OH(l) \rightleftharpoons CH_3COOC_2H_5(l) + H_2O(l)$

In each case, $$K_c = \frac{[CH_3COOC_2H_5][H_2O]}{[CH_3COOH][C_2H_5OH]}$$

On the assumption that the only water present is a result of the reaction, then in each case the equilibrium concentration of water is the same as that of $CH_3COOC_2H_5(l)$; that is, $[H_2O] = [CH_3COOC_2H_5]$.

first case: $K_c = \dfrac{(0.171)^2}{(1.00 - 0.171) \times (0.180 - 0.171)} = 3.9$

second case: $K_c = \dfrac{(0.667)^2}{(1.00 - 0.667) \times (1.00 - 0.667)} = 4.0$

third case: $\dfrac{(0.966)^2}{(1.00 - 0.966) \times (8.00 - 0.966)} = 3.9$

Equilibrium Table Calculations

14.35 (a) Initial concentration $Cl_2 = \dfrac{2.0 \times 10^{-3}\ \text{mol}}{2.0\ \text{L}} = 1.0 \times 10^{-3}$ mol/L

Concentration (mol/L)	$Cl_2(g)$ $\rightleftharpoons$	$2\ Cl(g)$	$K_c = 1.2 \times 10^{-7}$ at 1000 K
initial	1.0×10^{-3}	0	
change	$-x$	$+2x$	
equil.	$1.0 \times 10^{-3} - x$	$2x$	

so $K_c = \dfrac{[Cl]^2}{[Cl_2]} = \dfrac{(2x)^2}{(1.0 \times 10^{-3} - x)} = 1.2 \times 10^{-7}$

Assume in $(1.0 \times 10^{-3} - x)$, that x is negligible as a result of the small K_c, and so $4x^2 = (1.0 \times 10^{-3}) \times (1.2 \times 10^{-7})$.

Then $x = 5.5 \times 10^{-6}$ mol/L = amount/L decomposed.

equilibrium concentrations: $2x = [Cl] = 1.1 \times 10^{-5}$ mol/L

$[Cl_2] \approx 1.0 \times 10^{-3}$ mol/L, because x is small

% decomposition $= \dfrac{\text{amount/L decomposed}}{\text{initial amount/L}} \times 100\% = \dfrac{5.5 \times 10^{-6}\ \text{mol/L}}{1.0 \times 10^{-3}\ \text{mol/L}} \times 100\% = 0.55\%$

(b) Initial concentration of $F_2 = \dfrac{2.0 \times 10^{-3}\ \text{mol}}{2.0\ \text{L}} = 1.0 \times 10^{-3}$ mol/L

Concentration (mol/L)	$F_2(g)$ $\rightleftharpoons$	$2\ F(g)$	$K_c = 1.2 \times 10^{-4}$ at 1000 K
initial	1.0×10^{-3}	0	
change	$-x$	$+2x$	
equil.	$1.0 \times 10^{-3} - x$	$2x$	

so $K_c = \dfrac{[F]^2}{[F_2]} = \dfrac{(2x)^2}{(1.0 \times 10^{-3} - x)} = 1.2 \times 10^{-4}$

In this case, x is *not* negligible.

$4x^2 = (1.0 \times 10^{-3}) \times (1.2 \times 10^{-4}) - 1.2 \times 10^{-4}x$

$4x^2 + 1.2 \times 10^{-4}x - 1.2 \times 10^{-7} = 0$

Solve the quadratic: $x = 1.59 \times 10^{-4}$ mol/L = amount/L decomposed

equilibrium concentrations: $2x = [F] = 3.18 \times 10^{-4}$ mol/L

$[F_2] \approx 1.0 \times 10^{-3} - x = 8.41 \times 10^{-4}$ mol/L

$$\% \text{ decomposition} = \frac{1.59 \times 10^{-4} \text{ mol/L}}{8.41 \times 10^{-4} \text{ mol/L}} \times 100\% = 19\%$$

(c) Cl_2 is more stable at 1000 K than F_2 is because less decomposes.

14.37

Concentration (mol/L)	$2\,HBr(g)$	$\rightleftharpoons H_2(g)$	$+ Br_2(g)$	$K_c = 7.7 \times 10^{-11}$ at 500. K
initial	1.2×10^{-3}	0	0	
change	$-2x$	$+x$	$+x$	
equil.	$1.2 \times 10^{-3} - 2x$	x	x	

$$K_c = \frac{[H_2][Br_2]}{[HBr]^2} = \frac{(x)(x)}{(1.2 \times 10^{-3} - 2x)^2} = 7.7 \times 10^{-11}$$

In the term $(1.2 \times 10^{-3} - 2x)$, $2x$ is negligible.

$x^2 = 1.11 \times 10^{-16}$

$x = 1.10 \times 10^{-8} = [H_2] = [Br_2]$

$2x = 2.0 \times 10^{-8}$ mol/L = amount/L decomposed

$[HBr] = 1.2 \times 10^{-3} \text{ mol/L} - 2 \times (1.0 \times 10^{-8}) \approx 1.2 \times 10^{-3}$ mol/L

$$\% \text{ decomposition} = \frac{\text{amount/L decomposed}}{\text{initial concentration of reactant}} \times 100\% = \frac{2.0 \times 10^{-8} \text{ mol/L}}{1.2 \times 10^{-3} \text{ mol/L}} \times 100\% = 1.7 \times 10^{-3}\%$$

The percentage decomposition is close to negligible. At equilibrium, we find 1.2×10^{-3} mol/L HBr, 1.1×10^{-8} mol/L H_2, and 1×10^{-8} mol/L Br_2.

14.39 (a) $$[PCl_5] = \frac{n_{PCl_5}}{\text{Vol(L)}} = \frac{1.0 \text{ g } PCl_5/208.22 \text{ g/mol}}{0.250 \text{ L}} = 1.9 \times 10^{-2} \text{ mol/L}$$

= initial concentration

Concentration (mol/L)	$PCl_5(g)$	$\rightleftharpoons PCl_3(g)$	$+ Cl_2(g)$
initial	1.9×10^{-2}	0	0
change	$-x$	$+x$	$+x$
equil.	$1.9 \times 10^{-2} - x$	x	x

$$K_c = \frac{K_p}{(RT)^{\Delta n}} = \frac{0.36}{(0.082\,06) \times (400.)^1} = 1.1 \times 10^{-2} \quad \text{and, } K_c = \frac{x^2}{1.9 \times 10^{-2} - x}$$

$x^2 + 1.1 \times 10^{-2}x - 2.1 \times 10^{-4} = 0$

$x = [PCl_3] = [Cl_2] = 1.0 \times 10^{-2}$ mol/L

$[PCl_5] = 1.9 \times 10^{-2} - 1.0 \times 10^{-2}$

$= 9.0 \times 10^{-3}$ mol/L

At equilibrium, we find 1.0×10^{-2} mol/L PCl_3, 1.0×10^{-2} mol/L Cl_2, and 9.0×10^{-3} mol/L PCl_5.

(b) x = amount/L decomposed = 1.0×10^{-2} mol/L

$$\% \text{ decomposition} = \frac{1.0 \times 10^{-2} \text{ mol/L}}{1.9 \times 10^{-2} \text{ mol/L}} \times 100\% = 53\%$$

14.41 Initial $[NH_3] = \dfrac{0.400 \text{ mol}}{2.0 \text{ L}} = 0.20$ mol/L

Concentration (mol/L)	$NH_4HS(s)$ $\rightleftharpoons$	$NH_3(g)$	+ $H_2S(g)$
initial	—	0.20	0
change	—	$+x$	$+x$
equil.	—	$0.20 + x$	x

$K_c = [NH_3][H_2S] = 1.6 \times 10^{-4}$

$(0.200 + x) \times (x) = 1.6 \times 10^{-4}$

$x^2 + 0.20x - 1.6 \times 10^{-4} = 0$

equilibrium concentrations: $x = 8.0 \times 10^{-4}$ mol/L H_2S

$0.20 + 8.0 \times 10^{-4} = 0.20$ mol/L NH_3

14.43 Initial concentrations: $[PCl_5] = \dfrac{0.200 \text{ mol}}{4.00 \text{ L}} = 0.0500$ mol/L

$[PCl_3] = \dfrac{0.600 \text{ mol}}{4.00 \text{ L}} = 0.150$ mol/L

$[Cl_2] = 0$

Concentration (mol/L)	$PCl_5(g)$ $\rightleftharpoons$	$PCl_3(g)$	+ $Cl_2(g)$
initial	0.0500	0.150	0
change	$-x$	$+x$	$+x$
equil.	$0.0500 - x$	$0.150 + x$	x

$$K_c = \frac{[PCl_3][Cl_2]}{[PCl_5]} = 33.3 = \frac{(0.150 + x) \times (x)}{(0.0500 - x)}$$

$1.665 - 33.3x = x^2 + 0.15x$, and $x^2 + 33.45x - 1.665 = 0$

equilibrium concentrations: $x = 4.97 \times 10^{-2}$ mol/L Cl_2

$0.150 + 4.97 \times 10^{-2} = 2.00 \times 10^{-1}$ mol/L PCl_3

$0.0500 - 4.97 \times 10^{-2} = 3.00 \times 10^{-4}$ mol/L PCl_5

14.45

Concentration (mol/L)	$N_2(g)$	+	$O_2(g)$	$\rightleftharpoons$	$2\ NO(g)$
initial	0.114		0.114		0
change	$-x$		$-x$		$+2x$
equil.	$0.114 - x$		$0.114 - x$		$2x$

$$K_c = \frac{[NO]^2}{[N_2][O_2]} = 1.00 \times 10^{-5} = \frac{(2x)^2}{(0.114 - x) \times (0.114 - x)} = 1.00 \times 10^{-5}$$

$$4x^2 = (1.00 \times 10^{-5}) \times (x^2 - 0.228x + 0.013)$$

$$= 1.00 \times 10^{-5}x^2 - 2.28 \times 10^{-6}x + 1.3 \times 10^{-7}$$

$$4x^2 + 2.28 \times 10^{-6}x - 1.3 \times 10^{-7} = 0$$

Solving the quadratic equation, $x = 1.80 \times 10^{-4}$ mol/L

equilibrium concentrations: $2x = 3.60 \times 10^{-4}$ mol/L NO

$0.114 - x \approx 0.114$ mol/L N_2 and 0.114 mol/L O_2

14.47 $H_2(g) + I_2(g) \rightleftharpoons 2\ HI(g)$

0.400 mol H_2 $\times$ 0.60 = 0.240 mol H_2 reacted

Therefore, 0.240 mol I_2 reacted and 0.480 mol HI formed.

0.400 mol H_2 initial − 0.240 mol H_2 reacted = 0.160 mol H_2 at equilibrium

1.60 mol I_2 initial − 0.240 mol I_2 reacted = 1.36 mol I_2 at equilibrium

$$K_c = \frac{[HI]^2}{[H_2][I_2]} = \frac{\left(\frac{0.480\ \text{mol}}{3.00\ \text{L}}\right)^2}{\left(\frac{0.160\ \text{mol}}{3.00\ \text{L}}\right) \times \left(\frac{1.36\ \text{mol}}{3.00\ \text{L}}\right)} = 1.06$$

Effect of Added Reagents

14.49 (a) According to Le Chatelier's principle, an increase in the partial pressure of CO_2, a product, will favor a shift in the equilibrium to the left, thus decreasing the partial pressure of H_2.

(b) According to Le Chatelier's principle, a decrease in the partial pressure of CO, a reactant, will favor a shift in the equilibrium to the left, thus decreasing the partial pressure of CO_2.

(c) According to Le Chatelier's principle, an increase in the concentration of one of the reactants, CO, will favor the formation of products; thus the H_2 concentration will increase.

(d) Nothing; the equilibrium constant is a constant independent of concentrations.

14.51

Change	Quantity	Effect
add NO	amount of H_2O	decrease
add NO	amount of O_2	increase
remove H_2O	amount of NO	increase
remove O_2	amount of NH_3	increase
add NH_3	K_c	no change
remove NO	amount of NH_3	decrease
add NH_3	amount of O_2	decrease

Response to Pressure

14.53 General statement: If the number of moles of gaseous products and reactants are equal, no pressure effect will be observed. When the numbers of moles of gaseous products and reactants are not equal, an *increase* in pressure will favor the direction having the smaller number of moles of gas.
(a) reactants favored; smaller number of moles of gas
(b) reactants favored; smaller number of moles of gas
(c) reactants favored; smaller number of moles of gas
(d) no effect; same number of moles on each side
(e) reactants favored; smaller number of moles of gas

14.55 (a) The partial pressure of NH_3 increases when the partial pressure of NO increases. According to Le Chatelier's principle, an increase in partial pressure of a product shifts the equilibrium to the left, thereby favoring the production of reactant, NH_3.
(b) When the partial pressure of NH_3 decreases, the partial pressure of O_2 increases. According to Le Chatelier's principle, a decrease in the partial pressure of one of the reactants shifts the equilibrium to the left, thereby favoring the production of O_2.

Response to Temperature

14.57 General principle: According to Le Chatelier's principle, the reaction will shift in the direction of the substances having the higher internal energy when the temperature is increased. Therefore, for an exothermic reaction, an increase in temperature favors the reactants; whereas, for an endothermic reaction, the products are favored.

(a) products favored (endothermic reaction)
(b) reactants favored (exothermic reaction)
(c) reactants favored (exothermic reaction)

14.59 At 700. K, K_p is smaller than at 600. K. A smaller K_p means less product is formed, so the amount of ammonia will be less at 700. K than at 600. K.

SUPPLEMENTARY EXERCISES

14.61 More molecules of X_2 have decomposed to form X atoms. Because bond breaking is endothermic, (a) will lead to more X atoms dissociating. Neither (b), (c), nor (d) will lead to the change shown.

14.63 $2\ SO_2(g) + O_2(g) \rightleftharpoons 2\ SO_3(g) \quad (K_p = 3.0 \times 10^4,\ \ \Delta n = -1)$

Therefore, for $2\ SO_3(g) \rightleftharpoons 2\ SO_2(g) + O_2(g) \quad K_p = \dfrac{1}{3.0 \times 10^4}$

so, $SO_3(g) \rightleftharpoons SO_2(g) + \frac{1}{2}\ O_2(g)$

$$K_p = \sqrt{\frac{1}{3.0 \times 10^4}} = 5.8 \times 10^{-3}, \qquad \Delta n = \tfrac{1}{2}$$

(a) $K_c = \dfrac{K_p}{(RT)^{\Delta n}} = \dfrac{3.0 \times 10^4}{(0.082\ 06 \times 700.)^{-1}} = 1.7 \times 10^6$

(b) $K_c = \dfrac{K_p}{(RT)^{\Delta n}} = \dfrac{5.8 \times 10^{-3}}{\sqrt{0.082\ 06 \times 700.}} = 7.6 \times 10^{-4}$

14.65 $Q_c = \dfrac{[NH_3]^2}{[N_2][H_2]^3} = \dfrac{(0.500)^2}{(3.00) \times (2.00)^3} = 0.0104$ and $Q_c < K_c$

(because $0.0104 < 0.060$)

Therefore, the reaction is not at equilibrium; there is a tendency to form the product (NH_3) at the expense of the reactants (N_2 and H_2).

14.67 $Cl_2(g) + Br_2(g) \rightleftharpoons 2\ BrCl(g)$ and $K_c = 0.031 = \dfrac{[BrCl]^2}{[Cl_2][Br_2]}$

Let $x = [Br_2]$

$$0.031 = \frac{(0.097)^2}{(0.22) \times (x)}$$

$$x = \frac{(0.097)^2}{(0.22) \times (0.031)} = 1.4 \text{ mol/L } Br_2$$

14.69 Initial concentrations:

$$[CO] = \frac{2.00 \text{ mol}}{10.0 \text{ L}} = 0.200 \text{ mol/L}$$

$$[H_2] = \frac{3.00 \text{ mol}}{10.0 \text{ L}} = 0.300 \text{ mol/L}$$

equilibrium concentration: $[CH_4] = \dfrac{0.478 \text{ mol}}{10.0 \text{ L}} = 0.0478 \text{ mol/L}$

Concentration (mol/L)	$CO(g)$ +	$3\ H_2(g)$ $\rightleftharpoons$	$CH_4(g)$ +	$H_2O(g)$
initial	0.200	0.300	0	0
change	−0.0478	-3×0.0478	+0.0478	+0.0478
equil.	$0.152\overline{2}$	$0.156\overline{6}$	0.0478	0.0478

$$K_c = \frac{[CH_4][H_2O]}{[CO][H_2]^3} = \frac{(0.0478)^2}{0.152\overline{2} \times (0.156\overline{6})^3} = 3.91$$

14.71

Concentration (mol/L)	$SO_2(g)$ +	$NO_2(g)$ $\rightleftharpoons$	$NO(g)$ +	$SO_3(g)$
initial	$\dfrac{0.100 \text{ mol}}{5.00 \text{ L}}$	$\dfrac{0.200 \text{ mol}}{5.00 \text{ L}}$	$\dfrac{0.100 \text{ mol}}{5.00 \text{ L}}$	$\dfrac{0.150 \text{ mol}}{5.00 \text{ L}}$
change	$-x$	$-x$	$+x$	$+x$
equil.	$0.0200 - x$	$0.0400 - x$	$0.0200 + x$	$0.0300 + x$

$$K_c = 85.0 = \frac{[NO][SO_3]}{[SO_2][NO_2]} = \frac{(0.0200 + x) \times (0.0300 + x)}{(0.0200 - x) \times (0.0400 - x)}$$

$$85.0 = \frac{(0.0200 + x) \times (0.0300 + x)}{(0.0200 - x) \times (0.0400 - x)} = \frac{x^2 + 0.0500x + 6.00 \times 10^{-4}}{x^2 - 0.0600x + 8.00 \times 10^{-4}}$$

$$x^2 + 0.0500x + 6.00 \times 10^{-4} = 85.0x^2 - 5.10x + 6.80 \times 10^{-2}$$

$$84.0x^2 - 5.15x + 6.74 \times 10^{-2} = 0$$

Solving the quadratic yields $x = 0.0189$.

equilibrium concentrations:
$0.0200 + 0.0189 = 0.0389$ mol/L NO
$0.0300 + 0.0189 = 0.0489$ mol/L SO_3
$0.0200 - 0.0189 = 0.0011$ mol/L SO_2
$0.0400 - 0.0189 = 0.0211$ mol/L NO_2

14.73 $2\ N_2O(g) + 3\ O_2(g) \rightleftharpoons 4\ NO_2(g)$

(a) at equilibrium:

0.0200 mol/L NO_2

$$[N_2O] = 0.0200\ \text{mol/L} - (0.0200\ \text{mol/L}) \times \left(\frac{2\ \text{mol}\ N_2O}{4\ \text{mol}\ NO_2}\right)$$

$$= 0.0100\ \text{mol/L}\ N_2O$$

$$[O_2] = 0.0560\ \text{mol/L} - (0.0200\ \text{mol/L}) \times \left(\frac{3\ \text{mol}\ O_2}{4\ \text{mol}\ N_2O}\right)$$

$$= 0.0410\ \text{mol/L}\ O_2$$

(b) $$K_c = \frac{[NO_2]^4}{[N_2O]^2[O_2]^3} = \frac{(0.0200)^4}{(0.0100)^2(0.0410)^3} = 23.2$$

14.75

Pressure (atm)	$2\ HCl(g) \rightleftharpoons$	$H_2(g)$ +	$Cl_2(g)$
initial	0.22	0	0
change	$-2x$	$+x$	$+x$
equil.	$0.22 - 2x$	x	x

$$K_p = 3.2 \times 10^{-34} = \frac{P_{H_2}P_{Cl_2}}{P_{HCl}^{\ 2}} = \frac{(x) \times (x)}{(0.22 - x)^2}$$

Because K_p is very small, $(0.22 - x) \approx 0.22$

$$3.2 \times 10^{-34} = \frac{x^2}{(0.22)^2}$$

$$x^2 = 1.5 \times 10^{-35}$$

equilibrium partial pressures: $x = 3.9 \times 10^{-18}$ atm H_2 and 3.9×10^{-18} atm Cl_2

≈ 0.22 atm HCl

14.77 (a) $$n_{NOCl} = 30.1\ \text{g NOCl} \times \frac{1\ \text{mol NOCl}}{65.46\ \text{g NOCl}} = 0.459\overline{8}\ \text{mol NOCl}$$

$$\text{initial}\ P_{NOCl} = \frac{nRT}{V} = \frac{0.459\overline{8}\ \text{mol} \times 0.082\ 06\ \text{L·atm/K·mol} \times 500.\ \text{K}}{0.200\ \text{L}}$$

$$= 94.3\overline{3}\ \text{atm}$$

Pressure (atm)	$2\ NOCl(g) \rightleftharpoons$	$2\ NO(g)$ +	$Cl_2(g)$
initial	$94.3\overline{3}$	0	0
change	$-2x$	$+2x$	$+x$
equil.	$94.3\overline{3} - 2x$	$2x$	x

$$K_p = 1.13 \times 10^{-3} = \frac{P_{NO}^{\ 2}P_{Cl_2}}{P_{NOCl}^{\ 2}} = \frac{(2x)^2 \times x}{(94.3\overline{3} - 2x)^2}$$

Because K_p is small, we can approximate this expression by neglecting $2x$ in the denominator.

$$K_p = 1.13 \times 10^{-3} = \frac{4x^3}{(94.3\overline{3})^2}$$

$$x = \sqrt[3]{\frac{1.13 \times 10^{-3} \times (94.3\overline{3})^2}{4}} = 1.36 \text{ atm } Cl_2$$

$2x = 2 \times 1.36 \text{ atm} = 2.72 \text{ atm NO}$

$94.3\overline{3} \text{ atm} - 2x = 94.3\overline{3} \text{ atm} - 2.72 \text{ atm} = 91.6 \text{ atm NOCl}$

(b) $\% \text{ decomposition} = \frac{(94.3 - 91.6) \text{ atm}}{94.3 \text{ atm}} \times 100\% = 2.9\%$

Neglecting K_p does introduce some error. The Equilibrium option in the calculator tool on the CD will help you solve equations like this. Using this tool, we find (a) 0.846 atm Cl_2, 1.69 atm NO, 92.7 atm NOCl and (b) 1.8%.

14.79 Initial concentrations:

$$[CO] = \left(\frac{0.28 \text{ g CO}}{2.0 \text{ L}}\right) \times \left(\frac{1 \text{ mol CO}}{28.01 \text{ g CO}}\right) = 5.0 \times 10^{-3} \text{ mol/L}$$

$$[O_2] = \left(\frac{0.032 \text{ g } O_2}{2.0 \text{ L}}\right) \times \left(\frac{1 \text{ mol } O_2}{32.00 \text{ g } O_2}\right) = 5.0 \times 10^{-4} \text{ mol/L}$$

$[CO_2] = 0 \text{ mol/L}$

Concentration (mol/L)	$2\ CO(g)$	+	$O_2(g)$	$\rightleftharpoons$	$2\ CO_2(g)$
initial	5.0×10^{-3}		5.0×10^{-4}		0
change	$-2x$		$-x$		$+2x$
equil.	$5.0 \times 10^{-3} - 2x$		$5.0 \times 10^{-4} - x$		$2x$

$$K_c = \frac{(2x)^2}{(5.0 \times 10^{-3} - 2x)^2 \times (5.0 \times 10^{-4} - x)} = 0.66$$

This expression results in a cubic equation. Assume that x is small relative to the other values in the denominator. Then, an approximate quadratic expression is

$$0.66 = \frac{4x^2}{(5.0 \times 10^{-3})^2 \times (5.0 \times 10^{-4})}$$

$4x^2 = (0.66) \times (1.25 \times 10^{-8}) = 8.25 \times 10^{-9}$

$x^2 = 2.06 \times 10^{-9}$

$x = 4.5 \times 10^{-5} \text{ mol/L}$

equilibrium concentrations: $2x = 9.0 \times 10^{-5} \text{ mol/L } CO_2$

$\approx 4.9 \times 10^{-3} \text{ mol/L CO}$

$= 4.5 \times 10^{-4} \text{ mol/L } O_2$

It is instructive to compare these values to those obtained by solving the full cubic equation. This solution can be easily obtained from any of a number of mathematical software packages, as well as by many hand-held calculators. After multiplication and expansion of the terms in K_c above, the full cubic is

$2.64x^3 + 3.985x^2 + 2.31 \times 10^{-5}x - 8.25 \times 10^{-9} = 0$

The solution is $x = 4.2\overline{7} \times 10^{-5}$ mol/L.

Then, $2x = 8.5\bar{4} \times 10^{-5}$ mol/L CO_2
$5.0 \times 10^{-3} - 2x = 4.9\bar{1} \times 10^{-3}$ mol/L CO
$5.0 \times 10^{-4} - x = 4.5\bar{7} \times 10^{-4}$ mol/L
$= 4.6 \times 10^{-4}$ mol/L O_2

Comparing these results with the approximate one above, we see no measurable difference in [CO], very little in [O_2], but about 5% difference in [CO_2]. For [CO_2], the approximation may not be adequate, depending upon the accuracy required.

14.81

Concentration	$N_2O_4(g)$ $\rightleftharpoons$	$2\ NO_2(g)$
initial	$\left(\frac{2.50\ g}{2.00\ L}\right)\left(\frac{1\ mol}{92.02\ g}\right)$	0
change	$-x$	$+2x$
equil.	1.36×10^{-2} mol/L $- x$	$2x$

$$K_c = \frac{[NO_2]^2}{[N_2O_4]} = \frac{(2x)^2}{(1.36 \times 10^{-2} - x)} = 4.66 \times 10^{-3}$$

$4x^2 = 6.33 \times 10^{-5} - 4.66 \times 10^{-3}x$
$4x^2 + 4.66 \times 10^{-3}x - 6.33 \times 10^{-5} = 0$
Solving the quadratic yields $x = 3.44 \times 10^{-3}$
$2x = 6.88 \times 10^{-3}$ mol/L NO_2
$1.36 \times 10^{-2} - 3.44 \times 10^{-3} = 1.02 \times 10^{-2}$ mol/L N_2O_4

14.83 The reaction is $PCl_5(g) \rightleftharpoons PCl_3(g) + Cl_2(g)$ $(K_p = 4.96)$
The equilibrium constant expression is

$K_p = \frac{P_{PCl_3}P_{Cl_2}}{P_{PCl_5}}$, where P_{PCl_3}, P_{Cl_2}, and P_{PCl_5} are partial pressures.

Recall that partial pressure = mole fraction $\times P$ (P = total pressure)

Moles of species	P_{Cl_5}	P_{PCl_3}	P_{Cl_2}
initial	n		
change	αn	αn	αn
equilibrium	$n - \alpha n$	αn	αn

The mole fractions of each species are calculated thus:

$$\text{mole fraction } PCl_5 = \frac{n - \alpha n}{(n - \alpha n) + 2\alpha n} = \frac{n - \alpha n}{n + \alpha n} = \frac{1 - \alpha}{1 + \alpha}$$

$$\text{mole fraction } PCl_3 = \frac{\alpha n}{n + \alpha n} = \frac{\alpha}{1 + \alpha}$$

$$\text{mole fraction } Cl_2 = \frac{\alpha}{1 + \alpha}$$

$$K_p = \frac{P_{PCl_3}P_{Cl_2}}{P_{PCl_5}} = \frac{\left(\frac{\alpha}{1+\alpha}\right)P \times \left(\frac{\alpha}{1+\alpha}\right)P}{\left(\frac{1-\alpha}{1+\alpha}\right)P} = \frac{\left(\frac{\alpha}{1+\alpha}\right) \times \left(\frac{\alpha}{1+\alpha}\right)P}{\left(\frac{1-\alpha}{1+\alpha}\right)}$$

$$= \frac{\alpha^2 P}{(1+\alpha)(1-\alpha)} = \frac{\alpha^2 P}{1-\alpha^2} = K_p$$

$$\alpha^2 P = K_p - \alpha^2 K_p$$

$$\alpha^2 = \frac{K_p}{P} - \frac{\alpha^2 K_p}{P}$$

$$\alpha^2 + \frac{\alpha^2 K_p}{P} = \frac{K_p}{P}$$

$$\alpha^2\left(1 + \frac{K_p}{P}\right) = \frac{K_p}{P}$$

$$\alpha^2\left(\frac{P + K_p}{P}\right) = \frac{K_p}{P}$$

$$\alpha^2(P + K_p) = K_p$$

$$\alpha^2 = \frac{K_p}{K_p + P}$$

$$\alpha = \sqrt{\frac{K_p}{K_p + P}}$$

(a) $\alpha = \sqrt{\frac{4.96}{4.96 + 0.50}} = 0.95$ (b) $\alpha = \sqrt{\frac{4.96}{4.96 + 1.0}} = 0.91$

14.85 Choose any two temperatures; for example,

$T' = 2000.$ K, $T = 1000.$ K

Then $\ln K_p' = 2.5 + \frac{21\,700}{2000.} = -8.3\overline{5}$

$\ln K_p = 2.5 - \frac{21\,700}{1000.} = -19.2$

Then $\ln\left(\frac{K_p'}{K_p}\right) = 10.8\overline{5} = \frac{-\Delta H°}{R}\left(\frac{1}{1000.\text{ K}} - \frac{1}{2000.\text{ K}}\right) = \frac{\Delta H°}{R} \times \frac{1}{2000.}$

$\frac{\Delta H°}{R} = 10.8\overline{5} \times 2000. = 21\,700$ K

$\Delta H° = 8.314$ J/K·mol $\times$ 21 700 K $= 1.80 \times 10^5$ J/mol $= 180$ kJ/mol

Alternatively, we can subtract the simultaneous equations:

$\ln K_p' = 2.5 - 21\,700/(T'/\text{K})$

and $\ln K_p = 2.5 - 21\,700/(T/\text{K})$

which yields the van't Hoff equation:

$$\ln\left(\frac{K_p'}{K_p}\right) = -21\ 700\left(\frac{1}{T} - \frac{1}{T'}\right)$$

Thus, $\Delta H°/R = -21\ 700$ K and

$\Delta H° = 8.314$ J/K·mol × 21 700 K = 180. kJ/mol

APPLIED EXERCISES

14.87

Pressure (atm)	$N_2(g)$	$H_2(g)$	$NH_3(g)$
Initial	100.0	100.0	0
Change	$-x$	$-3x$	$+2x$
Equilibrium	$100.0 - x$	$100.0 - 3x$	$2x$

(a) $K_P = P_{NH3}^{\ 2}/P_{N2}P_{H2}^{\ 3}$

$$0.036 = \frac{(2x^2)}{(100.0 - x) \times (100.0 - 3x)^3}$$

Solution of this expression leads to a higher-order equation, and x cannot be neglected. Use the Equilibrium option in Calculator Tool of the CD. Then, $x = 30$ atm, and $P_{NH3} = 2x = 60.$ atm and $P_{N2} = 10.$ atm and $P_{H2} = 10.$ atm

(b)

Pressure (atm)	$N_2(g)$	$H_2(g)$	$NH_3(g)$
initial	70.	10.	0
change	$70. - x$	$10. - 3x$	$+2x$
equilibrium	$70. - x$	$10. - 3x$	$2x$

$$0.036 = \frac{(2x^2)}{(70. - x) \times (10.0 - 3x)^3}$$

Again, solution of this expression leads to a higher-order equation, and x cannot be neglected. Use the Equilibrium option in Calculator Tool of the CD. Then, $x = 3.9$ atm, and $P_{NH3} = 2x = 7.8$ atm and $P_{N2} = 66$ atm and $P_{H2} = 2$ atm

14.89 (c), (d), and (e) will shift the reaction toward the products. (a) and (f) will shift toward reactants; (b) will result in no change because there are the same number of moles of gas as products and as reactants.

INTEGRATED EXERCISES

14.91 (a)

Concentration (mol/L)	$CH_3COOH(l)$ +	$C_2H_5OH(l)$ $\rightleftharpoons$	$CH_3COOC_2H_5(l)$ +	$H_2O(l)$
initial	0.32	6.3	0	0
change	$-x$	$-x$	$+x$	$+x$
equil.	$0.32 - x$	$6.3 - x$	x	x

$$K_c = 4.0 = \frac{[CH_3COOC_2H_5][H_2O]}{[CH_3COOH][C_2H_5OH]} = \frac{(x)(x)}{(0.32 - x)(6.3 - x)}$$

$x^2 = 4.0(x^2 - 6.62x + 2.02) = 4x^2 - 26.5x + 8.08$

$3x^2 - 26.5x + 8.08 = 0$

Solving the quadratic yields $x = 0.316$ or ≈ 0.32 mol/L

equilibrium concentrations:

≈ 0.32 mol/L H_2O and 0.32 mol/L $CH_3COOC_2H_5$

0.32 mol/L − 0.32 mol/L $\approx$ 0 mol/L CH_3COOH

6.3 mol/L − 0.32 mol/L $\approx$ 6 mol/L C_2H_5OH

(b) $\Delta H° = \Delta H°\,(CH_3COOC_2H_5, l) + \Delta H°\,(H_2O, l)$

$- \Delta H°\,(CH_3COOH, l) - \Delta H°\,(C_2H_5OH, l)$

$= [-479.3 - 285.83 - (-484.5) - (-277.69)]$ kJ/mol

$= -2.9$ kJ/mol

The reaction is exothermic, so decreasing temperature will lead to more product formation.

14.93 (a) $K_c = [CrO_4^{2-}][Pb^{2+}]$

Chromates are insoluble, so K_c is small.

(b) $K_c = [SO_4^{2-}][Cs^+]^2$

Cs_2SO_4 is soluble, so K_c is large.

(c) $K_c = [Fe^{3+}][OH^-]^3$

$Fe(OH)_3$ is insoluble, so K_c is small.

CHAPTER 15
ACIDS AND BASES

EXERCISES

Unless stated otherwise, assume that all solutions are aqueous and that the temperature is 25°C.

Brønsted Acids and Bases

15.1 (a) amphiprotic (b) base (c) base (d) acid

15.3 (a) $CH_3NH_3^+$ (b) $NH_2NH_3^+$ (c) H_2CO_3 (d) CO_3^{2-} (e) $C_6H_5O^-$ (f) $CH_3CO_2^-$

15.5 (a) chloric acid, where ◯ = Cl, ○ = O

(b) nitrous acid, where ◯ = N, ○ = O

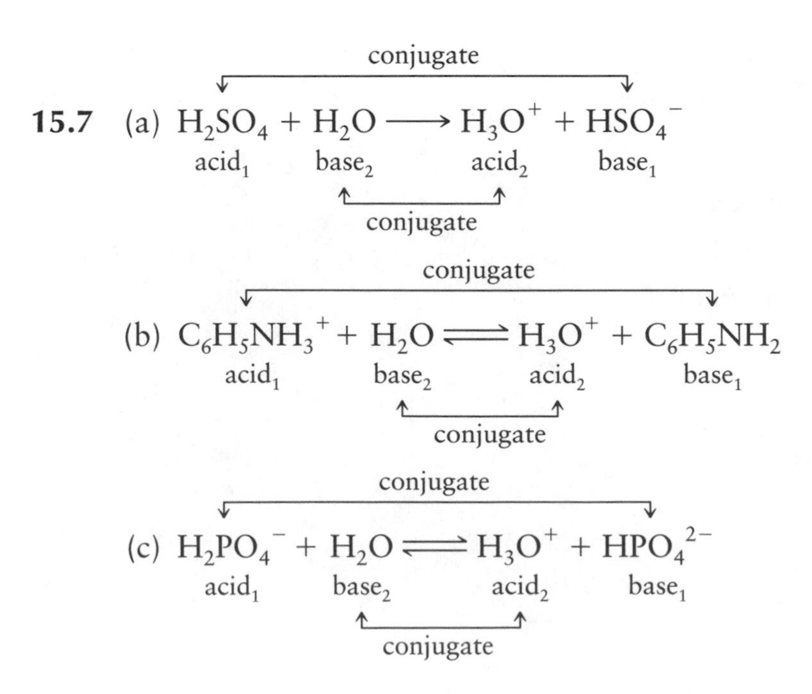

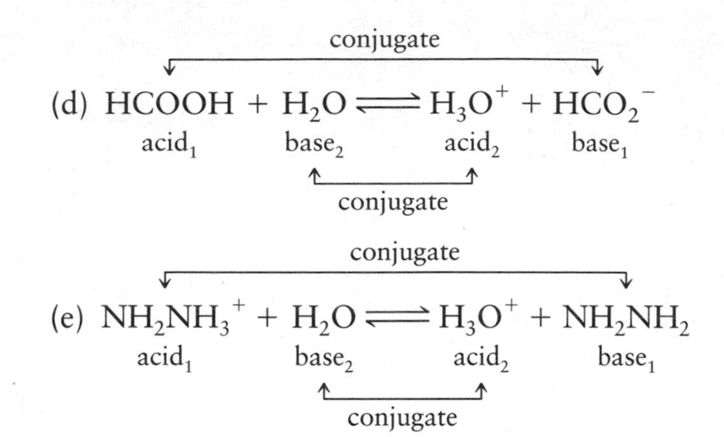

15.9 (a) HCO_3^-, as an acid:

conjugate

$$\underset{\text{acid}_1}{HCO_3^-} + \underset{\text{base}_2}{H_2O} \rightleftharpoons \underset{\text{acid}_2}{H_3O^+} + \underset{\text{base}_1}{CO_3^{2-}}$$

conjugate

HCO_3^-, as a base:

conjugate

$$\underset{\text{acid}_1}{H_2O} + \underset{\text{base}_2}{HCO_3^-} \rightleftharpoons \underset{\text{acid}_2}{H_2CO_3} + \underset{\text{base}_1}{OH^-}$$

conjugate

(b) HPO_4^{2-}, as an acid:

conjugate

$$\underset{\text{acid}_1}{HPO_4^{2-}} + \underset{\text{base}_2}{H_2O} \rightleftharpoons \underset{\text{acid}_2}{H_3O^+} + \underset{\text{base}_1}{PO_4^{3-}}$$

conjugate

HPO_4^{2-}, as a base:

conjugate

$$\underset{\text{acid}_1}{H_2O} + \underset{\text{base}_2}{HPO_4^{2-}} \rightleftharpoons \underset{\text{acid}_2}{H_2PO_4^-} + \underset{\text{base}_1}{OH^-}$$

conjugate

15.11 (a) Brønsted acid: HNO_3
Brønsted base: HPO_4^{2-}

(b) conjugate base to HNO_3: NO_3^-
conjugate acid to HPO_4^{2-}: $H_2PO_4^-$

Autoprotolysis of Water

15.13 In each case, use $K_w = [H_3O^+][OH^-] = 1.0 \times 10^{-14}$, then

$$[OH^-] = \frac{K_w}{[H_3O^+]} = \frac{1.0 \times 10^{-14}}{[H_3O^+]}$$

(a) $[OH^-] = \frac{1.0 \times 10^{-14}}{3.1 \times 10^{-2}} = 3.2 \times 10^{-13}$ mol/L

(b) $[OH^-] = \frac{1.0 \times 10^{-14}}{1.0 \times 10^{-4}} = 1.0 \times 10^{-10}$ mol/L

(c) $[OH^-] = \frac{1.0 \times 10^{-14}}{0.20} = 5.0 \times 10^{-14}$ mol/L

15.15 (a) $K_w = 2.5 \times 10^{-14} = [H_3O^+][OH^-] = x^2$, where $x = [H_3O^+] = [OH^-]$

$x = \sqrt{2.5 \times 10^{-14}} = 1.5\overline{8} \times 10^{-7} = 1.6 \times 10^{-7}$ mol/L

$pH = -\log[H_3O^+] = 6.80$

(b) $[OH^-] = [H_3O^+] = 1.5\overline{8} \times 10^{-7} = 1.6 \times 10^{-7}$ mol/L

$pOH = -\log[OH^-] = 6.80$

pH and pOH of Strong Acids and Bases

15.17 Because HCl is a strong acid, $[HCl]_0 = [H_3O^+] = [Cl^-]$,

where $[HCl]_0$ = nominal concentration of HCl

$$[HCl]_0 = \frac{0.48 \text{ mol}}{0.500 \text{ L}} = 0.96 \text{ mol/L} = [H_3O^+] = [Cl^-]$$

$$[OH^-] = \frac{K_w}{[H_3O^+]} = \frac{1.0 \times 10^{-14}}{0.96} = 1.0 \times 10^{-14} \text{ mol/L}$$

15.19 Because $Ba(OH)_2$ is a strong base, $Ba(OH)_2\ (aq) \longrightarrow Ba^{2+}(aq) + 2\ OH^-(aq)$, 100% Then, $[Ba(OH)_2]_0 = [Ba^{2+}]$, $[OH^-] = 2 \times [Ba(OH)_2]$,

where $[Ba(OH)_2]_0$ = nominal concentration of $Ba(OH)_2$

$$\text{amount (moles) of } Ba(OH)_2 = \frac{0.50 \text{ g}}{171.36 \text{ g/mol}} = 2.9 \times 10^{-3} \text{ mol}$$

$$[Ba(OH)_2]_0 = \frac{2.9 \times 10^{-3} \text{ mol}}{0.100 \text{ L}} = 2.9 \times 10^{-2} \text{ mol/L} = Ba^{2+}$$

$2 \times [Ba(OH)_2]_0 = 2 \times 2.9 \times 10^{-2}$ mol/L $= 5.8 \times 10^{-2}$ mol/L OH^-

$$\frac{K_w}{[OH-]} = \frac{1.0 \times 10^{-14}}{5.8 \times 10^{-2}} = 1.7 \times 10^{-13} \text{ mol/L } H_3O^+$$

15.21 Because $pH = -\log[H_3O^+]$, $\log[H_3O^+] = -pH$. Taking the antilogs of both sides gives $[H_3O^+] = 10^{-pH}$ mol/L

(a) $10^{-3.3} = 5 \times 10^{-4}$ mol/L H_3O^+

(b) $10^{-6.7}$ mol/L $= 2 \times 10^{-7}$ mol/L H_3O^+

(c) $10^{-4.4}$ mol/L $= 4 \times 10^{-5}$ mol/L H_3O^+

(d) $10^{-5.3}$ mol/L $= 5 \times 10^{-6}$ mol/L H_3O^+

(e) (b) < (d) < (c) < (a)

15.23 $pH = -\log[H_3O^+]$, $pOH = 14.00 - pH$

(a) $pH = -\log(2.0 \times 10^{-5}) = 4.70$,

$pOH = 14.00 - pH = 14.0 - 4.70 = 9.30$

(b) $pH = -\log(1.0) = 0.00$, $pOH = 14.00 - 0.00 = 14.00$

(c) $pH = -\log(5.0 \times 10^{-14}) = 13.30$, $pOH = 14.00 - 13.30 = 0.70$

(d) $pH = -\log(5.02 \times 10^{-5}) = 4.300$, $pOH = 14.000 - 4.300 = 9.700$

15.25 $[\text{acid}]_0$ = nominal concentration of acid

$[\text{base}]_0$ = nominal concentration of base

(a) $[HNO_3]_0 = [H_3O^+] = 0.010$ mol/L; HNO_3 is a strong acid.

$pH = -\log(0.010) = 2.00$, $pOH = 14.00 - 2.00 = 12.00$

(b) $[HCl]_0 = [H_3O^+] = 0.22$ mol/L; HCl is a strong acid.

$pH = -\log(0.22) = 0.66$, $pOH = 14.00 - 0.66 = 13.34$

(c) $[OH^-] = 2 \times [Ba(OH)_2] = 2 \times 1.0 \times 10^{-3}$ M $= 2.0 \times 10^{-3}$ mol/L

$pOH = -\log(2.0 \times 10^{-3}) = 2.70$, $pH = 14.00 - 2.70 = 11.30$

(d) $[NaOH]_0 = [OH^-]$; NaOH is a strong base.

$$\text{number of moles of NaOH} = \frac{0.0140 \text{ g}}{40.00 \text{ g/mol}} = 3.50 \times 10^{-4} \text{ mol}$$

$$[NaOH]_0 = \frac{3.50 \times 10^{-4} \text{ mol}}{0.250 \text{ L}} = 1.40 \times 10^{-3} \text{ mol/L } OH^-$$

$pOH = -\log(1.40 \times 10^{-3}) = 2.854$, $pH = 14.000 - 2.854 = 11.146$

Acidity and Basicity Constants

15.27

Name	K_a	pK_a
(a) formic acid	1.8×10^{-4}	3.75
(b) acetic acid	1.8×10^{-5}	4.75
(c) trichloroacetic acid	3.0×10^{-1}	0.52
(d) benzoic acid	6.5×10^{-5}	4.19

The larger the K_a, the stronger the acid; therefore, acetic acid < benzoic acid < formic acid < trichloroacetic acid.

15.29 $pK_{a1} = -\log K_{a1}$; therefore, after taking antilogs, $K_{a1} = 10^{-pK_{a1}}$

Acid	pK_{a1}	K_{a1}
(a) H_3PO_4	2.12	7.6×10^{-3}
(b) H_3PO_3	2.00	0.010
(c) H_3AsO_4	2.26	5.5×10^{-3}
(d) H_3AsO_3	9.29	5.1×10^{-10}

The larger the K_{a1}, the stronger the acid; therefore, $H_3AsO_3 < H_3PO_4 < H_3AsO_4 < H_3PO_3$.

15.31 The weakest acid has the strongest conjugate base and vice versa.
(a) $H_2PO_3^-$, dihydrogen phosphite ion, strongest base
$H_2AsO_3^-$, dihydrogen arsenite ion, weakest base
(b) $H_2PO_3^-$, $K_b = \frac{K_w}{K_a} = \frac{1.0 \times 10^{-14}}{0.010} = 1.0 \times 10^{-12}$

$H_2AsO_3^-$, $K_b = \frac{1.0 \times 10^{-14}}{5.13 \times 10^{-10}} = 1.9 \times 10^{-5}$

(c) $HAsO_3^{2-}$, hydrogen arsenite ion, strongest base corresponds to highest pH.

Structures and Strengths of Acids

15.33 For oxoacids, the greater the number of highly electronegative O atoms attached to the central atom, the stronger the acid. This effect is related to the increased oxidation number of the central atom as the number of O atoms increases. Therefore, HIO_3 is the stronger acid, with the lower pK_a.

15.35 (a) HCl is the stronger acid, because its bond strength is much weaker than the bond in HF, and bond strength is the dominant factor in determining the strength of binary acids.
(b) $HClO_2$ is stronger; there is one more O atom attached to the Cl atom in $HClO_2$ than in HClO. The additional O in $HClO_2$ helps to pull the electron of the H atom out of the H—O bond. The oxidation state of Cl is higher in $HClO_2$ than in HClO.
(c) $HClO_2$ is stronger; Cl has a greater electronegativity than Br, making the H—O bond $HClO_2$ more polar than in $HBrO_2$.

(d) $HClO_4$ is stronger; Cl has a greater electronegativity than P (See Toolbox 15.3 in the text).

(e) HNO_3 is stronger; the explanation is the same as that for part (b). HNO_3 has one more O atom.

(f) H_2CO_3 is stronger; C has greater electronegativity than Ge. See part (c).

15.37 (a) The $—CCl_3$ group that is bonded to the carboxyl group, —COOH, in trichloroacetic acid, is more electron withdrawing than the $—CH_3$ group in acetic acid. Therefore, trichloroacetic acid is the stronger acid.

(b) The $—CH_3$ group in acetic acid has electron-donating properties, which means that it is less electron withdrawing than the —H attached to the carboxyl group in formic acid, HCOOH. Thus, formic acid is a slightly stronger acid than acetic acid. However, it is not nearly as strong as trichloroacetic acid. The order is

$CCl_3COOH \gg HCOOH > CH_3COOH$

15.39 The larger the K_a, the stronger the corresponding acid. 2,4,6-Trichlorophenol is the stronger acid because the chlorines have a greater electron-withdrawing power than the hydrogens they replaced in the unsubstituted phenol.

15.41 The larger the pK_a of an acid, the stronger the corresponding conjugate base; the order is aniline $<$ ammonia $<$ methylamine $<$ ethylamine. Although one should not draw conclusions from such a small data set, we might suggest the possibility that

(1) arylamines $<$ ammonia $<$ alkylamines

(2) methyl $<$ ethyl $<$ etc.

(Arylamines are amines in which the nitrogen of the amine is attached to a benzene ring.)

Weak Acid and Weak Base Calculations

Refer to Tables 15.3 and 15.6 for the appropriate K_a and K_b values for the following exercises.

15.43 (a)

Concentration (mol/L)	C_6H_5COOH +	$H_2O \rightleftharpoons$	H_3O^+ +	$C_6H_5CO_2^-$
initial	0.20	—	0	0
change	$-x$	—	$+x$	$+x$
equilibrium	$0.20 - x$	—	x	x

$$K_a = \frac{[H_3O^+][C_6H_5CO_2^-]}{[C_6H_5COOH]} = \frac{x \cdot x}{0.20 - x} \approx \frac{x^2}{0.20} = 6.5 \times 10^{-5}$$

$x = \sqrt{0.20 \times 6.5 \times 10^{-5}} = 3.6 \times 10^{-3}$ mol/L H_3O^+

$\frac{K_w}{[H_3O^+]} = \frac{1.0 \times 10^{-14}}{3.6 \times 10^{-3}} = 2.8 \times 10^{-12}$ mol/L OH^-

(b) Concentration (mol/L) $H_2O + NH_2NH_2 \rightleftharpoons NH_2NH_3^+ + OH^-$

	H_2O	NH_2NH_2	$NH_2NH_3^+$	OH^-
initial	—	0.20	0	0
change	—	$-x$	$+x$	$+x$
equilibrium	—	$0.20 - x$	x	x

$$K_b = \frac{[NH_2NH_3^-][OH^-]}{[NH_2NH_2]} = \frac{x \cdot x}{0.20 - x} \approx \frac{x^2}{0.20} = 1.7 \times 10^{-6}$$

$x = \sqrt{0.20 \times 1.7 \times 10^{-6}} = 5.8 \times 10^{-4}$ mol/L OH^-

$\frac{K_w}{[OH^-]} = \frac{1.0 \times 10^{-14}}{5.8 \times 10^{-4}} = 1.7 \times 10^{-11}$ mol/L H_3O^+

(c) Concentration (mol/L) $H_2O + (CH_3)_3N \rightleftharpoons (CH_3)_3NH^+ + OH^-$

	H_2O	$(CH_3)_3N$	$(CH_3)_3NH^+$	OH^-
initial	—	0.20	0	0
change	—	$-x$	$+x$	$+x$
equilibrium	—	$0.20 - x$	x	x

$$K_b = \frac{[(CH_3)_3NH^+][OH^-]}{[(CH_3)_3N]} = \frac{x \cdot x}{0.20 - x} \approx \frac{x^2}{0.20} = 6.5 \times 10^{-5}$$

$x = \sqrt{0.20 \times 6.5 \times 10^{-5}} = 3.6 \times 10^{-3}$ mol/L OH^-

$\frac{1.0 \times 10^{-14}}{3.6 \times 10^{-3}} = 2.8 \times 10^{-12}$ mol/L H_3O^+

15.45 (a) Concentration (mol/L) $HCOOH + H_2O \rightleftharpoons H_3O^+ + HCO_2^-$

	$HCOOH$	H_2O	H_3O^+	HCO_2^-
initial	0.20	—	0	0
change	$-x$	—	$+x$	$+x$
equilibrium	$0.20 - x$	—	x	x

$$K_a = \frac{[H_3O^+][HCO_2^-]}{[HCOOH]} = \frac{x \cdot x}{0.20 - x} \approx \frac{x^2}{0.20} = 1.8 \times 10^{-4}$$

$x = \sqrt{0.20 \times 1.8 \times 10^{-4}} = 6.0 \times 10^{-3}$ mol/L H_3O^+

$pH = -\log(6.0 \times 10^{-3}) = 2.22$

(b) Concentration (mol/L) $H_2O + NH_2NH_2 \rightleftharpoons NH_2NH_3^+ + OH^-$

	H_2O	NH_2NH_2	$NH_2NH_3^+$	OH^-
initial	—	0.12	0	0
change	—	$-x$	$+x$	$+x$
equilibrium	—	$0.12 - x$	x	x

$$K_b = \frac{[NH_2NH_3^+][OH^-]}{[NH_2NH_2]} = \frac{x \cdot x}{0.12 - x} \approx \frac{x^2}{0.12} = 1.7 \times 10^{-6}$$

$x = \sqrt{0.12 \times 1.7 \times 10^{-6}} = 4.5 \times 10^{-4}$ mol/L OH^-

$pOH = -\log(4.5 \times 10^{-4}) = 3.35, \quad pH = 14.00 - 3.35 = 10.65$

(c)

Concentration (mol/L)	C_6H_5COOH +	H_2O $\rightleftharpoons$	H_3O^+ +	$C_6H_5CO_2^-$
initial	0.15	—	0	0
change	$-x$	—	$+x$	$+x$
equilibrium	$0.15 - x$	—	x	x

$$K_a = \frac{[H_3O^+][C_6H_5CO_2^-]}{[C_6H_5COOH]} = \frac{x^2}{0.15 - x} \approx \frac{x^2}{0.15} = 6.5 \times 10^{-5}$$

$x = \sqrt{0.15 \times 6.5 \times 10^{-5}} = 3.1 \times 10^{-3}$ mol/L H_3O^+

$pH = -\log(3.1 \times 10^{-3}) = 2.51$

(d)

Concentration (mol/L)	H_2O +	$C_{10}H_{14}N_2$ $\rightleftharpoons$	$C_{10}H_{14}N_2H^+$ +	OH^-
initial	—	0.0034	0	0
change	—	$-x$	$+x$	$+x$
equilibrium	—	$0.0058 - x$	x	x

$$K_b = \frac{[C_{10}H_{14}N_2H^+][OH^-]}{[C_{10}H_{14}N_2]} = \frac{x^2}{0.0034 - x} \approx \frac{x^2}{0.0034} = 1.0 \times 10^{-6}$$

$x = \sqrt{0.0034 \times 1.0 \times 10^{-6}} = 5.8 \times 10^{-5}$ mol/L OH^-

$pOH = -\log(5.8 \times 10^{-5}) = 4.24, \quad pH = 14.00 - 4.24 = 9.76$

15.47 (a)

Concentration (mol/L)	CH_3COOH +	H_2O $\rightleftharpoons$	H_3O^+ +	$CH_3CO_2^-$
initial	0.15	—	0	0
change	$-x$	—	$+x$	$+x$
equilibrium	$0.15 - x$	—	x	x

$$K_a = 1.8 \times 10^{-5} = \frac{[H_3O^+][CH_3CO_2^-]}{[CH_3COOH]} = \frac{x^2}{0.15 - x} \approx \frac{x^2}{0.15}$$

$x = 1.6 \times 10^{-3}$ mol/L H_3O^+

$pH = -\log(1.6 \times 10^{-3}) = 2.78, \quad pOH = 14.00 - 2.78 = 11.22$

(b) The equilibrium table for (b) is similar to that for (a).

$$K_a = 3.0 \times 10^{-1} = \frac{[H_3O^+][CCl_3CO_2^-]}{[CCl_3COOH]} = \frac{x^2}{0.15 - x}$$

or $x^2 + 3.0 \times 10^{-1}x - 0.045 = 0$

$$x = \frac{-3.0 \times 10^{-1} \pm \sqrt{(3.0 \times 10^{-1})^2 - (4)(-0.045)}}{2} = 0.11\bar{0}, -0.41\bar{0}$$

The negative root is not possible and can be eliminated.

$x = 0.11\bar{0}$ mol/L H_3O^+

$pH = -\log(0.11\bar{0}) = 0.96, \quad pOH = 14.00 - 0.96 = 13.04$

(c)

Concentration (mol/L)	$HCOOH$ +	H_2O ⇌	H_3O^+ +	HCO_2^-
initial	0.15	—	0	0
change	$-x$	—	$+x$	$+x$
equilibrium	$0.15 - x$	—	x	x

$$K_a = \frac{[H_3O^+][HCO_2^-]}{[HCOOH]} = \frac{x \cdot x}{0.15 - x} \approx \frac{x^2}{0.15} = 1.8 \times 10^{-4}$$

$x = \sqrt{0.15 \times 1.8 \times 10^{-4}} = 5.2 \times 10^{-3}$ mol/L H_3O^+

$pH = -\log(5.2 \times 10^{-3}) = 2.28, \quad pOH = 14.00 - 2.28 = 11.72$

15.49 (a)

Concentration (mol/L)	H_2O +	NH_3 ⇌	NH_4^+ +	OH^-
initial	—	0.10	0	0
change	—	$-x$	$+x$	$+x$
equilibrium	—	$0.10 - x$	x	x

$$K_b = \frac{[NH_4^+][OH^-]}{[NH_3]} = \frac{x \cdot x}{0.10 - x} \approx \frac{x^2}{0.10} = 1.8 \times 10^{-5}$$

$x = \sqrt{0.10 \times 1.8 \times 10^{-5}} = 1.3 \times 10^{-3}$ mol/L OH^-

$pOH = -\log(1.3 \times 10^{-3}) = 2.87, \quad pH = 14.00 - 2.87 = 11.13$

$$\text{percentage protonation} = \frac{1.3 \times 10^{-3}}{0.10} \times 100\% = 1.3\%$$

(b)

Concentration (mol/L)	NH_2OH +	H_2O ⇌	$^+NH_3OH$ +	OH^-
initial	0.017	—	0	0
change	$-x$	—	$+x$	$+x$
equilibrium	$0.017 - x$	—	x	x

$$K_b = 1.1 \times 10^{-8} = \frac{x^2}{0.017 - x} \approx \frac{x^2}{0.017}$$

$x = 1.3\bar{7} \times 10^{-5}$ mol/L OH^-

$pOH = -\log(1.3\bar{7} \times 10^{-5}) = 4.86, \quad pH = 14.00 - 4.86 = 9.14$

$$\text{percentage protonation} = \frac{1.3\bar{7} \times 10^{-5}}{0.017} \times 100\% = 0.080\%$$

(c) See the solution to Exercise 15.43 (c), which involves the same aqueous solution of $(CH_3)_3N$.

$[OH^-] = 3.6 \times 10^{-3}$ mol/L

$pOH = -\log(3.6 \times 10^{-3}) = 2.44, \quad pH = 14.00 - 2.44 = 11.56$

$$\text{percentage protonation} = \frac{3.6 \times 10^{-3}}{0.20} \times 100\% = 1.8\%$$

(d) $pK_b = 14.00 - pK_a = 14.00 - 8.21 = 5.79, \quad K_b = 1.6 \times 10^{-6}$

$\text{codeine} + H_2O \rightleftharpoons \text{codeineH}^+ + OH^-$

$$K_b = 1.6 \times 10^{-6} = \frac{x^2}{0.020 - x} \approx \frac{x^2}{0.020}$$

$x = 1.7\overline{9} \times 10^{-4}$ mol/L OH^-

$pOH = -\log(1.7\overline{9} \times 10^{-4}) = 3.75, \quad pH = 14.00 - 3.75 = 10.25$

$$\text{percentage protonation} = \frac{1.7\overline{9} \times 10^{-4}}{0.020} \times 100\% = 0.90\%$$

15.51 (a) $HClO_2 + H_2O \rightleftharpoons H_3O^+ + ClO_2^-$

$[H_3O^+] = [ClO_2^-] = 10^{-pH} = 10^{-1.2} = 0.06\overline{3}$ mol/L

$$K_a = \frac{[H_3O^+][ClO_2^-]}{[HClO_2]} = \frac{(0.06\overline{3})^2}{0.10 - 0.063} = 0.1\overline{1} = 0.1 \text{ (1 sf)}$$

$pK_a = -\log(0.1\overline{1}) = 1.0$

(b) $C_3H_7NH_2 + H_2O \rightleftharpoons C_3H_7NH_3^+ + OH^-$

$pOH = 14.00 - 11.86 = 2.14$

$[C_3H_7NH_3^+] = [OH^-] = 10^{-2.14} = 7.2\overline{4} \times 10^{-3}$ mol/L

$$K_b = \frac{[C_3H_7NH_3^+][OH^-]}{[C_3H_7NH_2]} = \frac{(7.2\overline{4} \times 10^{-3})^2}{0.10 - 7.2\overline{4} \times 10^{-3}} = 5.7 \times 10^{-4}$$

$pK_b = -\log(5.7 \times 10^{-4}) = 3.25$

15.53 (a) $pH = 4.6, \quad [H_3O^+] = 10^{-pH} = 10^{-4.6} = 2.\overline{5} \times 10^{-5}$ mol/L

Let x = nominal concentration of HClO, then

Concentration	$HClO$ +	$H_2O \rightleftharpoons$	H_3O^+ +	ClO^-
nominal conc.	x	—	0	0
at equil.	$x - 2.\overline{5} \times 10^{-5}$	—	$2.\overline{5} \times 10^{-5}$	$2.\overline{5} \times 10^{-5}$

$$K_a = 3.0 \times 10^{-8} = \frac{(2.\overline{5} \times 10^{-5})^2}{x - 2.\overline{5} \times 10^{-5}}$$

$$\text{Solve for } x;\ x = \frac{(2.5 \times 10^{-5})^2 + 2.\overline{5} \times 10^{-5} \times 3.0 \times 10^{-8}}{3.0 \times 10^{-8}}$$

$= 2 \times 10^{-2}$ mol/L $= 0.02$ mol/L

(b) $pOH = 14.00 - pH = 14.00 - 10.2 = 3.8$

$[OH^-] = 10^{-pOH} = 10^{-3.8} = 1.\overline{6} \times 10^{-4}$ mol/L

Let x = nominal concentration of NH_2NH_2, then

Concentration	NH_2NH_2	+ $H_2O \rightleftharpoons$	$NH_2NH_3^+$	+ OH^-
nominal conc.	x	—	0	0
at equil.	$x - 1.\bar{6} \times 10^{-4}$	—	$1.\bar{6} \times 10^{-4}$	$1.\bar{6} \times 10^{-4}$

$$K_b = 1.7 \times 10^{-6} = \frac{(1.\bar{6} \times 10^{-4})^2}{x - 1.\bar{6} \times 10^{-4}}$$

Solve for x; $x = 1.\bar{5} \times 10^{-2}$ mol/L = 0.02 mol/L

15.55

Concentration (mol/L)	C_6H_5COOH	+ $H_2O \rightleftharpoons$	H_3O^+	+ $C_6H_5CO_2^-$
initial	0.110	—	0	0
change	$-x$	—	$+x$	$+x$
equilibrium	$0.110 - x$	—	x	x

To find the concentration of deprotonated benzoic acid and the concentration of hydronium ion, we multiply the percent deprotonation (written in decimal form) by the initial molarity. Thus

$x = 0.024 \times 0.110$ mol/L $= [H_3O^+] = [C_6H_5CO_2^-]$

$$K_a = \frac{[H_3O^+][C_6H_5COO^-]}{[C_6H_5COOH]} = \frac{(0.024 \times 0.110)^2}{(1 - 0.024) \times 0.110}$$

$$= \frac{(2.6\bar{4} \times 10^{-3})^2}{0.107\bar{4}} = 6.5 \times 10^{-5}$$

$\text{pH} = -\log(2.6\bar{4} \times 10^{-3}) = 2.58$

15.57 H_2O + octylamine $\rightleftharpoons$ octylamineH$^+$ + OH^-

The change in the concentration of octylamine is the percent deprotonation (as a decimal) times the initial molarity, or $x = 0.067 \times 0.10 = 0.0067$ mol/L. So the equilibrium table is

Concentration (mol/L)	H_2O	+ octylamine	$\rightleftharpoons$ octylamineH$^+$	+ OH^-
initial	—	0.10	0	0
change	—	−0.0067	+0.0067	+0.0067
equilibrium	—	0.10 − 0.0067	0.0067	0.0067

The equilibrium concentrations are

[octylamine] $= 0.10 - 0.067 \times 0.10 = 0.09\bar{3}$ mol/L

$[OH^-]$ = [octylamineH$^+$] = 0.0067 mol/L

$\text{pOH} = -\log(0.0067) = 2.17, \quad \text{pH} = 14.00 - 2.17 = 11.83$

$$K_b = \frac{[\text{octylamineH}^+][OH^-]}{[\text{octylamine}]} = \frac{(6.7 \times 10^{-3})^2}{0.09\bar{3}} = 5 \times 10^{-4}$$

Polyprotic Acids

15.59 (a) $H_2SO_4 + H_2O \longrightarrow H_3O^+ + HSO_4^-$

$HSO_4^- + H_2O \rightleftharpoons H_3O^+ + SO_4^{2-}$

(b) $H_3AsO_4 + H_2O \rightleftharpoons H_3O^+ + H_2AsO_4^-$

$H_2AsO_4^- + H_2O \rightleftharpoons H_3O^+ + HAsO_4^{2-}$

$HAsO_4^{2-} + H_2O \rightleftharpoons H_3O^+ + AsO_4^{3-}$

(c) $C_6H_4(COOH)_2 + H_2O \rightleftharpoons H_3O^+ + C_6H_4(COOH)CO_2^-$

$C_6H_4(COOH)CO_2^- + H_2O \rightleftharpoons H_3O^+ + C_6H_4(CO_2)_2^{2-}$

15.61 The initial concentrations of HSO_4^- and H_3O^+ are both 0.15 mol/L as a result of the complete ionization of H_2SO_4 in the first step. The second ionization is incomplete.

Concentration (mol/L)	HSO_4^-	+ H_2O $\rightleftharpoons$	H_3O^+	+ SO_4^{2-}
initial	0.15	—	0.15	0
change	$-x$	—	$+x$	$+x$
equilibrium	$0.15 - x$	—	$0.15 + x$	x

$$K_{a2} = 1.2 \times 10^{-2} = \frac{[H_3O^+][SO_4^{2-}]}{[HSO_4^-]} = \frac{(0.15 + x)(x)}{0.15 - x}$$

$$x^2 + 0.162x - 1.8 \times 10^{-3} = 0$$

$$x = \frac{-0.162 + \sqrt{(0.162)^2 + (4)(1.8 \times 10^{-3})}}{2} = 0.010\bar{4} \text{ mol/L}$$

$$[H_3O^+] = 0.15 + x = (0.15 + 0.010\bar{4})\text{mol/L} = 0.16 \text{ mol/L}$$

$$pH = -\log(0.16) = 0.80$$

15.63 (a)

Concentration (mol/L)	H_2CO_3	+ H_2O $\rightleftharpoons$	H_3O^+	+ HCO_3^-
initial	0.0010	—	0	0
change	$-x$	—	$+x$	$+x$
equilibrium	$0.0010 - x$	—	x	x

$$K_{a1} = \frac{[H_3O^+][HCO_3^-]}{[H_2CO_3]} = \frac{x^2}{0.0010 - x} \approx \frac{x^2}{0.0010} = 4.3 \times 10^{-7}$$

$$x = [H_3O^+] = 2.1 \times 10^{-5} \text{ mol/L}$$

$$pH = -\log(2.1 \times 10^{-5}) = 4.68$$

(b)

Concentration (mol/L)	H_2S	+ H_2O $\rightleftharpoons$	H_3O^+	+ HS^-
initial	0.20	—	0	0
change	$-x$	—	$+x$	$+x$
equilibrium	$0.20 - x$	—	x	x

$$K_{a1} = 1.3 \times 10^{-7} = \frac{[H_3O^+][HS^-]}{[H_2S]} = \frac{x^2}{0.20 - x}$$

$x = 1.6 \times 10^{-4}$ mol/L H_3O^+

$pH = -\log(1.6 \times 10^{-4}) = 3.79$

SUPPLEMENTARY EXERCISES

15.65 $[H_3O^+] = [OH^-]$

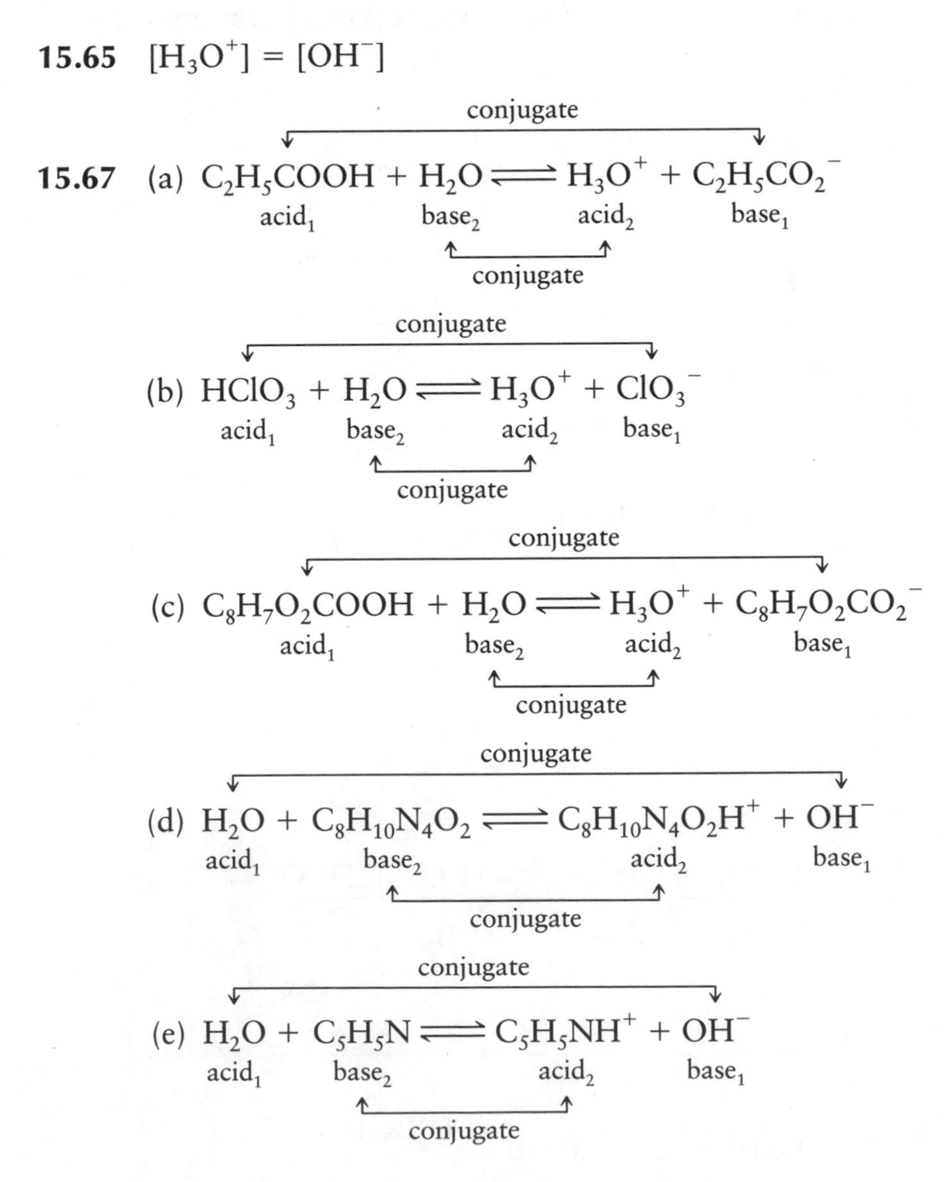

15.69 (a) If a solution is a concentrated solution of an acid with $[H_3O^+] > 1$ M, the pH will be negative.

(b) If a solution is very basic, with $[OH^-] > 1$ M (which means $[H_3O^+] < 10^{-14}$ M), the pH will be greater than 14.

15.71 $pH = -\log[H_3O^+]$; therefore, $[H_3O^+] = 10^{-pH}$ mol/L

(a) $10^{-9.33}$ mol/L $= 4.7 \times 10^{-10}$ mol/L

(b) $10^{-7.95}$ mol/L $= 1.1 \times 10^{-8}$ mol/L

(c) $10^{-0.01}$ mol/L $= 0.98$ mol/L

(d) $10^{-4.33}$ mol/L $= 4.7 \times 10^{-5}$ mol/L

(e) $10^{-1.99}$ mol/L $= 0.010$ mol/L

(f) $10^{-11.95}$ mol/L $= 1.1 \times 10^{-12}$ mol/L

15.73 (a) $C_6H_5OH + H_2O \rightleftharpoons H_3O^+ + C_6H_5O^-$

$NH_4^+ + H_2O \rightleftharpoons H_3O^+ + NH_3$

(b) $pK_a(C_6H_5OH) = -\log(1.3 \times 10^{-10}) = 9.89$

$pK_a(NH_4^+) = -\log(5.6 \times 10^{-10}) = 9.25$

(c) NH_4^+ is the stronger acid; it has the smaller pK_a and the larger K_a.

15.75 (a) $HBrO_4$, because Br is slightly more electronegative than I.

(b) HI, because H—I is a weaker bond than H—F.

(c) HIO_3, because I has more oxygens attached to it in HIO_3 than in HIO_2.

(d) H_2SeO_4, because Se is more electronegative than As.

15.77 Let B = base; $B + H_2O \rightleftharpoons BH^+ + OH^-$

$pOH = 14.00 - 10.05 = 3.95$

The equilibrium concentrations are

$[OH^-] = [BH^+] = 10^{-pOH}$ mol/L $= 10^{-3.95}$ mol/L $= 1.1\bar{2} \times 10^{-4}$ mol/L

initial concentration of B: $[B]_0 = \left(\dfrac{0.150 \text{ g}}{0.0500 \text{ L}}\right) \times \left(\dfrac{1 \text{ mol}}{31.06 \text{ g}}\right) = 0.0966$ mol/L

equilibrium concentration of B: $[B] = (0.0966 - 1.1\bar{2} \times 10^{-4})$ mol/L $= 0.0965$ mol/L

Concentration (mol/L)	H_2O	+	B	$\rightleftharpoons$	BH^+	+	OH^-
initial	—		0.0966		0		0
change	—		$-1.1\bar{2} \times 10^{-4}$		$+1.1\bar{2} \times 10^{-4}$		$+1.1\bar{2} \times 10^{-4}$
equilibrium	—		$0.0966 - 1.1\bar{2} \times 10^{-4}$		$1.1\bar{2} \times 10^{-4}$		$1.1\bar{2} \times 10^{-4}$

$$K_b = \frac{[BH^+][OH^-]}{[B]} = \frac{(1.1\bar{2} \times 10^{-4})^2}{0.0965} = 1.3 \times 10^{-7}$$

$$pK_b = -\log(1.3 \times 10^{-7}) = 6.89, \quad pK_a(BH^+) = 14.00 - 6.89 = 7.11$$

$$\text{percentage protonation} = \frac{1.1\bar{2} \times 10^{-4}}{0.0965} \times 100\% = 0.12\%$$

15.79 (a) $NH_3 + NH_3 \rightleftharpoons NH_4^+ + NH_2^-$

(b) acid = NH_4^+, base = NH_2^-

(c) $pK_{am} = -\log K_{am} = -\log(1 \times 10^{-33}) = 33.0$

(d) $pK_{am} = [NH_4^+][NH_2^-] = x^2 \quad (x = [NH_4^+] = [NH_2^-])$

$x = \sqrt{1 \times 10^{-33}} = 3.\bar{2} \times 10^{-17}$ mol/L NH_4^+

(e) $pNH_4^+ = pNH_2^- = -\log(3.\bar{2} \times 10^{-17}) = 16$ (assuming 2 sfs)

(f) $pNH_4^+ + pNH_2^- = pK_{am} \approx 33.0$

15.81 (a) $D_2O + D_2O \rightleftharpoons D_3O^+ + OD^-$

(b) $K_w = [D_3O^+][OD^-] = 1.35 \times 10^{-15}, \quad pK_w = -\log K_w = 14.870$

(c) $[D_3O^+] = [OD^-] = \sqrt{1.35 \times 10^{-15}} = 3.67 \times 10^{-8}$ mol/L

(d) $pD = -\log(3.67 \times 10^{-8}) = 7.435 = pOD$

(e) $pD + pOD = pK_w(D_2O) = 14.870$

APPLIED EXERCISES

15.83 (a) $S(s) + O_2(g) \longrightarrow SO_2(g)$

$$\text{mass } SO_2 = (0.025 \times 1000) \text{ kg S} \times \left(\frac{1 \text{ mol S}}{0.032\,06 \text{ kg S}}\right) \times \left(\frac{1 \text{ mol } SO_2}{1 \text{ mol S}}\right) \times \left(\frac{0.064\,06 \text{ kg } SO_2}{1 \text{ mol } SO_2}\right) = 50. \text{ kg } SO_2$$

(b) $$\text{volume of water} = (0.020 \text{ m}) \times (2.6 \text{ km}^2) \times \left(\frac{10^3 \text{ m}}{1 \text{ km}}\right)^2 \times \left(\frac{10^3 \text{ L}}{1 \text{ m}^3}\right) = 5.2 \times 10^7 \text{ L}$$

$$\text{amount (moles) of } SO_2 = (50. \text{ kg } SO_2) \times \left(\frac{10^3 \text{ g } SO_2}{1 \text{ kg } SO_2}\right) \times \left(\frac{1 \text{ mol } SO_2}{64.06 \text{ g } SO_2}\right) = 7.8 \times 10^2 \text{ mol } SO_2$$

$$\text{molar concentration of } SO_2 = [SO_2]_0 = \frac{7.8 \times 10^2 \text{ mol } SO_2}{5.2 \times 10^7 \text{ L}} = 1.5 \times 10^{-5} \text{ mol/L}$$

When $SO_2(g)$ dissolves in water, the following reactions occur:

$SO_2 + H_2O(l) \longrightarrow H_2SO_3(aq)$

Concentration (mol/L)	$H_2SO_3(aq)$	+ $H_2O(l)$ $\rightleftharpoons$	$H_3O^+(aq)$ +	$HSO_3^-(aq)$
initial	1.5×10^{-5}	—	0	0
change	$-x$	—	$+x$	$+x$
equil.	$1.5 \times 10^{-5} - x$	—	x	x

$$K_{a1} = 10^{-pK_{a1}} \text{ mol/L} = 10^{-1.82} \text{ mol/L} = 1.5 \times 10^{-2} \text{ mol/L}$$

$$= \frac{[H_3O^+][HSO_3^-]}{[SO_2]} = 1.5 \times 10^{-2} \text{ mol/L}$$

Ignoring the further ionization of HSO_3^-, we can justifiably assume $[H_3O^+] = [HSO_3^-] = x$.

$$K_{a1} = \frac{x^2}{[SO_2]_0 - x}$$

Because of the relatively large value of K_{a1}, the exact quadratic equation is needed here. Rearranging into standard quadratic form gives

$$x^2 + K_{a1}x - K_{a1}[SO_2]_0 = 0$$

$$x = \frac{-K_{a1} \pm \sqrt{K_{a1}^2 + 4K_{a1}[SO_2]_0}}{2}$$

$$x = \frac{-1.5 \times 10^{-2} \pm \sqrt{(1.5 \times 10^{-2})^2 + 4 \times 1.5 \times 10^{-2} \times 1.5 \times 10^{-5}}}{2} \text{ mol/L}$$

$$= 1.5 \times 10^{-5} \text{ mol/L } H_3O^+$$

Essentially all of the SO_2 has been converted to HSO_3^-.

$pH = -\log[H_3O^+] = -\log(1.5 \times 10^{-5}) = 4.82$

(c) When SO_2 is first oxidized to SO_3 and then dissolved in water to form $H_2SO_4(aq)$, the first ionization of H_2SO_4 is complete, but the second is neither complete nor negligible, so it cannot be ignored. All SO_2 is converted to SO_3 by the reaction $SO_2(g) + \frac{1}{2} O_2(g) \longrightarrow SO_3(g)$; and $SO_3(g) + H_2O(l) \longrightarrow H_2SO_4(aq)$. The reactions with water are now

(1) $H_2SO_4(aq) + H_2O(l) \longrightarrow H_3O^+(aq) + HSO_4^-(aq)$

(2) $HSO_4^-(aq) + H_2O(l) \rightleftharpoons H_3O^+(aq) + SO_4^{2-}(aq)$

Because (1) is complete, $[H_3O^+]_0 = [HSO_4^-]_0 = [SO_3]_0 = [SO_2]_0$

Concentration (mol/L)	$HSO_4^-(aq)$	+ $H_2O(l)$	$\rightleftharpoons$ $H_3O^+(aq)$	+ $SO_4^{2-}(aq)$
initial	1.5×10^{-5}	—	1.5×10^{-5}	0
change	$-x$	—	$+x$	$+x$
equil.	$1.5 \times 10^{-5} - x$	—	$1.5 \times 10^{-5} + x$	x

$$K_{a2} = 1.20 \times 10^{-2} = \frac{[H_3O^+][SO_4^{2-}]}{[HSO_4^-]} = \frac{(1.5 \times 10^{-5} + x)(x)}{1.5 \times 10^{-5} - x}$$

which leads to the quadratic $x^2 + 1.20 \times 10^{-2}\, x - 1.80 \times 10^{-7} = 0$

The solution for x is $1.4\overline{98} \times 10^{-5}$, which is very close to 1.5×10^{-5}. Thus,

$2 \times 1.5 \times 10^{-5} \text{ mol/L} = 3.0 \times 10^{-5} \text{ mol/L } H_3O^+$

$pH = -\log(3.0 \times 10^{-7}) = 4.52$

The concentration of HSO_4^- is essentially 0; all of it has been converted to SO_4^{2-} and H_3O^+, decreasing the pH to 4.52.

15.85 When $CO_2(g)$ dissolves in water, the following reaction takes place:
$CO_2(g) + H_2O(l) \longrightarrow H_2CO_3(aq)$ and

Concentration	$H_2CO_3(aq)$	+ $H_2O(l)$ $\rightleftharpoons$	$H_3O^+(aq)$ +	$HCO_3^-(aq)$
initial	$[H_2CO_3]_0$	—	0	0
change	$-x$	—	$+x$	$+x$
equilibrium	$[H_2CO_3]_0 - x$	—	x	x

We can justifiably ignore the second ionization of H_2CO_3, because $K_{a2} \ll K_{a1}$. The initial concentration of H_2CO_3 can be calculated from the Henry's law constant for CO_2:

$[H_2CO_3]_0 = [CO_2] = 2.3 \times 10^{-2}$ mol/L·atm $\times 3.04 \times 10^{-4}$ atm
$= 6.9\overline{9} \times 10^{-6}$ mol/L

$K_{a1} = 10^{-pK_a}$ mol/L $= 10^{-6.37}$ mol/L $= 4.2\overline{6} \times 10^{-7}$ mol/L

$$K_{a1} = 4.2\overline{6} \times 10^{-7} = \frac{[H_3O^+][HCO_3^-]}{[H_2CO_3]} = \frac{x^2}{[H_2CO_3]_0 - x} = \frac{x^2}{6.9\overline{9} \times 10^{-6} - x}$$

Because $[H_2CO_3]_0$ is so small, we cannot neglect x relative to it. This leads to the quadratic:

$x^2 + 4.2\overline{6} \times 10^{-7}x - 2.9\overline{8} \times 10^{-12} = 0$

which has the solution:

$x = 1.5\overline{2} \times 10^{-6}$ mol/L H_3O^+ and

$pH = -\log[H_3O^+] = -\log(1.5\overline{2} \times 10^{-6}) = 5.82$

which is close to 5.7.

There are two possible explanations for this small but not negligible difference. The first is that we have not included in the calculation the contribution to $[H_3O^+]$ from the autoprotolysis of water. However, when we do so, the pH changes by only 0.03 to 5.79 (~5.8). The remaining difference, assuming it is real (namely, outside of experimental error), could be accounted for by the fact that the Henry's law constant is sensitive to temperature changes. An average temperature for rain, for all seasons and at all elevations, may not correspond to the temperature (20°C) of the Henry's law constant given in the case.

15.87 (a) $M_{dil} = \dfrac{(0.022\ \text{M})(10.0\ \text{mL})}{250.\ \text{mL}} = 8.8 \times 10^{-4}\ \text{M}$. KOH is a strong base and dissociates completely, so $[OH^-] = 8.8 \times 10^{-4}$ mol/L.

$pOH = -\log(8.8 \times 10^{-4}) = 3.06$

$pH = 14.00 - 3.06 = 10.94$

(b) $M_{dil} = \dfrac{(0.00043\ \text{M})(50.0\ \text{mL})}{250.\ \text{mL}} = 8.6 \times 10^{-5}$ mol/L. HBr is a strong acid and dissociates completely, so $[H^+] = 8.6 \times 10^{-5}$ mol/L. The dissociation of water, where $[H^+] = 1.0 \times 10^{-7}$, is still negligible.

$pH = -\log(8.6 \times 10^{-5}) = 4.06$

$pOH = 14.00 - 4.06 = 9.94$

15.89 Recall that

$$\ln\left(\frac{K_p'}{K_p}\right) = \frac{\Delta H°}{R}\left(\frac{1}{T} - \frac{1}{T'}\right),\ \text{where}\ \Delta H° = \frac{R\ln\left(\dfrac{K_p'}{K_p}\right)}{\left(\dfrac{1}{T} - \dfrac{1}{T'}\right)}$$

$$\Delta H° = \frac{8.314\ \text{J/mol·K} \times \ln(1.768 \times 10^{-4}/1.765 \times 10^{-4})}{\left(\dfrac{1}{293\ \text{K}} - \dfrac{1}{303\ \text{K}}\right)} = 125\ \text{J/mol}$$

CHAPTER 16
AQUEOUS EQUILIBRIA

The values for the acidity and basicity constants are listed in Tables 15.3 and 15.6. Unless stated otherwise, take the solutions to be aqueous and at 25°C. The calculated pH and pOH values are, in most cases, accurate to one fewer significant figure than given.

EXERCISES

Ions as Acids and Bases

16.1 (a) pH < 7, acidic; $NH_4^+(aq) + H_2O(l) \rightleftharpoons H_3O^+(aq) + NH_3(aq)$

(b) pH > 7, basic; $H_2O(l) + CO_3^{2-}(aq) \rightleftharpoons HCO_3^-(aq) + OH^-(aq)$

(c) pH > 7, basic; $H_2O(l) + F^-(aq) \rightleftharpoons HF(aq) + OH^-(aq)$

(d) pH = 7, neutral; K^+ is not an acid, Br^- is not a base

(e) pH < 7, acidic; $Al(H_2O)_6^{3+}(aq) + H_2O(l) \rightleftharpoons H_3O^+(aq) + Al(H_2O)_5OH^{2+}(aq)$

(f) pH < 7, acidic; $Cu(H_2O)_6^{2+}(aq) + H_2O(l) \rightleftharpoons H_3O^+(aq) + Cu(H_2O)_5OH^+(aq)$

16.3 In each case, determine whether the ion is a weak acid or a weak base. Its acidity or basicity constant is then calculated from the K_b or K_a of its conjugate base or acid, using $K_a \times K_b = K_w$. K_b and K_a data are from Table 15.3.

(a) $K_a \times K_b = K_w$; conjugate base is NH_3.

$$K_a = \frac{K_w}{K_b} = \frac{1.00 \times 10^{-14}}{1.8 \times 10^{-5}} = 5.6 \times 10^{-10}$$

(b) $K_a \times K_b = K_w$; conjugate acid is HCO_3^-.

$$K_b = \frac{K_w}{K_{a2}} = \frac{1.00 \times 10^{-14}}{5.6 \times 10^{-11}} = 1.8 \times 10^{-4}$$

(c) $K_a \times K_b = K_w$; conjugate acid is HF.

$$K_b = \frac{K_w}{K_a} = \frac{1.00 \times 10^{-14}}{3.5 \times 10^{-4}} = 2.8 \times 10^{-11}$$

(d) $K_a \times K_b = K_w$; conjugate acid is HClO.

$$K_b = \frac{K_w}{K_a} = \frac{1.00 \times 10^{-14}}{3.0 \times 10^{-8}} = 3.3 \times 10^{-7}$$

(e) $K_a \times K_b = K_w$; conjugate acid is H_2CO_3.

$$K_b = \frac{K_w}{K_a} = \frac{1.00 \times 10^{-14}}{4.3 \times 10^{-7}} = 2.3 \times 10^{-8}$$

(f) $K_a \times K_b = K_w$; conjugate base is $(CH_3)_3N$.

$$K_a = \frac{K_w}{K_b} = \frac{1.00 \times 10^{-14}}{6.5 \times 10^{-5}} = 1.5 \times 10^{-10}$$

16.5 (a) $K_b = \frac{K_w}{K_a} = \frac{1.00 \times 10^{-14}}{1.8 \times 10^{-5}} = 5.6 \times 10^{-10}$

Concentration (mol/L)	$CH_3CO_2^-(aq)$ +	$H_2O(l) \rightleftharpoons$	$HCH_3CO_2(aq)$ +	$OH^-(aq)$
initial	0.20	—	0	0
change	$-x$	—	$+x$	$+x$
equilibrium	$0.20 - x$	—	x	x

$$K_b = \frac{[HCH_3CO_2][OH^-]}{[CH_3CO_2^-]} = 5.6 \times 10^{-10} = \frac{x^2}{0.20 - x} \approx \frac{x^2}{0.20}$$

$x = 1.0 \times 10^{-5}$ mol/L OH^-, $pOH = -\log(1.0 \times 10^{-5}) = 5.00$

$pH = 14.00 - pOH = 14.00 - 5.00 = 9.00$

(b) $K_a = \frac{K_w}{K_b} = \frac{1.00 \times 10^{-14}}{1.8 \times 10^{-5}} = 5.6 \times 10^{-10}$

Concentration (mol/L)	$NH_4^+(aq)$ +	$H_2O(l) \rightleftharpoons$	$H_3O^+(aq)$ +	$NH_3(aq)$
initial	0.10	—	0	0
change	$-x$	—	$+x$	$+x$
equilibrium	$0.10 - x$	—	x	x

$$K_a = \frac{[H_3O^+][NH_3]}{[NH_4Cl]} = 5.6 \times 10^{-10} = \frac{x^2}{0.10 - x} \approx \frac{x^2}{0.10}$$

$x^2 = 5.6 \times 10^{-11}$

$x = 7.4\overline{8} \times 10^{-6}$ mol/L H_3O^+

$pH = -\log(7.4\overline{8} \times 10^{-6}) = 5.13$

(c)

Concentration (mol/L)	$Al(H_2O)_6^{3+}(aq)$ +	$H_2O(l) \rightleftharpoons$	$H_3O^+(aq)$ +	$Al(H_2O)_5OH^{2+}(aq)$
initial	0.10	—	0	0
change	$-x$	—	$+x$	$+x$
equilibrium	$0.10 - x$	—	x	x

$$K_a = \frac{[H_3O^+][Al(H_2O)_5OH^{2+}]}{[Al(H_2O)_6^{3+}]} = 1.4 \times 10^{-5} = \frac{x^2}{0.10 - x} \approx \frac{x^2}{0.10}$$

$x^2 = 1.42 \times 10^{-6}$

$x = \sqrt{1.4 \times 10^{-6}} = 1.2 \times 10^{-3}$ mol/L H_3O^+

$pH = -\log(1.2 \times 10^{-3}) = 2.92$

(d)

Concentration (mol/L)	$H_2O(l)$ +	$CN^-(aq)$ $\rightleftharpoons$	$HCN(aq)$ +	$OH^-(aq)$
initial	—	0.15	0	0
change	—	$-x$	$+x$	$+x$
equilibrium	—	$0.15 - x$	x	x

$$K_b = \frac{K_w}{K_a} = \frac{1.00 \times 10^{-14}}{4.9 \times 10^{-10}} = 2.0 \times 10^{-5} = \frac{[HCN][OH^-]}{[CN^-]} = \frac{x^2}{0.15 - x} \approx \frac{x^2}{0.15}$$

$x^2 = (0.15) \times (2.0 \times 10^{-5}) = 3.0 \times 10^{-6}$

$x = 1.7 \times 10^{-3}$ mol/L OH^-

$pOH = -\log(1.7 \times 10^{-3}) = 2.77, \quad pH = 11.23$

16.7 (a) 250. mL of solution contains 10.0 g $KC_2H_3O_2$, molar mass = 98.14 g/mol

$$(10.0 \text{ g } KC_2H_3O_2) \times \left(\frac{1 \text{ mol } KC_2H_3O_2}{98.14 \text{ g } KC_2H_3O_2}\right) \times \left(\frac{1}{0.250 \text{ L}}\right) = 0.408 \text{ M } KC_2H_3O_2$$

Concentration (mol/L)	$H_2O(l)$ +	$C_2H_3O_2^-(aq)$ $\rightleftharpoons$	$HC_2H_3O_2(aq)$ +	$OH^-(aq)$
initial	—	0.408	0	0
change	—	$-x$	$+x$	$+x$
equilibrium	—	$0.408 - x$	x	x

$$\frac{1.0 \times 10^{-14}}{1.8 \times 10^{-5}} = \frac{x^2}{0.408 - x} \approx \frac{x^2}{0.408}$$

$[OH^-] = 1.5 \times 10^{-5}$ mol/L

$[H_3O^+] = 6.7 \times 10^{-10}$ mol/L

$pH = -\log(6.7 \times 10^{-10}) = 9.17$

(b) 100. mL of solution contains 5.75 g NH_4Br, molar mass = 97.95 g/mol

$$(5.75 \text{ g } NH_4Br) \times \left(\frac{1 \text{ mol } NH_4Br}{97.95 \text{ g } NH_4Br}\right) \times \left(\frac{1}{0.100 \text{ L}}\right) = 0.587 \text{ M } NH_4Br$$

Concentration (mol/L)	$NH_4^+(aq)$ +	$H_2O(l)$ $\rightleftharpoons$	$NH_3(aq)$ +	$H_3O^+(aq)$
initial	0.587	—	0	0
change	$-x$	—	$+x$	$+x$
equilibrium	$0.587 - x$	—	x	x

$$\frac{1.0 \times 10^{-14}}{1.8 \times 10^{-5}} = \frac{x^2}{0.587 - x} \approx \frac{x^2}{0.587}$$

$x = 1.8 \times 10^{-5}$ mol/L H_3O^+

$pH = -\log(1.8 \times 10^{-5}) = 4.74$

16.9 (a) $\frac{0.200\text{ mol/L NaCH}_3\text{CO}_2 \times 0.200\text{ L}}{0.500\text{ L}} = 0.0800\text{ mol/L}$

Concentration (mol/L)	$H_2O(l)$ +	$CH_3CO_2^-(aq)$ ⇌	$CH_3COOH(aq)$ +	$OH^-(aq)$
initial	—	0.0800	0	0
change	—	$-x$	$+x$	$+x$
equilibrium	—	$0.0800 - x$	x	x

$$K_b = \frac{K_w}{K_a} = \frac{1.00 \times 10^{-14}}{1.8 \times 10^{-5}} = 5.6 \times 10^{-10} = \frac{[CH_3COOH][OH^-]}{[CH_3CO_2^-]}$$

$$5.6 \times 10^{-10} = \frac{x^2}{0.0800 - x} \approx \frac{x^2}{0.0800}$$

$x^2 = 4.4 \times 10^{-11}$

$x = 6.7 \times 10^{-6}$ mol/L CH_3COOH

(b) $\left(\frac{5.75\text{ g NH}_4\text{Br}}{400.\text{ mL}}\right) \times \left(\frac{1\text{ mL}}{10^{-3}\text{ L}}\right) \times \left(\frac{1\text{ mol NH}_4\text{Br}}{97.95\text{ g NH}_4\text{Br}}\right) = 0.147\text{ (mol NH}_4\text{Br)/L}$

Concentration (mol/L)	$NH_4^+(aq)$ +	$H_2O(l)$ ⇌	$H_3O^+(aq)$ +	$NH_3(aq)$
initial	0.147	—	0	0
change	$-x$	—	$+x$	$+x$
equilibrium	$0.147 - x$	—	x	x

$$K_a = \frac{K_w}{K_b} = \frac{1.00 \times 10^{-14}}{1.8 \times 10^{-5}} = 5.6 \times 10^{-10} = \frac{[NH_3][H_3O^+]}{[NH_4^+]}$$

$$5.6 \times 10^{-10} = \frac{x^2}{0.147 - x} \approx \frac{x^2}{0.147}$$

$x^2 = 8.2 \times 10^{-11}$

$x = 9.1 \times 10^{-6}$ mol/L H_3O^+ and pH $= -\log(9.1 \times 10^{-6}) = 5.04$

Mixed Solutions

16.11 (a) When solid sodium acetate is added to an acetic acid solution, the concentration of H_3O^+ decreases because the equilibrium

$HC_2H_3O_2(aq) + H_2O(l) \rightleftharpoons H_3O^+(aq) + C_2H_3O_2^-(aq)$

shifts to the left to relieve the stress imposed by the increase of $[C_2H_3O_2^-]$ (Le Chatelier's principle).

(b) When HCl is added to a benzoic acid solution, the percentage of benzoic acid that is deprotonated decreases because the equilibrium

$C_6H_5COOH(aq) + H_2O(l) \rightleftharpoons H_3O^+(aq) + C_6H_5CO_2^-(aq)$

shifts to the left to relieve the stress imposed by the increased $[H_3O^+]$ (Le Chatelier's principle).

(c) When solid NH_4Cl is added to an ammonia solution, the pH decreases because the equilibrium

$NH_3(aq) + H_2O(l) \rightleftharpoons NH_4^+(aq) + OH^-(aq)$

shifts to the left to relieve the stress imposed by the increased $[NH_4^+]$ (Le Chatelier's principle). Because $[OH^-]$ decreases, $[H_3O^+]$ increases and pH decreases.

16.13 (a) $\text{pH} = \text{p}K_a + \log\left(\frac{[\text{base}]_0}{[\text{acid}]_0}\right) = \text{p}K_a + \log\left(\frac{[\text{lactate ion}]}{[\text{lactic acid}]}\right)$

$\text{pH} = \text{p}K_a = 3.08, \quad K_a = 8.3 \times 10^{-4}$

(b) Let x = [lactate ion] = $[L^{-1}]$ and $y = [H_3O^+]$.

Concentration (mol/L)	$HL(aq)$	$+ H_2O(l) \rightleftharpoons$	$H_3O^+(aq)$	$+ L^-(aq)$
initial	$2x$	—	—	x
change	$-y$	—	$+y$	$+y$
equilibrium	$2x - y$	—	y	$y + x$

$$K_a = \frac{[H_3O^+][L^-]}{[HL]} = \frac{(y)(y + x)}{(2x - y)} \approx \frac{(y)(x)}{(2x)} = 8.3 \times 10^{-4}$$

$y = 2(8.3 \times 10^{-4}) \approx 1.66 \times 10^{-3}\ \text{mol/L} \approx [H_3O^+]$

$\text{pH} \approx 2.78$

16.15 (a)

Concentration (mol/L)	$HBrO(aq)$	$+ H_2O(l) \rightleftharpoons$	$H_3O^+(aq)$	$+ BrO^-(aq)$
initial	0.20	—	0	0.10
change	$-x$	—	$+x$	$+x$
equilibrium	$0.20 - x$	—	x	$0.10 + x$

$$K_a = \frac{[H_3O^+][BrO^-]}{[HBrO]} = \frac{(x)(0.10 + x)}{(0.20 - x)} = 2.0 \times 10^{-9} \approx \frac{[H_3O^+](0.10)}{(0.20)}$$

$x \approx 4.0 \times 10^{-9}\ \text{mol/L}\ H_3O^+$

(b)

Concentration (mol/L)	$(CH_3)_2NH(aq)$	$+ H_2O(l) \rightleftharpoons$	$(CH_3)_2NH_2^+(aq)$	$+ OH^-(aq)$
initial	0.010	—	0.150	0
change	$-x$	—	$+x$	$+x$
equilibrium	$0.010 - x$	—	$0.150 + x$	x

$$K_b = \frac{[(CH_3)_2NH_2^+][OH^-]}{[(CH_3)_2NH]} = \frac{(0.150 + x)(x)}{(0.010 - x)} = 5.4 \times 10^{-4} \approx \frac{(0.150)[OH^-]}{(0.010)}$$

$x \approx 3.6 \times 10^{-5}$ mol/L OH^-

$$\text{Because } [H_3O^+] = \frac{K_w}{[OH^-]} = \frac{1.00 \times 10^{-14}}{3.6 \times 10^{-5}} = 2.8 \times 10^{-10} \text{ mol/L}$$

(c)

Concentration (mol/L)	$HBrO(aq)$ +	$H_2O(l) \rightleftharpoons$	$H_3O^+(aq)$ +	$BrO^-(aq)$
initial	0.10	—	0	0.20
change	$-x$	—	$+x$	$+x$
equilibrium	$0.10 - x$	—	x	$0.20 + x$

$$K_a = \frac{[H_3O^+][BrO^-]}{[HBrO]} = \frac{(x)(0.20 + x)}{(0.10 - x)} = 2.0 \times 10^{-9} \approx \frac{[H_3O^+](0.20)}{0.10}$$

$x = 1.0 \times 10^{-9}$ mol/L H_3O^+

(d)

Concentration (mol/L)	$(CH_3)_2NH(aq)$ +	$H_2O(l) \rightleftharpoons$	$(CH_3)_2NH_2^+(aq)$ +	$OH^-(aq)$
initial	0.020	—	0.030	0
change	$-x$	—	$+x$	$+x$
equilibrium	$0.020 - x$	—	$0.030 + x$	x

$$K_b = \frac{[(CH_3)_2NH_2^+][OH^-]}{[(CH_3)_2NH]} = \frac{(0.030 + x)(x)}{(0.020 - x)} = 5.4 \times 10^{-4} \approx \frac{(0.030)[OH^-]}{(0.020)}$$

$\approx 3.6 \times 10^{-4}$ mol/L OH^-

$$[H_3O^+] = \frac{K_w}{[OH^-]} = \frac{1.00 \times 10^{-14}}{3.6 \times 10^{-4}} = 2.8 \times 10^{-11} \text{ mol/L}$$

16.17 In each case, the equilibrium involved is

$HSO_4^-(aq) + H_2O(l) \rightleftharpoons H_3O^+(aq) + SO_4^{2-}(aq)$

$HSO_4^-(aq)$ and $SO_4^{2-}(aq)$ are conjugate acid and base; therefore, the pH calculation is most easily performed with the Henderson-Hasselbalch equation:

$$pH = pK_a + \log\left(\frac{[\text{base}]}{[\text{acid}]}\right) = pK_a + \log\left(\frac{[SO_4^{2-}]}{[HSO_4^-]}\right)$$

(a) $pH = 1.92 + \log\left(\frac{0.80 \text{ mol/L}}{0.40 \text{ mol/L}}\right) = 2.22, \quad pOH = 14.00 - 2.22 = 11.78$

(b) $pH = 1.92 + \log\left(\frac{0.20 \text{ mol/L}}{0.40 \text{ mol/L}}\right) = 1.62, \quad pOH = 12.38$

(c) $pH = pK_a = 1.92, \quad pOH = 12.08$

16.19 (a) $HCN(aq) + H_2O(l) \rightleftharpoons H_3O^+(aq) + CN^-(aq)$

total volume = 100.0 mL = 0.1000 L

moles of HCN = 0.0200 L × 0.050 mol/L = 1.0×10^{-3} mol HCN

moles of NaCN = 0.0800 L × 0.030 mol/L = 2.4×10^{-3} mol NaCN

$$\text{initial } [\text{HCN}]_0 = \frac{1.0 \times 10^{-3}\ \text{mol}}{0.1000\ \text{L}} = 1.0 \times 10^{-2}\ \text{mol/L}$$

$$\text{initial } [\text{CN}^-]_0 = \frac{2.4 \times 10^{-3}\ \text{mol}}{0.1000\ \text{L}} = 2.4 \times 10^{-2}\ \text{mol/L}$$

Concentration (mol/L)	$HCN(aq)$ +	$H_2O(l) \rightleftharpoons$	$H_3O^+(aq)$ +	$CN^-(aq)$
initial	1.0×10^{-2}	—	0	2.4×10^{-2}
change	$-x$	—	$+x$	$+x$
equilibrium	$1.0 \times 10^{-2} - x$	—	x	$2.4 \times 10^{-2} + x$

$$K_a = \frac{[H_3O^+][CN^-]}{[HCN]} = \frac{(x)(2.4 \times 10^{-2} + x)}{(1.0 \times 10^{-2} - x)} \approx \frac{(x)(2.4 \times 10^{-2})}{(1.0 \times 10^{-2})} = 4.9 \times 10^{-10}$$

$x \approx 2.0 \times 10^{-10}$ mol/L H_3O^+

$pH = -\log[H_3O^+] = -\log(2.0 \times 10^{-10}) = 9.70$

(b) The solution here is the same as for part (a), except for the initial concentrations:

$$[\text{HCN}]_0 = \frac{0.0800\ \text{L} \times 0.030\ \text{mol/L}}{0.1000\ \text{L}} = 2.4 \times 10^{-2}\ \text{mol/L}$$

$$[\text{CN}^-]_0 = \frac{0.0200\ \text{L} \times 0.050\ \text{mol/L}}{0.100\ \text{L}} = 1.0 \times 10^{-2}\ \text{mol/L}$$

$$K_a = 4.9 \times 10^{-10} = \frac{(x)(1.0 \times 10^{-2})}{(2.4 \times 10^{-2})}$$

$x = 1.1\overline{8} \times 10^{-9}$ mol/L H_3O^+

$pH = -\log(1.1\overline{8} \times 10^{-9}) = 8.93$

(c) $[HCN]_0 = [NaCN]_0$ after mixing; therefore,

$$K_a = 4.9 \times 10^{-10} = \frac{(x)[\text{NaCN}]_0}{[\text{HCN}]_0} = x = [H_3O^+]$$

$pH = pK_a = -\log(4.9 \times 10^{-10}) = 9.31$

Strong Acid-Strong Base Titrations

16.21 In each case, the net ionic reaction is $H_3O^+(aq) + OH^-(aq) \longrightarrow 2\ H_2O(l)$; therefore, the neutralization requires 1 mol H_3O^+ = 1 mol OH^-

(a) number of moles of H_3O^+ (from the acid) = (0.0250 L) × (0.30 mol/L) = 7.5×10^{-3} mol H_3O^+

number of moles of OH^- (from the base) = (0.0250 L) × (0.20 mol/L) = 5.0×10^{-3} mol OH^-

excess $H_3O^+ = (7.5 \times 10^{-3} - 5.0 \times 10^{-3})$ mol $H_3O^+ = 2.5 \times 10^{-3}$ mol H_3O^+

$$[H_3O^+] = \frac{2.5 \times 10^{-3} \text{ mol } H_3O^+}{0.0500 \text{ L}} = 0.050 \text{ mol/L}$$

$pH = -\log(0.050) = 1.30$

(b) number of moles of H_3O^+ (from the acid) $= (0.0250 \text{ L}) \times (0.15 \text{ mol/L})$

$= 3.7\overline{5} \times 10^{-3}$ mol H_3O^+

number of moles of OH^- (from the base) $= (0.0500 \text{ L}) \times (0.15 \text{ mol/L})$

$= 7.5 \times 10^{-3}$ mol OH^-

excess $OH^- = (7.5 \times 10^{-3} - 3.7\overline{5} \times 10^{-3})$ mol $OH^- = 3.7\overline{5} \times 10^{-3}$ mol OH^-

$$\frac{3.7\overline{5} \times 10^{-3} \text{ mol } OH^-}{0.0750 \text{ L}} = 0.050 \text{ mol/L } OH^-$$

$pOH = -\log(0.050) = 1.30, \quad pH = 14.00 - 1.30 = 12.70$

(c) moles of H_3O^+ (from the acid) $= 0.0217 \text{ L} \times 0.27$ mol/L

$= 5.8\overline{6} \times 10^{-3}$ mol H_3O^+

moles of OH^- (from the base) $= 0.0100 \text{ L} \times 0.30$ mol/L

$= 3.0\overline{0} \times 10^{-3}$ mol OH^-

excess $H_3O^+ = (5.8\overline{6} \times 10^{-3} - 3.0\overline{0} \times 10^{-3})$ mol $= 2.9 \times 10^{-3}$ mol

$$\frac{2.9 \times 10^{-3} \text{ mol } H_3O^+}{0.0317 \text{ L}} = 0.091 \text{ mol/L } H_3O^+$$

$pH = -\log(0.091) = 1.04$

16.23 $[NaOH]_0 = \left(\dfrac{14.0 \text{ g NaOH}}{0.250 \text{ L}}\right) \times \left(\dfrac{1 \text{ mol NaOH}}{40.00 \text{ g NaOH}}\right) = 1.40 \text{ mol/L} = [OH^-]_0$

moles of OH^- (from the base) $= (0.0250 \text{ L}) \times (1.40 \text{ (mol } OH^-\text{)/L})$

$= 0.0350$ mol OH^-

moles of H_3O^+ (from the acid) $= (0.0500 \text{ L}) \times (0.20 \text{ mol } H_3O^+/\text{L})$

$= 0.010$ mol H_3O^+

excess $OH^- = (0.0350 - 0.010)$ mol $OH^- = 0.025$ mol OH^-

$$\frac{0.025 \text{ mol } OH^-}{(0.0500 + 0.0250) \text{ L}} = 0.33 \text{ mol/L } OH^-$$

$pOH = -\log(0.33) = 0.48, \quad pH = 14.00 - 0.48 = 13.52$

16.25 1 mol H_3O^+ = 1 mol OH^- in neutralization, thus 1 mol HCl = 1 mol KOH

moles of HCl required $= (0.0250 \text{ L KOH}) \times \left(\dfrac{0.0497 \text{ mol KOH}}{1 \text{ L KOH}}\right) \times \left(\dfrac{1 \text{ mol HCl}}{1 \text{ mol KOH}}\right)$

$= 1.24 \times 10^{-3}$ mol HCl

$$\text{volume of HCl required} = \frac{1.24 \times 10^{-3}\ \text{mol HCl}}{0.0631\ \text{mol/L}} = 0.0197\ \text{L HCl} = 19.7\ \text{mL HCl}$$

16.27 initial pH = $-\log(0.10) = 1.00$

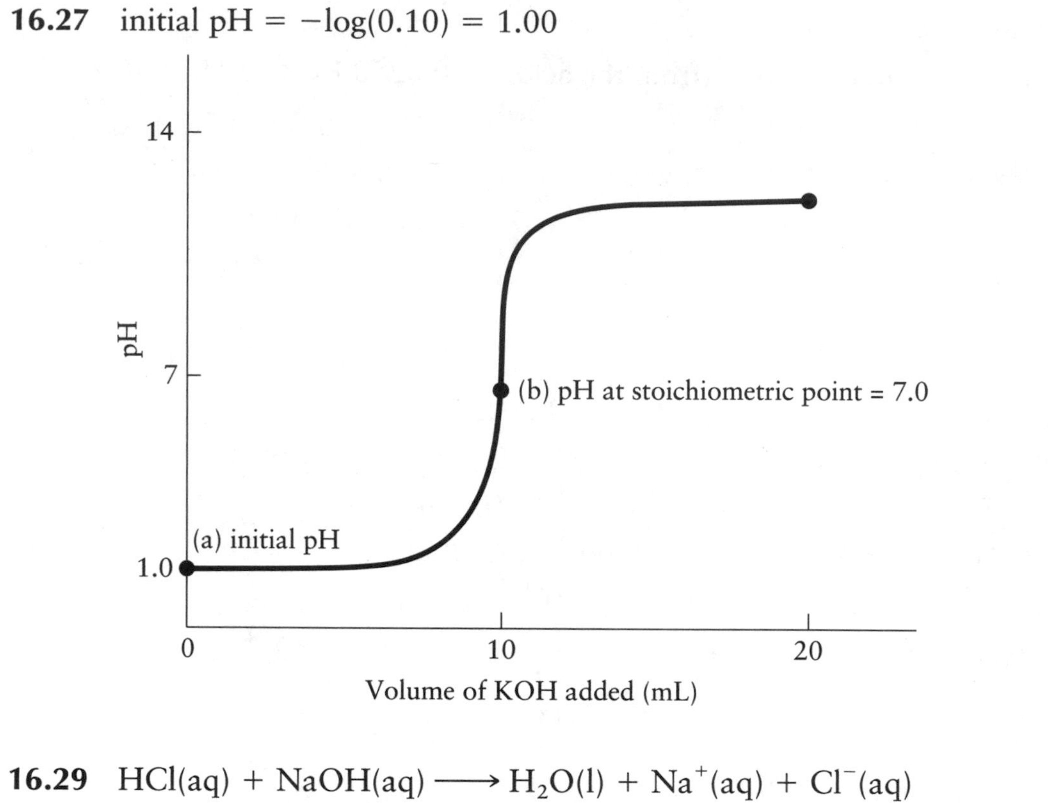

16.29 $HCl(aq) + NaOH(aq) \longrightarrow H_2O(l) + Na^+(aq) + Cl^-(aq)$

(a) $$\text{volume of HCl} = (\tfrac{1}{2}) \times (25.0\ \text{mL}) \times \left(\frac{10^{-3}\ \text{L}}{1\ \text{mL}}\right) \times \left(\frac{0.110\ \text{mol NaOH}}{1\ \text{L}}\right)$$
$$\times \left(\frac{1\ \text{mol HCl}}{1\ \text{mol NaOH}}\right) \times \left(\frac{1\ \text{L HCl}}{0.150\ \text{mol HCl}}\right) = 9.17 \times 10^{-3}\ \text{L HCl}$$

(b) $2 \times 9.17 \times 10^{-3}\ \text{L} = 0.0183\ \text{L}$

(c) volume = $(0.0250 + 0.0183)\ \text{L} = 0.0433\ \text{L}$

$$[Na^+] = (0.0250\ \text{L}) \times \left(\frac{0.110\ \text{mol NaOH}}{1\ \text{L}}\right) \times \left(\frac{1\ \text{mol Na}^+}{1\ \text{mol NaOH}}\right) \times \left(\frac{1}{0.0433\ \text{L}}\right)$$
$$= 0.0635\ \text{mol/L}$$

(d) $$\text{number of moles of } H_3O^+ \text{ (from acid)} = (0.0200\ \text{L}) \times \left(\frac{0.150\ \text{mol}}{1\ \text{L}}\right)$$
$$= 3.00 \times 10^{-3}\ \text{mol } H_3O^+$$

$$\text{number of moles of } OH^- \text{ (from base)} = (0.0250\ \text{L}) \times \left(\frac{0.110\ \text{mol Na}^+}{1\ \text{L}}\right)$$
$$= 2.75 \times 10^{-3}\ \text{mol } OH^-$$

$$\text{excess } H_3O^+ = (3.00 - 2.75) \times 10^{-3}\ \text{mol} = 2.5 \times 10^{-4}\ \text{mol } H_3O^+$$

$$= \frac{2.5 \times 10^{-4}\ \text{mol}}{0.0450\ \text{L}} = 5.5\overline{6} \times 10^{-3}\ \text{mol/L } H_3O^+$$

$$\text{pH} = -\log(5.5\overline{6} \times 10^{-3}) = 2.25$$

16.31 $HCl(aq) + NaOH(aq) \longrightarrow H_2O(l) + Na^+(aq) + Cl^-(aq)$

(a) $\text{moles of HCl} = (2.54\text{ g NaOH}) \times \left(\frac{1\text{ mol NaOH}}{40.00\text{ g NaOH}}\right) \times \left(\frac{1\text{ mol HCl}}{1\text{ mol NaOH}}\right)$

$= 6.35 \times 10^{-2}\text{ mol HCl}$

$\text{volume of HCl} = \frac{6.35 \times 10^{-2}\text{ mol}}{0.150\text{ mol/L}} = 0.423\text{ L} = 423\text{ mL}$

(b) $\text{moles of Cl}^- = 0.423\text{ L} \times \left(\frac{0.150\text{ mol HCl}}{1\text{ L}}\right) \times \left(\frac{1\text{ mol Cl}^-}{1\text{ mol HCl}}\right)$

$= 6.34 \times 10^{-2}\text{ mol Cl}^-$

$= \frac{6.34 \times 10^{-2}\text{ mol Cl}^-}{(0.423 + 0.0250)\text{ L}} = 0.142\text{ mol/L Cl}^-$

16.33 (a) $\text{pOH} = -\log(0.110) = 0.959, \quad \text{pH} = 14.00 - 0.959 = 13.04$

(b) $\text{initial moles of OH}^-\text{ (from base)} = (0.0250\text{ L}) \times \left(\frac{0.110\text{ mol}}{1\text{ L}}\right)$

$= 2.75 \times 10^{-3}\text{ mol OH}^-$

$\text{moles of H}_3\text{O}^+\text{ added} = (0.0050\text{ L}) \times \left(\frac{0.150\text{ mol}}{1\text{ L}}\right) = 7.5 \times 10^{-4}\text{ mol H}_3\text{O}^+$

$\text{excess OH}^- = (2.75 - 0.75) \times 10^{-3}\text{ mol} = 2.00 \times 10^{-3}\text{ mol OH}^-$

$= \frac{2.00 \times 10^{-3}\text{ mol}}{0.030\text{ L}} = 0.066\overline{7}\text{ mol/L OH}^-$

$\text{pOH} = -\log(0.066\overline{7}) = 1.18, \quad \text{pH} = 14.00 - 1.18 = 12.82$

(c) $\text{moles of H}_3\text{O}^+\text{ added} = 2 \times 7.5 \times 10^{-4}\text{ mol} = 1.50 \times 10^{-3}\text{ mol H}_3\text{O}^+$

$\text{excess OH}^- = (2.75 - 1.50) \times 10^{-3}\text{ mol} = 1.25 \times 10^{-3}\text{ mol OH}^-$

$= \frac{1.25 \times 10^{-3}\text{ mol}}{0.035\text{ L}} = 0.035\overline{7}\text{ mol/L OH}^-$

$\text{pOH} = -\log(0.035\overline{7}) = 1.45, \quad \text{pH} = 14.00 - 1.45 = 12.55$

(d) $\text{pH} = 7.00$

volume of HCl at stoichiometric point =

$(2.75 \times 10^{-3}\text{ mol NaOH}) \times \left(\frac{1\text{ mol HCl}}{1\text{ mol NaOH}}\right) \times \left(\frac{1\text{ L HCl}}{0.150\text{ mol HCl}}\right) = 0.0183\text{ L}$

(e) $(0.0050\text{ L}) \times \left(\frac{0.150\text{ mol}}{1\text{ L}}\right) \times \left(\frac{1}{(0.0250 + 0.0183 + 0.0050)\text{ L}}\right)$

$= 0.015\overline{5}\text{ mol/L H}_3\text{O}^+$

$\text{pH} = -\log(0.015\overline{5}) = 1.81$

(f) $\left(\frac{0.010\text{ L}}{0.0533\text{ L}}\right) \times \left(\frac{0.150\text{ mol}}{1\text{ L}}\right) = 0.028\overline{1}\text{ mol/L H}_3\text{O}^+$

$\text{pH} = -\log(0.028\overline{1}) = 1.55$

16.35 (a)

Concentration (mol/L)	$CH_3COOH(aq)$	$+ H_2O(l) \rightleftharpoons$	$H_3O^+(aq)$	$+ CH_3CO_2^-(aq)$
initial	0.10	—	0	0
change	$-x$	—	$+x$	$+x$
equilibrium	$0.10 - x$	—	x	x

$$K_a = \frac{[H_3O^+][CH_3CO_2^-]}{[CH_3COOH]} = 1.8 \times 10^{-5} = \frac{x^2}{0.10 - x} \approx \frac{x^2}{0.10}$$

$x^2 = 1.8 \times 10^{-6}$

$x = 1.3\bar{4} \times 10^{-3}$ mol/L H_3O^+

initial pH $= -\log(1.3\bar{4} \times 10^{-3}) = 2.87$

(b) moles of $CH_3COOH = (0.0250\text{ L}) \times (0.10\text{ M})$

$= 2.50 \times 10^{-3}$ mol CH_3COOH

moles of NaOH $= (0.0100\text{ L}) \times (0.10\text{ M}) = 1.0 \times 10^{-3}$ mol OH^-

So, $\dfrac{1.50 \times 10^{-3}\text{ mol } CH_3COOH}{0.0350\text{ L}} = 4.28 \times 10^{-2}$ mol/L

and $\dfrac{1.0 \times 10^{-3}\text{ mol } CH_3CO_2^-}{0.0350\text{ L}} = 2.86 \times 10^{-2}$ mol/L

Then, consider equilibrium, $K_a = \dfrac{[H_3O^+][CH_3CO_2^-]}{[CH_3COOH]}$

Concentration (mol/L)	$CH_3COOH(aq)$	$+ H_2O(l) \rightleftharpoons$	$H_3O^+(aq)$	$+ CH_3CO_2^-(aq)$
initial	4.28×10^{-2}	—	0	2.86×10^{-2}
change	$-x$	—	$+x$	$+x$
equilibrium	$4.28 \times 10^{-2} - x$	—	x	$2.86 \times 10^{-2} + x$

$1.8 \times 10^{-5} = \dfrac{(x)(x + 2.86 \times 10^{-2})}{(4.29 \times 10^{-2} - x)}$; assume $+x$ and $-x$ negligible

$x = 2.7 \times 10^{-5}$ mol/L H_3O^+ and pH $= -\log(2.7 \times 10^{-5}) = 4.57$

(c) Because acid and base concentrations are equal, their volumes are equal at the stoichiometric point. Therefore, 25.0 mL NaOH are required to reach the stoichiometric point and 12.5 mL NaOH are required to reach half the stoichiometric point.

(d) At the half stoichiometric point, pH $= pK_a$ and pH $= 4.74$

(e) 25.0 mL; see part (c)

(f) The final pH is that of 0.050 M $NaCH_3CO_2$.

Concentration (mol/L)	$H_2O(l)$ +	$CH_3CO_2^-(aq)$ ⇌	$CH_3COOH(aq)$ +	$OH^-(aq)$
initial	—	0.050	0	0
change	—	$-x$	$+x$	$+x$
equilibrium	—	$0.050 - x$	x	x

$$K_b = \frac{K_w}{K_a} = \frac{1.00 \times 10^{-14}}{1.8 \times 10^{-5}} = 5.6 \times 10^{-10} = \frac{x^2}{0.050 - x} \approx \frac{x^2}{0.050}$$

$x^2 = 2.8 \times 10^{-11}$

$x = 5.3 \times 10^{-6}$ mol/L OH^-

pOH = 5.28, pH = 14.00 − 5.28 = 8.72

(g) thymol blue; $pK_{in} = 8.9$ or phenolphthalein; $pK_{in} = 9.4$ } close to stoichiometric point pH of 8.72

16.37 (a) $K_b = \dfrac{[NH_4^+][OH^-]}{[NH_3]} = 1.8 \times 10^{-5}$

Concentration (mol/L)	$H_2O(l)$ +	$NH_3(aq)$ ⇌	$NH_4^+(aq)$ +	$OH^-(aq)$
initial	—	0.15	0	0
change	—	$-x$	$+x$	$+x$
equilibrium	—	$0.15 - x$	x	x

$$1.8 \times 10^{-5} = \frac{x^2}{0.15 - x} \approx \frac{x^2}{0.15}$$

$x = 1.6 \times 10^{-3}$ mol/L OH^-

pOH = 2.80, initial pH = 14.00 − 2.80 = 11.20

(b) moles of NH_3 = (0.0150 L) × (0.15 mol/L) = $2.2\overline{5} \times 10^{-3}$ mol NH_3

moles of HCl = (0.0150 L) × (0.10 mol/L) = 1.5×10^{-3} mol HCl

$$\frac{(2.25 \times 10^{-3} - 1.5 \times 10^{-3}) \text{ mol } NH_3}{0.0300 \text{ L}} = 2.5 \times 10^{-2} \text{ (mol } NH_3\text{)/L}$$

$$\frac{1.5 \times 10^{-3} \text{ mol HCl}}{0.0300 \text{ L}} = 5.0 \times 10^{-2} \text{ mol HCl/L} \approx 5.0 \times 10^{-2} \text{ (mol } NH_4^+\text{)/L}$$

Then, consider the equilibrium.

Concentration (mol/L)	$H_2O(l)$ +	$NH_3(aq)$ ⇌	$NH_4^+(aq)$ +	$OH^-(aq)$
initial	—	2.5×10^{-2}	5.0×10^{-2}	0
change	—	$-x$	$+x$	$+x$
equilibrium	—	$2.5 \times 10^{-2} - x$	$5.0 \times 10^{-2} + x$	x

$$K_b = \frac{[NH_4^+][OH^-]}{[NH_3]} = 1.8 \times 10^{-5}$$

$$= \frac{(x)(5.0 \times 10^{-2} + x)}{(2.5 \times 10^{-2} - x)}; \text{ assume } +x \text{ and } -x \text{ are negligible}$$

$x = 9.0 \times 10^{-6}$ mol/L and pOH = 5.04 OH^-

Therefore, pH = 14.00 − 5.04 = 8.96

(c) At the stoichiometric point, moles NH_3 = moles HCl.

$$\text{volume HCl added} = \frac{(0.15 \text{ mol } NH_3/L)(0.0150 \text{ L})}{0.10 \text{ mol HCl/L}} = 0.0225 \text{ L HCl}$$

Therefore, halfway to the stoichiometric point,

volume HCl added = 22.5/2 = 11.25 mL or 11.2 mL

(d) At half stoichiometric point, pOH = pK_b and pOH = 4.75.

Therefore, pH = 14.00 − 4.75 = 9.25.

(e) 22.5 mL; see part (c)

(f) $NH_4^+(aq) + H_2O(l) \rightleftharpoons H_3O^+(aq) + NH_3(aq)$

The initial moles of NH_3 have now been converted to moles NH_4^+ in a (15 + 22.5 = 37.5) mL volume

$$[NH_4^+] = \frac{2.25 \times 10^{-3} \text{ mol}}{0.0375 \text{ L}} = 0.060 \text{ mol/L}$$

$$K_a = \frac{K_w}{K_b} = \frac{1.00 \times 10^{-14}}{1.8 \times 10^{-5}} = 5.6 \times 10^{-10}$$

Concentration (mol/L)	$NH_4^+(aq)$	+ $H_2O(l)$ $\rightleftharpoons$	$H_3O^+(aq)$	+ $NH_3(aq)$
initial	0.060	—	0	0
change	$-x$	—	$+x$	$+x$
equilibrium	$0.060 - x$	—	x	x

$$K_a = 5.6 \times 10^{-10} = \frac{x^2}{0.060 - x} \approx \frac{x^2}{0.060}$$

$x = 5.8 \times 10^{-6}$ mol/L H_3O^+

pH = $-\log(5.8 \times 10^{-6}) = 5.24$

(g) methyl red or bromocresol green

16.39 (a) Let HL = lactic acid; then, $HL(aq) + H_2O(l) \rightleftharpoons H_3O^+(aq) + L^-(aq)$

$$K_a = \frac{[H_3O^+][L^-]}{[HL]} = 8.4 \times 10^{-4}$$

Concentration (mol/L)	$HL(aq)$	+ $H_2O(l)$ $\rightleftharpoons$	$H_3O^+(aq)$	+ $L^-(aq)$
initial	0.110	—	0	0
change	$-x$	—	$+x$	$+x$
equilibrium	$0.110 - x$	—	x	x

$$8.4 \times 10^{-4} = \frac{x^2}{0.110 - x} \approx \frac{x^2}{0.110}$$

$x = 9.6 \times 10^{-3}$ mol/L H_3O^+

initial pH $= -\log(9.6 \times 10^{-3}) = 2.02$

(b) moles of HL $= (0.0250 \text{ L}) \times (0.110 \text{ M}) = 2.75 \times 10^{-3}$ mol HL

moles of NaOH $= (0.0050 \text{ L}) \times (0.150 \text{ M}) = 7.5 \times 10^{-4}$ mol NaOH

$$[\text{HL}] = \frac{2.00 \times 10^{-3} \text{ mol HL}}{0.0300 \text{ L}} = 6.67 \times 10^{-2} \text{ mol HL/L}$$

$$[\text{L}^-] = \frac{7.5 \times 10^{-4} \text{ mol L}^-}{0.0300 \text{ L}} = 2.5 \times 10^{-2} \text{ mol L}^-\text{/L}$$

Then, consider the equilibrium:

Concentration (mol/L)	$HL(aq)$ +	$H_2O(l) \rightleftharpoons$	$H_3O^+(aq)$ +	$L^-(aq)$
initial	6.67×10^{-2}	—	0	2.5×10^{-2}
change	$-x$	—	$+x$	$+x$
equilibrium	$6.67 \times 10^{-2} - x$	—	x	$2.5 \times 10^{-2} + x$

$$K_a = \frac{[H_3O^+][L^-]}{[HL]} = 8.4 \times 10^{-4} = \frac{(x)(x + 2.5 \times 10^{-2})}{(6.67 \times 10^{-2} - x)}; \ +x \text{ and } -x \text{ negligible}$$

$x = 2.2\bar{4} \times 10^{-3}$ mol/L H_3O^+

pH $= -\log(2.2\bar{4} \times 10^{-3}) = 2.6$

(c) moles of HL $= (0.0250 \text{ L}) \times (0.110 \text{ mol/L}) = 2.75 \times 10^{-3}$ mol HL

moles of NaOH $= (0.0100 \text{ L}) \times (0.150 \text{ mol/L}) = 1.50 \times 10^{-3}$ mol NaOH

$$[\text{HL}] = \frac{1.25 \times 10^{-3} \text{ mol HL}}{0.0350 \text{ L}} = 3.57 \times 10^{-2} \text{ mol HL/L}$$

$$[\text{L}^-] = \frac{1.50 \times 10^{-3} \text{ mol L}^-}{0.0350 \text{ L}} = 4.28 \times 10^{-2} \text{ mol L}^-\text{/L}$$

Then, consider the equilibrium:

Concentration (mol/L)	$HL(aq)$ +	$H_2O(l) \rightleftharpoons$	$H_3O^+(aq)$ +	$L^-(aq)$
initial	3.57×10^{-2}	—	0	4.28×10^{-2}
change	$-x$	—	$+x$	$+x$
equilibrium	$3.57 \times 10^{-2} - x$	—	x	$4.28 \times 10^{-2} + x$

$$K_a = \frac{[H_3O^+][L^-]}{[HL]} = 8.4 \times 10^{-4} = \frac{(x)(x + 4.28 \times 10^{-2})}{(3.57 \times 10^{-2} - x)}; \ +x \text{ and } -x \text{ negligible}$$

$x = 7.01 \times 10^{-4}$ mol/L H_3O^+

pH $= -\log(7.01 \times 10^{-4}) = 3.15$

(d) At the stoichiometric point, moles acid = moles base = moles L^-

$$\text{volume base} = \frac{(0.110 \text{ mol acid/L}) \times (0.0250 \text{ L acid})}{0.150 \text{ mol base/L}} = 0.0183 \text{ L base}$$

$$K_b = \frac{K_w}{K_a} = \frac{1.00 \times 10^{-14}}{8.4 \times 10^{-4}} = 1.2 \times 10^{-11}$$

$$[L^-] = \frac{2.75 \times 10^{-3} \text{ mol/L}}{(0.025 + 0.0183) \text{ L}} = 6.35 \times 10^{-2} \text{ mol } L^-/\text{L}$$

Concentration (mol/L)	$H_2O(l)$	+ $L^-(aq)$	$\rightleftharpoons$ $HL(aq)$	+ $OH^-(aq)$
initial	—	6.35×10^{-2}	0	0
change	—	$-x$	$+x$	$+x$
equilibrium	—	$6.35 \times 10^{-2} - x$	x	x

$$1.2 \times 10^{-11} = \frac{[HL][OH^-]}{[L^-]} = \frac{x^2}{6.35 \times 10^{-2} - x} \approx \frac{x^2}{6.34 \times 10^{-2}}$$

$x = 8.7 \times 10^{-7}$ mol/L OH^-

$\text{pOH} = -\log(8.7 \times 10^{-7}) = 6.1, \quad \text{pH} = 14.00 - 6.1 = 7.9$

(e) At 5.0 mL base beyond the stoichiometric point, consider only strong base.

$$\text{molar concentration of } OH^- = \frac{(0.0050 \text{ L}) \times (0.150 \text{ mol/L})}{(0.0250 \text{ L} + 0.0183 \text{ L} + 0.0050 \text{ L})} = 1.5\bar{5} \times 10^{-2} \text{ mol/L}$$

$\text{pOH} = -\log(1.5\bar{5} \times 10^{-2}) = 1.8, \quad \text{pH} = 14.00 - 1.8 = 12.2$

(f) At 10.0 mL base beyond stoichiometric point, consider only strong base.

$$\text{molar concentration of } OH^- = \frac{(0.0100 \text{ L}) \times (0.150 \text{ mol/L})}{(0.0250 \text{ L} + 0.0183 \text{ L} + 0.0100 \text{ L})} = 2.81 \times 10^{-2} \text{ mol/L}$$

$\text{pOH} = -\log(2.81 \times 10^{-2}) = 1.55, \quad \text{pH} = 14.00 - 1.55 = 12.45$

(g) phenol red, $pK_{in} = 7.9$ (close to pH = 7.9)

Indicators

For the pH ranges over which common indicators change color, see Table 16.3.

16.41 (a) methyl orange, $pK_{in} = 3.4$; 3.2–4.4

(b) litmus, $pK_{in} = 6.5$; 5.0–8.0

(c) methyl red, $pK_{in} = 5.0$; 4.8–6.0

(d) phenolphthalein, $pK_{in} = 9.4$; 8.2–10.0

Values are from Table 16.3.

16.43 At the stoichiometric point, the volume of solution will have doubled; therefore, the concentration of $CH_3CO_2^-$ will be 0.10 M. The equilibrium is

Concentration (mol/L)	$CH_3CO_2^-(aq)$	+ $H_2O(l)$ $\rightleftharpoons$	$HCH_3CO_2(aq)$	+ $OH^-(aq)$
initial	0.10	—	0	0
change	$-x$	—	$+x$	$+x$
equilibrium	$0.10 - x$	—	x	x

$$K_b = \frac{K_w}{K_a} = \frac{1.00 \times 10^{-14}}{1.8 \times 10^{-5}} = 5.6 \times 10^{-10}$$

$$K_b = \frac{[HCH_3CO_2][OH^-]}{[CH_3CO_2^-]} = \frac{x^2}{0.10 - x} \approx \frac{x^2}{0.10} = 5.6 \times 10^{-10}$$

$x = 7.5 \times 10^{-6}$ mol/L OH^-

$pOH = -\log(7.5 \times 10^{-6}) = 5.12$, $pH = 14.00 - 5.12 = 8.88$

From Table 16.3, we see that this pH value lies within the range for phenolphthalein, so that indicator would be suitable; the others would not.

Buffers

16.45 (a) not a buffer; strong acid/salt of strong acid and strong base

(b) weak acid/conjugate base; thus, a buffer:

$HClO(aq) + H_2O(l) \rightleftharpoons H_3O^+(aq) + ClO^-(aq)$

(c) weak base/conjugate acid; thus, a buffer:

$(CH_3)_3N(aq) + H_2O(l) \rightleftharpoons (CH_3)_3NH^+(aq) + OH^-(aq)$

(d) After partial neutralization, we have a weak acid (CH_3COOH) and its salt ($NaCH_3CO_2$) in solution. Thus, weak acid/conjugate base, and a buffer:

$CH_3COOH(aq) + H_2O(l) \rightleftharpoons H_3O^+(aq) + CH_3CO_2^-(aq)$

(e) not a buffer; complete neutralization results in a salt and water

16.47 In a solution containing $HClO(aq)$ and $ClO^-(aq)$, the following equilibrium occurs:

$HClO(aq) + H_2O(l) \rightleftharpoons H_3O^+(aq) + ClO^-(aq)$

The ratio $[ClO^-]/[HClO]$ is related to pH, as given by the Henderson-Hasselbalch equation:

$$pH = pK_a + \log\left(\frac{[ClO^-]}{[HClO]}\right), \text{ or}$$

$$\log\left(\frac{[ClO^-]}{[HClO]}\right) = pH - pK_a = 6.50 - 7.53 = -1.03$$

$$\frac{[ClO^-]}{[HClO]} = 9.3 \times 10^{-2}$$

16.49 The rule of thumb we use is that the effective range of a buffer is roughly within plus or minus one pH unit of the pK_a of the acid. Therefore,

(a) $pK_a = 3.08$; pH range, 2–4

(b) $pK_a = 4.19$; pH range, 3–5

(c) $pK_{a3} = 12.68$; pH range, 11.5–13.5

(d) $pK_{a2} = 7.21$; pH range, 6–8

(e) $pK_b = 7.97$, $pK_a = 6.03$; pH range, 5–7

16.51 Choose a buffer system in which the conjugate acid has a pK_a close to the desired pH. Therefore,

(a) $HClO_2$ and $NaClO_2$, $pK_a = 2.00$

(b) NaH_2PO_4 and Na_2HPO_4, $pK_{a2} = 7.21$

(c) $CH_2ClCOOH$ and $NaCH_2ClCO_2$, $pK_a = 2.85$ or $CH_3CH(OH)COOH$ and $NaCH_3CH(OH)CO_2$, $pK_a = 3.08$

(d) Na_2HPO_4 and Na_3PO_4, $pK_a = 12.68$

16.53 (a) $HCO_3^-(aq) + H_2O(l) \rightleftharpoons CO_3^{2-}(aq) + H_3O^+(aq)$

$$K_{a2} = \frac{[H_3O^+][CO_3^{2-}]}{[HCO_3^-]}, \quad pK_{a2} = 10.25$$

$$pH = pK_{a2} + \log\left(\frac{[CO_3^{2-}]}{[HCO_3^-]}\right)$$

$$\log\left(\frac{[CO_3^{2-}]}{[HCO_3^-]}\right) = pH - pK_{a2} = 11.0 - 10.25 = 0.75$$

$$\frac{[CO_3^{2-}]}{[HCO_3^-]} = 5.6$$

(b) $[CO_3^{2-}] = 5.6 \times [HCO_3^-] = 5.6 \times 0.100 \text{ mol/L} = 0.56 \text{ mol/L}$

moles of CO_3^{2-} = moles of K_2CO_3 = $0.56 \text{ mol/L} \times 1 \text{ L} = 0.56 \text{ mol}$

$$\text{mass of } K_2CO_3 = 0.56 \text{ mol} \times \left(\frac{138.21 \text{ g } K_2CO_3}{1 \text{ mol } K_2CO_3}\right) = 77 \text{ g } K_2CO_3$$

$$\text{(c) } [HCO_3^-] = \frac{[CO_3^{2-}]}{5.6} = \frac{0.100 \text{ mol/L}}{5.6} = 1.7\overline{8} \times 10^{-2} \text{ mol/L}$$

moles of HCO_3^- = moles of $KHCO_3$ = $1.7\overline{8} \times 10^{-2} \text{ mol/L} \times 1 \text{ L}$
$= 1.7\overline{8} \times 10^{-2} \text{ mol}$

mass $KHCO_3$ = $1.7\overline{8} \times 10^{-2} \text{ mol} \times 100.12 \text{ g/mol} = 1.8 \text{ g } KHCO_3$

(d) $[CO_3^{2-}] = 5.6 \times [HCO_3^-]$

moles HCO_3^- = moles $KHCO_3$ = $0.100 \text{ mol/L} \times 0.100 \text{ L} = 1.00 \times 10^{-2} \text{ mol}$

Because the final total volume is the same for both $KHCO_3$ and K_2CO_3, the number of moles of K_2CO_3 required is $5.6 \times 1.00 \times 10^{-2} \text{ mol} = 5.6 \times 10^{-2} \text{ mol}$.

Thus,

$$\text{volume of } K_2CO_3 \text{ solution} = \frac{5.6 \times 10^{-2}\ \text{mol}}{0.200\ \text{mol/L}} = 0.28\ \text{L} = 2.8 \times 10^{2}\ \text{mL}$$

16.55 $CH_3COOH(aq) + H_2O(l) \rightleftharpoons H_3O^+(aq) + CH_3CO_2^-(aq)$

$$K_a = \frac{[H_3O^+][CH_3CO_2^-]}{[CH_3COOH]}$$

$$pH = pK_a + \log\left(\frac{[CH_3CO_2^-]}{[CH_3COOH]}\right)$$

(a) $pH = pK_a + \log\left(\frac{0.10}{0.10}\right) = pK_a = 4.75$

(b) 3.0 mmol NaOH = 3.0×10^{-3} mol NaOH (strong base) produces 3.0×10^{-3} mol $CH_3CO_2^-$ from CH_3COOH (assume no volume change)

0.10 mol/L × 0.100 L = 1.0×10^{-2} mol CH_3COOH initially

0.10 mol/L × 0.100 L = 1.0×10^{-2} mol $CH_3CO_2^-$ initially

after adding NaOH:

$$[CH_3COOH] = \frac{(1.0 \times 10^{-2} - 3.0 \times 10^{-3})\ \text{mol}}{0.100\ \text{L}} = \frac{7.0 \times 10^{-3}\ \text{mol}}{0.100\ \text{L}}$$

$$= 7.0 \times 10^{-2}\ \text{mol/L}$$

$$[CH_3CO_2^-] = \frac{(1.0 \times 10^{-2} + 3.0 \times 10^{-3})\ \text{mol}}{0.100\ \text{L}} = \frac{1.3 \times 10^{-2}\ \text{mol}}{0.100\ \text{L}}$$

$$= 1.3 \times 10^{-1}\ \text{mol/L}$$

$$pH = 4.75 + \log\left(\frac{1.3 \times 10^{-1}}{7.0 \times 10^{-2}}\right)$$

$pH = 4.75 + 0.27 = 5.02$ ($\Delta pH = 0.27$)

(c) 6.0 mmol HNO_3 = 6.0×10^{-3} mol HNO_3 (strong acid) produces 6.0×10^{-3} mol CH_3COOH from $CH_3CO_2^-$ (assume no volume change)

after adding HNO_3:

$$[CH_3COOH] = \frac{(1.0 \times 10^{-2} + 6.0 \times 10^{-3})\ \text{mol}}{0.100\ \text{L}} = \frac{1.6 \times 10^{-2}\ \text{mol}}{0.100\ \text{L}}$$

$$= 1.6 \times 10^{-1}\ \text{mol/L}$$

$$[CH_3CO_2^-] = \frac{(1.0 \times 10^{-2} - 6.0 \times 10^{-3})\ \text{mol}}{0.100\ \text{L}} = \frac{4.0 \times 10^{-3}\ \text{mol}}{0.100\ \text{L}}$$

$$= 4.0 \times 10^{-2}\ \text{mol/L}$$

$$pH = 4.75 + \log\left(\frac{4.0 \times 10^{-2}}{1.6 \times 10^{-1}}\right)$$

$pH = 4.75 - 0.60 = 4.15$ ($\Delta pH = -0.60$)

16.57 (a) $pH = pK_a + \log\left(\dfrac{[CH_3CO_2^-]}{[CH_3COOH]}\right)$ (see Exercise 16.55)

$$pH = pK_a + \log\left(\frac{0.10}{0.10}\right) = 4.75 \text{ (initial pH)}$$

final pH: $(0.0100\text{ L})(0.950\text{ mol/L}) = 9.5 \times 10^{-3}$ mol NaOH (strong base) produces 9.5×10^{-3} mol $CH_3CO_2^-$ from CH_3COOH

$0.10\text{ mol/L} \times 0.100\text{ L} = 1.0 \times 10^{-2}$ mol CH_3COOH initially

$0.10\text{ mol/L} \times 0.100\text{ L} = 1.0 \times 10^{-2}$ mol $CH_3CO_2^-$ initially

after adding NaOH:

$$[CH_3COOH] = \frac{(1.0 \times 10^{-2} - 9.5 \times 10^{-3})\text{ mol}}{0.110\text{ L}} = 4.54 \times 10^{-3}\text{ mol/L}$$

$$[CH_3CO_2^-] = \frac{(1.0 \times 10^{-2} + 9.5 \times 10^{-3})\text{ mol}}{0.110\text{ L}} = 1.77 \times 10^{-1}\text{ mol/L}$$

$$pH = 4.75 + \log\left(\frac{1.77 \times 10^{-1}\text{ mol/L}}{4.54 \times 10^{-3}\text{ mol/L}}\right) = 4.75 + 1.59 = 6.34$$

$\Delta pH = 1.59$

(b) $(0.0200\text{ L}) \times (0.10\text{ mol/L}) = 2.00 \times 10^{-3}$ mol HNO_3 (strong acid) produces 2.00×10^{-3} mol CH_3COOH from $CH_3CO_2^-$

after adding HNO_3 [see part (a) of this exercise]:

$$[CH_3COOH] = \frac{(1.0 \times 10^{-2} + 2.00 \times 10^{-3})\text{ mol}}{0.120\text{ L}} = 1.00 \times 10^{-1}\text{ mol/L}$$

$$[CH_3CO_2^-] = \frac{(1.0 \times 10^{-2} - 2.00 \times 10^{-3})\text{ mol}}{0.120\text{ L}} = 6.67 \times 10^{-2}\text{ mol/L}$$

$$pH = 4.75 + \log\left(\frac{6.67 \times 10^{-2}\text{ mol/L}}{1.00 \times 10^{-1}\text{ mol/L}}\right) = 4.75 - 0.18 = 4.57$$

$\Delta pH = -0.18$

Solubility Products

The values for the solubility products of various sparingly soluble salts are listed in Table 16.5.

16.59 (a) $K_{sp} = [Ag^+][Br^-]$

(b) $K_{sp} = [Ag^+]^2[S^{2-}]$

(c) $K_{sp} = [Ca^{2+}][OH^-]^2$

(d) $K_{sp} = [Ag^+]^2[CrO_4^{2-}]$

16.61 (a) The solubility equilibrium is $AgBr(s) \rightleftharpoons Ag^+(aq) + Br^-(aq)$.

$[Ag^+] = [Br^-] = 8.8 \times 10^{-7}\text{ mol/L} = S =$ solubility

$K_{sp} = [Ag^+][Br^-] = (8.8 \times 10^{-7})(8.8 \times 10^{-7}) = 7.7 \times 10^{-13}$

(b) The solubility equilibrium is $PbCrO_4(s) \rightleftharpoons Pb^{2+}(aq) + CrO_4^{2-}(aq)$

$[Pb^{2+}] = 1.3 \times 10^{-7}$ mol/L $= S$, $[CrO_4^{2-}] = 1.3 \times 10^{-7}$ mol/L $= S$

$K_{sp} = [Pb^{2+}][CrO_4^{2-}] = (1.3 \times 10^{-7})(1.3 \times 10^{-7}) = 1.7 \times 10^{-14}$

(c) The solubility equilibrium is $Ba(OH)_2(s) \rightleftharpoons Ba^{2+}(aq) + 2\ OH^-(aq)$

$[Ba^{2+}] = 0.11$ mol/L $= S$, $[OH^-] = 0.22$ mol/L $= 2S$

$K_{sp} = [Ba^{2+}][OH^-]^2 = (0.11)(0.22)^2 = 5.3 \times 10^{-3}$

(d) The solubility equilibrium is $MgF_2(s) \rightleftharpoons Mg^{2+}(aq) + 2\ F^-(aq)$

$[Mg^{2+}] = 1.2 \times 10^{-3}$ mol/L $= S$, $[F^-] = 2.4 \times 10^{-3}$ mol/L $= 2S$

$K_{sp} = [Mg^{2+}][F^-]^2 = (1.2 \times 10^{-3})(2.4 \times 10^{-3})^2 = 6.9 \times 10^{-9}$

16.63 (a) Equilibrium equation: $MgF_2(s) \rightleftharpoons Mg^{2+}(aq) + 2\ F^-(aq)$

$K_{sp} = [Mg^{2+}][F^-]^2 = S \times (2S)^2 = 4S^3 = 6.4 \times 10^{-9}$

$S = 1.2 \times 10^{-3}$ mol/L

(b) Equilibrium equation: $BaSO_4(s) \rightleftharpoons Ba^{2+}(aq) + SO_4^{2-}(aq)$

$K_{sp} = [Ba^{2+}][SO_4^{2-}] = S \times S = S^2 = 1.1 \times 10^{-10}$

$S = 1.0 \times 10^{-5}$ mol/L

(c) Equilibrium equation: $CuI(s) \rightleftharpoons Cu^+(aq) + I^-(aq)$

$K_{sp} = [Cu^+][I^-] = S \times S = S^2 = 5.1 \times 10^{-2}$

$S = 0.23$ mol/L

16.65 $Tl_2CrO_4(s) \rightleftharpoons 2\ Tl^+(aq) + CrO_4^{2-}(aq)$

$[CrO_4^{2-}] = S = 6.3 \times 10^{-5}$ mol/L

$[Tl^+] = 2S = 2(6.3 \times 10^{-5})$ mol/L

$K_{sp} = [Tl^+]^2[CrO_4^{2-}] = (2S)^2 \times (S)$

$K_{sp} = [2(6.3 \times 10^{-5})]^2 \times (6.3 \times 10^{-5}) = 1.0 \times 10^{-12}$

Common-Ion Effect

16.67 (a)

Concentration (mol/L)	$AgCl(s) \rightleftharpoons$	$Ag^+(aq)$ +	$Cl^-(aq)$
initial	—	0	0.20
change	—	$+S$	$+S$
equilibrium	—	S	$S + 0.20$

$K_{sp} = [Ag^+][Cl^-] = (S) \times (S + 0.20) = 1.6 \times 10^{-10}$

Assume S in $S + 0.20$ is negligible, so $0.20\ S = 1.6 \times 10^{-10}$

$S = 8.0 \times 10^{-10}$ mol/L Ag^+ = molar solubility of AgCl in 0.20 M NaCl

(b)

Concentration (mol/L)	$Hg_2Cl_2(s) \rightleftharpoons$	$Hg_2^{2+}(aq)$ +	$2\ Cl^-(aq)$
initial	—	0	0.10
change	—	$+S$	$+2S$
equilibrium	—	S	$0.10 + 2S$

$K_{sp} = [Hg_2^{2+}][Cl^-]^2 = (S) \times (2S + 0.10)^2 = 1.3 \times 10^{-18}$

Assume $2S$ in $2S + 0.10$ is negligible, so $0.010S = 1.3 \times 10^{-18}$

$S = 1.3 \times 10^{-16}$ mol/L Hg_2^{2+} = molar solubility of Hg_2Cl_2 in 0.10 M NaCl

(c)

Concentration (mol/L)	$PbCl_2(s) \rightleftharpoons$	$Pb^{2+}(aq)$ +	$2\ Cl^-(aq)$
initial	—	0	$2 \times 0.10 = 0.20$
change	—	$+S$	$+S$
equilibrium	—	S	$S + 0.20$

$K_{sp} = [Pb^{2+}][Cl^-]^2 = S \times (S + 0.20)^2 = 1.6 \times 10^{-5}$

S may not be negligible relative to 0.20, so the full cubic form may be required. We do it both ways:

For $S^3 + 0.40\ S^2 + 4 \times 10^{-2}S - 1.6 \times 10^{-5} = 0$, the solution by standard methods is $S = 4.0 \times 10^{-4}$ mol/L.

If S had been neglected, the answer would have been the same, 4.0×10^{-4}, to within two significant figures.

(d)

Concentration (mol/L)	$Fe(OH)_2(s) \rightleftharpoons$	$Fe^{2+}(aq)$ +	$2\ OH^-(aq)$
initial	—	1.0×10^{-4}	0
change	—	$+S$	$+2S$
equilibrium	—	$1.0 \times 10^{-4} + S$	$2S$

$K_{sp} = [Fe^{2+}][OH^-]^2 = (S + 1.0 \times 10^{-4}) \times (2S)^2 = 1.6 \times 10^{-14}$

Assume S in $S + 1.0 \times 10^{-4}$ is negligible, so $4S^2 \times 1.0 \times 10^{-4} = 1.6 \times 10^{-14}$

$S^2 = 4 \times 10^{-11}$

$S = 6.3 \times 10^{-6}$ mol/L = molar solubility of $Fe(OH)_2$ in 1.0×10^{-4} M $FeCl_2$

16.69 (a) $Ag^+(aq) + Cl^-(aq) \rightleftharpoons AgCl(s)$

Concentration (mol/L)	Ag^+	Cl^-
initial	0	1.0×10^{-5}
change	$+x$	0
equilibrium	x	1.0×10^{-5}

$K_{sp} = [Ag^+][Cl^-] = 1.6 \times 10^{-10} = (x)(1.0 \times 10^{-5})$
$x = 1.6 \times 10^{-5}$ mol/L Ag^+

(b) $\text{mass } AgNO_3 = \left(\frac{1.6 \times 10^{-5} \text{ mol } AgNO_3}{1 \text{ L}}\right) \times \left(\frac{10^{-3} \text{ L}}{1 \text{ mL}}\right) \times (100. \text{ mL})$
$\times \left(\frac{169.88 \text{ g } AgNO_3}{1 \text{ mol } AgNO_3}\right) \times \left(\frac{1 \ \mu\text{g}}{10^{-6} \text{ g}}\right) = 2.7 \times 10^2 \ \mu\text{g } AgNO_3$

16.71 $Ni^{2+}(aq) + 2\ OH^-(aq) \rightleftharpoons Ni(OH)_2(s)$

Concentration (mol/L)	Ni^{2+}	OH^-
initial	0.010	0
change	0	$+x$
equilibrium	0.010	x

$K_{sp} = [Ni^{2+}][OH^-]^2 = 6.5 \times 10^{-18} = (0.010)(x)^2$
$x = \sqrt{\frac{6.5 \times 10^{-18}}{0.010}} = 2.5 \times 10^{-8}$ mol/L OH^-
$\text{pOH} = -\log(2.5 \times 10^{-8}) = 7.60, \quad \text{pH} = 14.00 - 7.60 = 6.40$

Predicting Precipitation Reactions

16.73 (a) $Ag^+(aq) + Cl^-(aq) \rightleftharpoons AgCl(s), \quad [Ag^+][Cl^-] = K_{sp}$
$Q_{sp} = \left[\frac{(0.073 \text{ L})(0.0040 \text{ mol/L})}{0.100 \text{ L}}\right] \times \left[\frac{(0.027 \text{ L})(0.0010 \text{ mol/L})}{0.100 \text{ L}}\right]$
$= (2.9 \times 10^{-3}) \times (2.7 \times 10^{-4}) = 7.9 \times 10^{-7}$
will precipitate, because $Q_{sp}\ (7.8 \times 10^{-7}) > K_{sp}\ (1.6 \times 10^{-10})$
(b) $Ca^{2+}(aq) + SO_4^{\ 2-}(aq) \rightleftharpoons CaSO_4(s), \quad [Ca^{2+}][SO_4^{\ 2-}] = K_{sp}$
$Q_{sp} = \left[\frac{(0.0100 \text{ L}) \times (0.0030 \text{ mol/L})}{0.111 \text{ L}}\right] \times \left[\frac{(0.0010 \text{ L}) \times (1.0 \text{ mol/L})}{0.111 \text{ L}}\right]$
$= (2.7 \times 10^{-4}) \times (9.0 \times 10^{-3}) = 2.4 \times 10^{-6}$
will not precipitate, because $Q_{sp}\ (2.4 \times 10^{-6}) < K_{sp}(2.4 \times 10^{-5})$

16.75 $\left(\frac{1 \text{ mL}}{20 \text{ drops}}\right) \times 1 \text{ drop} = 0.05 \text{ mL} = 0.05 \times 10^{-3} \text{ L} = 5 \times 10^{-5} \text{ L}$
and $(5 \times 10^{-5} \text{ L})(0.010 \text{ mol/L}) = 5 \times 10^{-7}$ mol NaCl $= 5 \times 10^{-7}$ mol Cl^-
(a) $Ag^+(aq) + Cl^-(aq) \rightleftharpoons AgCl(s), \quad [Ag^+][Cl^-] = K_{sp}$
$Q_{sp} = \left[\frac{(0.010 \text{ L}) \times (0.0040 \text{ mol/L})}{0.010 \text{ L}}\right] \times \left[\frac{5 \times 10^{-7} \text{ mol}}{0.010 \text{ L}}\right]$
$= (4.0 \times 10^{-3}) \times (5 \times 10^{-5}) = 2 \times 10^{-7}$

will precipitate, because Q_{sp} $(2 \times 10^{-7}) > K_{sp}$ (1.6×10^{-10})

(b) $Pb^{2+}(aq) + 2\ Cl^-(aq) \rightleftharpoons PbCl_2(s)$, $[Pb^{2+}][Cl^-]^2 = K_{sp}$

$$Q_{sp} = \left[\frac{(0.0100\ L) \times (0.0040\ mol/L)}{0.010\ L}\right] \times \left[\frac{5 \times 10^{-7}\ mol}{0.010\ L}\right]^2$$

$$= (4.0 \times 10^{-3}) \times (2.\bar{5} \times 10^{-9}) = 1 \times 10^{-11}$$

will not precipitate, because Q_{sp} $(1 \times 10^{-11}) < K_{sp}$ (1.6×10^{-5})

16.77 Because the sulfides of interest have the same general formula (MS), each forms two ions in solution. Because the number of ions in solution is the same, $K_{sp} = [M^{2+}][S^{2-}]$ and the values of K_{sp} can be compared directly. CuS precipitates first, followed by FeS, then ZnS.

16.79 (a) $K_{sp}[Ni(OH)_2] < K_{sp}[Mg(OH)_2] < K_{sp}[Ca(OH)_2]$

This is the order for the solubility products of these hydroxides. Thus, order of precipitation is (first to last): $Ni(OH)_2$, $Mg(OH)_2$, $Ca(OH)_2$

(b) $K_{sp}[Ni(OH)_2] = 6.5 \times 10^{-18} = [Ni^{2+}][OH^-]^2$

$$[OH^-]^2 = \frac{6.5 \times 10^{-18}}{0.0010} = 6.5 \times 10^{-15}$$

$$= \sqrt{\frac{6.5 \times 10^{-18}}{0.0010}} = 8.1 \times 10^{-8}\ mol/L\ OH^-$$

$pOH = -\log[OH^-] = 7.1, \quad pH = 6.9$

$K_{sp}[Mg(OH)_2] = 1.1 \times 10^{-11} = [Mg^{2+}][OH^-]^2$

$$= \sqrt{\frac{1.1 \times 10^{-11}}{0.0010}} = 1.0\bar{5} \times 10^{-4}\ mol/L\ OH^-$$

$pOH = -\log(1.0\bar{5} \times 10^{-4}) = 3.98 = 4.0, \quad pH = 14.0 - 4.0 = 10.0$

$K_{sp}[Ca(OH)^2] = 5.5 \times 10^{-6} = [Ca^{2+}][OH^-]^2$

$$= \sqrt{\frac{5.5 \times 10^{-6}}{0.0010}} = 7.4 \times 10^{-2}\ mol/L\ OH^-$$

$pOH = -\log(7.4 \times 10^{-2}) = 1.1, \quad pH = 14.0 - 1.1 = 12.9$

Dissolving Precipitates and Qualitative Analysis

16.81

$AgBr(s) \rightleftharpoons Ag^+(aq) + Br^-(aq)$	$K_{sp} = 7.7 \times 10^{-13}$
$Ag^+(aq) + 2\ CN^-(aq) \rightleftharpoons Ag(CN)_2^-(aq)$	$K_f = 5.6 \times 10^8$
$AgBr(s) + 2\ CN^-(aq) \rightleftharpoons Ag(CN)_2^-(aq) + Br^-(aq)$	$K = 4.3 \times 10^{-4}$

$$\text{thus, } K = \frac{[Ag(CN)_2^-][Br^-]}{[CN^-]^2} = 4.3 \times 10^{-4}$$

Concentration (mol/L)	$AgBr(s)$ +	$2\ CN^-(aq) \rightleftharpoons$	$Ag(CN)_2^-(aq)$ +	$Br^-(aq)$
initial	—	0.10	0	0
change	—	$-2S$	$+S$	$+S$
equilibrium	—	$0.10 - 2S$	S	S

$$\frac{[Ag(CN)_2^-][Br^-]}{[CN^-]^2} = \frac{S^2}{(0.10 - 2S)^2} = 4.3 \times 10^{-4}$$

$$\frac{S}{0.10 - 2S} = \sqrt{4.3 \times 10^{-4}} = 2.1 \times 10^{-2}$$

$S = 2.1 \times 10^{-3} - 4.2 \times 10^{-2}S$

$1.04S = 2.1 \times 10^{-3}$

$S = 2.0 \times 10^{-3}$ mol/L = molar solubility of AgBr

16.83 (a) pH = 7.0; $[OH^-] = 1.0 \times 10^{-7}$ mol/L

$Al^{3+}(aq) + 3\ OH^-(aq) \rightleftharpoons Al(OH)_3(s)$

$[Al^{3+}][OH^-]^3 = K_{sp} = 1.0 \times 10^{-33}$

$S \times (10^{-7})^3 = 1.0 \times 10^{-33}$

$$S = \frac{1.0 \times 10^{-33}}{1 \times 10^{-21}} = 1.0 \times 10^{-12} \text{ mol/L } Al^{3+}$$

= molar solubility of $Al(OH)_3$ at pH = 7.0

(b) pH = 4.5; pOH = 9.5; $[OH^-] = 3.2 \times 10^{-10}$ mol/L

$[Al^{3+}][OH^-]^3 = K_{sp} = 1.0 \times 10^{-33}$

$S \times (3.2 \times 10^{-10})^3 = 1.0 \times 10^{-33}$

$$S = \frac{1.0 \times 10^{-33}}{3.3 \times 10^{-29}} = 3.0 \times 10^{-5} \text{ mol/L } Al^{3+}$$

= molar solubility of $Al(OH)_3$ at pH = 4.5

(c) pH = 7.0; $[OH^-] = 1.0 \times 10^{-7}$ mol/L

$Zn^{2+}(aq) + 2\ OH^-(aq) \rightleftharpoons Zn(OH)_2(s)$

$[Zn^{2+}][OH^-]^2 = K_{sp} = 2.0 \times 10^{-17}$

$S \times (1.0 \times 10^{-7})^2 = 2.0 \times 10^{-17}$

$$S = \frac{2.0 \times 10^{-17}}{1.0 \times 10^{-14}} = 2.0 \times 10^{-3} \text{ mol/L } Zn^{2+}$$

= molar solubility of $Zn(OH)_2$ at pH = 7.0

(d) pH = 6.0; pOH = 8.0; $[OH^-] = 1.0 \times 10^{-8}$ mol/L

$[Zn^{2+}][OH^-]^2 = 2.0 \times 10^{-17} = K_{sp}$

$S \times (1.0 \times 10^{-8})^2 = 2.0 \times 10^{-17}$

$$S = \frac{2.0 \times 10^{-17}}{1.0 \times 10^{-16}} = 2.0 \times 10^{-1} = 0.20 \text{ mol/L } Zn^{2+}$$

= molar solubility of $Zn(OH)_2$ at pH = 6.0

16.85 (a) Multiply the second equilibrium equation by 2 and add to the first equilibrium. Then

$CaF_2(s) + 2\ H_2O(l) \rightleftharpoons Ca^{2+}(aq) + 2\ HF(aq) + 2\ OH^-(aq)$

$K = K_{sp}K_b^{\,2} = 4.0 \times 10^{-11} \times (2.9 \times 10^{-11})^2 = 3.3\overline{6} \times 10^{-32} = 3.4 \times 10^{-32}$

(b) $K = [Ca^{2+}][HF]^2[OH^-]^2$

$3.4 \times 10^{-32} = (S)(2S)^2(1.0 \times 10^{-7})^2$

$3.4 \times 10^{-18} = 4S^3$

$S = 9.5 \times 10^{-7}$ mol/L Ca^{2+} = molar solubility of CaF_2 at pH = 7.0

(c) $K = [Ca^{2+}][HF]^2[OH^-]^2$

$3.4 \times 10^{-32} = (S)(2S)^2(1.0 \times 10^{-9})^2$

$3.4 \times 10^{-14} = 4S^3$

$S = 2.0 \times 10^{-5}$ mol/L Ca^{2+} = molar solubility of CaF_2 at pH = 5.0

SUPPLEMENTARY EXERCISES

16.87 (a) KI is neutral, because neither K^+ or I^- is acidic or basic.

(b) CsF is basic, because F^- is basic.

(c) CrI_3 is acidic, because $Cr(H_2O)_6^{\,3+}$ is an acid.

(d) $C_6H_5NH_3Cl$ is acidic, because $C_6H_5NH_3^{\,+}$ is the conjugate acid of $C_6H_5NH_2$.

(e) Na_2CO_3 is basic, because $CO_3^{\,2-}$ is a base.

(f) $Cu(NO_3)_2$ is acidic, because $Cu(H_2O)_6^{\,2+}$ is an acid.

16.89 The presence of K^+ and Cl^- ions does not affect the pH. Total volume of solution = 200. mL + 150. mL = 350. mL.

moles of Na_3PO_4 = 0.200 L × 0.27 mol/L = 5.4×10^{-2} mol

molar concentration of $Na_3PO_4 = \dfrac{5.4 \times 10^{-2}\ \text{mol}}{0.350\ \text{L}} = 0.154$ mol/L

Concentration (mol/L)	$H_2O(l)$ +	$PO_4^{\,3-}(aq)$ $\rightleftharpoons$	$HPO_4^{\,2-}(aq)$ +	$OH^-(aq)$
initial	—	0.154	0	0
change	—	$-x$	$+x$	$+x$
equilibrium	—	$0.154 - x$	x	x

$$K_b = \frac{K_w}{K_a} = \frac{[HPO_4^{\,2-}][OH^-]}{[PO_4^{\,3-}]} = \frac{x^2}{0.154 - x}$$

$$\frac{1.0 \times 10^{-14}}{2.1 \times 10^{-13}} = \frac{x^2}{0.154 - x} = 4.8 \times 10^{-2} \quad (x \text{ cannot be neglected})$$

$x^2 + 4.8 \times 10^{-2}x - 7.4 \times 10^{-3} = 0$

Solving this quadratic yields $x = 6.5 \times 10^{-2}$ mol/L OH^-

pOH = $-\log(6.5 \times 10^{-2}) = 1.19$, pH = 14.00 − 1.19 = 12.81.

16.91 The end point of an acid-base titration is the pH at which the indicator color change is observed. The stoichiometric point is the pH at which an exactly equivalent amount of titrant (in moles) has been added to the solution being titrated.

16.93 $HNO_3 + NaOH \longrightarrow NaNO_3 + H_2O$

$$\text{moles of } HNO_3 = (0.0210\text{ L}) \times \left(\frac{3.0\text{ mol } HNO_3}{1\text{ L}}\right) = 0.063$$

$$\text{moles of NaOH} = (0.0252\text{ L}) \times \left(\frac{2.5\text{ mol NaOH}}{1\text{ L}}\right) = 0.063$$

To within experimental error, there is complete neutralization; therefore, pH = 7.00.

16.95 (a) $\text{pH} = -\log(0.020) = 1.70$

(b) initial moles of $H_3O^+ = 0.020\text{ L} \times 0.020\text{ mol/L} = 4.0 \times 10^{-4}\text{ mol}$

moles of OH^- added $= 0.005\,00\text{ L} \times 0.035\text{ mol/L} = 1.7\bar{5} \times 10^{-4}\text{ mol}$

excess $H_3O^+ = (4.0 \times 10^{-4} - 1.7\bar{5} \times 10^{-4})\text{ mol} = 2.2\bar{5} \times 10^{-4}\text{ mol}$

$$[H_3O^+] = \frac{2.2\bar{5} \times 10^{-4}\text{ mol}}{(0.020 + 0.0050)\text{ L}} = 9.0 \times 10^{-3}\text{ mol/L}$$

$\text{pH} = -\log(9.0 \times 10^{-3}) = 2.04$

(c) moles of OH^- added $= 2 \times 1.7\bar{5} \times 10^{-4}\text{ mol} = 3.5\bar{0} \times 10^{-4}\text{ mol}$

excess $H_3O^+ = (4.0 \times 10^{-4} - 3.5\bar{0} \times 10^{-4})\text{ mol} = 5.\bar{0} \times 10^{-5}\text{ mol}$

$$[H_3O^+] = \frac{5.\bar{0} \times 10^{-5}\text{ mol}}{0.030\text{ L}} = 1.\bar{7} \times 10^{-3}\text{ mol/L}$$

$\text{pH} = -\log(1.\bar{7} \times 10^{-3}) = 2.8$

(d) moles of OH^- added $= 3 \times 1.7\bar{5} \times 10^{-4} = 5.25 \times 10^{-4}\text{ mol}$

excess $OH^- = (5.25 \times 10^{-4} - 4.0 \times 10^{-4})\text{ mol} = 1.25 \times 10^{-4}\text{ mol}$

$$[OH^-] = \frac{1.25 \times 10^{-4}\text{ mol}}{0.035\text{L}} = 3.5\bar{7} \times 10^{-3}\text{ mol/L}$$

$\text{pOH} = -\log(3.5\bar{7} \times 10^{-3}) = 2.4, \quad \text{pH} = 14.0 - 2.4 = 11.6$

(e) moles of OH^- added $= 4 \times 1.75 \times 10^{-4}\text{ mol} = 7.00 \times 10^{-4}\text{ mol}$

excess $OH^- = (7.00 \times 10^{-4} - 4.0 \times 10^{-4})\text{ mol} = 3.0 \times 10^{-4}\text{ mol}$

$$[OH^-] = \frac{3.0 \times 10^{-4}\text{ mol}}{0.040\text{ L}} = 7.5 \times 10^{-3}\text{ mol/L}$$

$\text{pOH} = -\log(7.5 \times 10^{-3}) = 2.12, \quad \text{pH} = 14.00 - 2.12 = 11.88$

(f)
$$\text{volume of KOH} = (4.0 \times 10^{-4}\text{ mol HCl}) \times \left(\frac{1\text{ mol KOH}}{1\text{ mol HCl}}\right) \times \left(\frac{1\text{ L KOH}}{0.035\text{ mol KOH}}\right) = 0.011\bar{4}\text{ L KOH} = 11\text{ mL KOH}$$

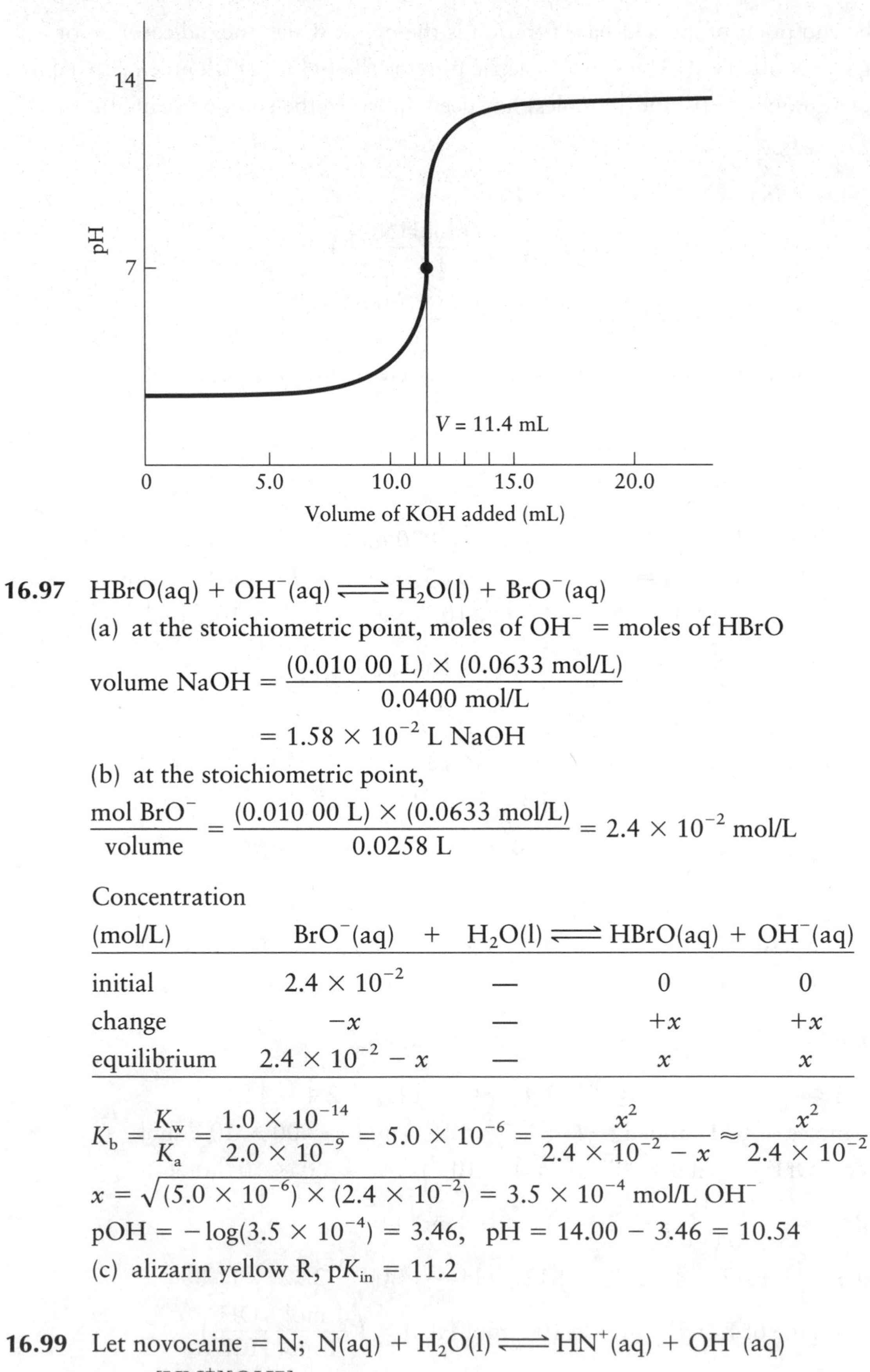

16.97 $HBrO(aq) + OH^-(aq) \rightleftharpoons H_2O(l) + BrO^-(aq)$

(a) at the stoichiometric point, moles of OH^- = moles of HBrO

$$\text{volume NaOH} = \frac{(0.010\ 00\ \text{L}) \times (0.0633\ \text{mol/L})}{0.0400\ \text{mol/L}}$$

$$= 1.58 \times 10^{-2}\ \text{L NaOH}$$

(b) at the stoichiometric point,

$$\frac{\text{mol BrO}^-}{\text{volume}} = \frac{(0.010\ 00\ \text{L}) \times (0.0633\ \text{mol/L})}{0.0258\ \text{L}} = 2.4 \times 10^{-2}\ \text{mol/L}$$

Concentration (mol/L)	$BrO^-(aq)$	+ $H_2O(l)$ ⇌	$HBrO(aq)$	+ $OH^-(aq)$
initial	2.4×10^{-2}	—	0	0
change	$-x$	—	$+x$	$+x$
equilibrium	$2.4 \times 10^{-2} - x$	—	x	x

$$K_b = \frac{K_w}{K_a} = \frac{1.0 \times 10^{-14}}{2.0 \times 10^{-9}} = 5.0 \times 10^{-6} = \frac{x^2}{2.4 \times 10^{-2} - x} \approx \frac{x^2}{2.4 \times 10^{-2}}$$

$x = \sqrt{(5.0 \times 10^{-6}) \times (2.4 \times 10^{-2})} = 3.5 \times 10^{-4}$ mol/L OH^-

$pOH = -\log(3.5 \times 10^{-4}) = 3.46$, $pH = 14.00 - 3.46 = 10.54$

(c) alizarin yellow R, $pK_{in} = 11.2$

16.99 Let novocaine = N; $N(aq) + H_2O(l) \rightleftharpoons HN^+(aq) + OH^-(aq)$

$$K_b = \frac{[HN^+][OH^-]}{[N]}$$

$pK_a = pK_w - pK_b = 14.00 - 5.05 = 8.95$

$$pH = pK_a + \log\left(\frac{[N]}{[HN^+]}\right)$$

$$\log\left(\frac{[N]}{[HN^+]}\right) = pH - pK_a = 7.4 - 8.95 = -1.5\overline{5}$$

Therefore, the ratio of the concentrations of novocaine and its conjugate acid is $[N]/[HN^+] = 10^{-1.55} = 2.8 \times 10^{-2}$

16.101 Equilibrium equation: $HCO_3^-(aq) + H_2O(l) \rightleftharpoons H_3O^+(aq) + CO_3^{2-}(aq)$

$$K_a = \frac{[H_3O^+][CO_3^{2-}]}{[HCO_3^-]}$$

$$pH = pK_a + \log\left(\frac{[CO_3^{2-}]}{[HCO_3^-]}\right)$$

$$10 = -\log(5.6 \times 10^{-11}) + \log\left(\frac{[CO_3^{2-}]}{[HCO_3^-]}\right)$$

$$\log\left(\frac{[CO_3^{2-}]}{[HCO_3^-]}\right) = 10 - 10.25 = -0.25$$

$$\frac{[CO_3^{2-}]}{[HCO_3^-]} = 0.56$$

Prepare a solution of Na_2CO_3 and $NaHCO_3$ in a molar ratio of 0.56 : 1.

16.103 (a) $BaSO_4(s) \rightleftharpoons Ba^{2+}(aq) + SO_4^{2-}(aq)$

$K_{sp} = [Ba^{2+}][SO_4^{2-}] = S \times S = S^2$

$1.1 \times 10^{-10} = S^2$

$S = 1.0 \times 10^{-5}$ mol/L = molar solubility of $BaSO_4$

(b) $1.1 \times 10^{-10} = (S + 2.0 \times 10^{-4}) \times (S) \approx (2.0 \times 10^{-4})S$

$S = 5.5 \times 10^{-7}$ mol/L

= molar solubility of $BaSo_4$ in 2.0×10^{-4} M $Ba(NO_3)_2$

(c) mass of $BaSO_4 = 5.5 \times 10^{-7}\ \text{mol/L} \times 10.0\ \text{L} \times \left(\frac{233.39\ \text{g}\ BaSO_4}{1\ \text{mol}\ BaSO_4}\right)$

$= 1.3 \times 10^{-3}$ g or 1.3 mg $BaSO_4$

16.105 moles of $Ag^+ = (0.1000\ \text{L}) \times (1.0 \times 10^{-4}\ \text{mol/L})$

$= 1.0 \times 10^{-5}$ mol Ag^+ in 0.200 L

$$\text{moles of } CO_3^{2-} = (0.100 \text{ L}) \times (1.0 \times 10^{-4} \text{ mol/L})$$
$$= 1.0 \times 10^{-5} \text{ mol } CO_3^{2-} \text{ in } 0.200 \text{ L}$$

$$Ag_2CO_3(s) \rightleftharpoons 2\,Ag^+(aq) + CO_3^{2-}(aq)$$

$$[Ag^+]^2[CO_3^{2-}] = Q_{sp}$$

$$\left(\frac{1.0 \times 10^{-5}}{0.200}\right)^2\left(\frac{1.0 \times 10^{-5}}{0.200}\right) = 1.25 \times 10^{-13} = Q_{sp}$$

Because Q_{sp} (1.25×10^{-13}) is less than K_{sp} (6.2×10^{-12}), no precipitate will form.

16.107 (1) First stoichiometric point: $OH^-(aq) + H_2SO_4(aq) \longrightarrow H_2O(l) + HSO_4^-(aq)$. Then, because the volume of the solution has doubled, $[HSO_4^-] = 0.10$ M,

Concentration (mol/L)	$HSO_4^-(aq)$ +	$H_2O(l) \rightleftharpoons$	$SO_4^{2-}(aq)$ +	$H_3O^+(aq)$
initial	0.10	—	0	0
change	$-x$	—	$+x$	$+x$
equilibrium	$0.10 - x$	—	x	x

$$K_{a2} = 0.012 = \frac{x^2}{0.10 - x}$$

$$0.0012 - 0.012x = x^2$$

$$x^2 + 0.012x - 0.0012 = 0$$

$$x = \frac{-0.012 + \sqrt{(0.012)^2 + (4) \times (0.0012)}}{2}$$

$$x = H_3O^+ \; 0.029 \text{ mol/L } H_3O^+$$

$$pH = -\log(0.029) = 1.54$$

(2) Second stoichiometric point: $HSO_4^-(aq) + OH^-(aq) \longrightarrow SO_4^{2-}(aq) + H_2O(l)$. Then, because the volume of the solution has increased by an equal amount again,

Concentration (mol/L)	$SO_4^{2-}(aq)$ +	$H_2O(l) \rightleftharpoons$	$HSO_4^-(aq)$ +	$OH^-(aq)$
initial	0.067	—	0	1×10^{-7}
change	$-x$	—	$+x$	$+x$
equilibrium	$0.067 - x$	—	x	$1 \times 10^{-7} + x$

$$K_b = 8.3 \times 10^{-13} = \frac{(x)(1 \times 10^{-7} + x)}{0.067 - x}$$

$$(5.6 \times 10^{-14}) - (8.3 \times 10^{-13})x = (1 \times 10^{-7})x + x^2$$

$$x^2 + (1 \times 10^{-7})x - (5.6 \times 10^{-14}) = 0$$

$$x = \frac{-(1 \times 10^{-7}) + \sqrt{(1 \times 10^{-7})^2 + (4) \times (5.6 \times 10^{-14})}}{2}$$

$x = 1.9 \times 10^{-7}$ mol/L HSO_4^-

$(1.0 \times 10^{-7}) + (1.9 \times 10^{-7}) = 2.9 \times 10^{-7}$ mol/L OH^-

$pOH = -\log(2.9 \times 10^{-7}) = 6.54, \quad pH = 14.00 - 6.54 = 7.46$

16.109 (a) $K_{sp} = 1.1 \times 10^{-10}$ for $BaSO_4$

$[Ba^{2+}][SO_4^{2-}] = 1.1 \times 10^{-10} = S \times S$

$S = 1.0 \times 10^{-5}$ mol/L SO_4^{2-} for $BaSO_4$ to precipitate.

(b) $K_{sp} = 1.6 \times 10^{-8}$ for $PbSO_4$

$[Pb^{2+}][SO_4^{2-}] = 1.6 \times 10^{-8} = S \times S$

$S = 1.3 \times 10^{-4}$ mol/L SO_4^{2-} for $PbSO_4$ to precipitate.

(c) $[Ba^{2+}][1.3 \times 10^{-4} \text{ mol/L}] = 1.1 \times 10^{-10}$

$$[Ba^{2+}] = \frac{1.1 \times 10^{-10}}{1.3 \times 10^{-4} \text{ mol/L}} = 8.5 \times 10^{-7} \text{ mol/L}$$

APPLIED EXERCISES

16.111 (a) The equilibria involved are

$CO_2(g) + H_2O(l) \rightleftharpoons H_2CO_3(aq)$

$H_2CO_3(aq) + H_2O(l) \rightleftharpoons H_3O^+(aq) + HCO_3^-(aq)$

Hyperventilation decreases the amount of $CO_2(g)$, thus driving the first equilibrium to the left, which in turn drives the second equilibrium to the left. Note that $[HCO_3^-]$ is controlled by excretion in urine, but this mechanism is slow; thus we will ignore it here. Therefore, both $[HCO_3^-]$ and $[H_2CO_3]$ are diminished, but it is not immediately apparent how their ratio is affected. We can analyze the effect on the ratio through the use of the Henderson-Hasselbalch equation:

$$pH = -\log[H_3O^+] = pK_{a1} + \log\left(\frac{[HCO_3^-]}{[H_2CO_3]}\right)$$

With loss of CO_2 during hyperventilation, there is a decrease in $[H_3O^+]$, which corresponds to an increased pH, which in turn corresponds to an increased $[HCO_3^-]/[H_2CO_3]$ ratio, or a decreased $[H_2CO_3]/[HCO_3^-]$ ratio.

(b) Paramedics should anticipate alkalosis.

16.113 From the Henderson-Hasselbalch equation, the effective buffer range in water is $pH = pK_a, \pm \log 10 = 6.37 \pm 1.00$, or from 5.37 to 7.37.

16.115 $CaF_2(s) \rightleftharpoons Ca^{2+}(aq) + 2\, F^-(aq)$

$[Ca^{2+}][F^-]^2 = Q_{sp}$

$(2 \times 10^{-4}) \times (5 \times 10^{-5})^2 = 5 \times 10^{-13} = Q_{sp}$

Because Q_{sp} (5×10^{-13}) is less than K_{sp} (4.0×10^{-11}), no precipitate will form.

INTEGRATED EXERCISES

16.117 (a) $H_2SO_4(aq) + Ba(OH)_2(aq) \longrightarrow BaSO_4(s) + 2\ H_2O(l)$

(b) The bulb goes out when the $BaSO_4$ has precipitated.

$$0.02500\ L \times \left(\frac{0.010\ mol\ H_2SO_4}{L}\right) \times \left(\frac{1\ mol\ Ba(OH)_2}{1\ mol\ H_2SO_4}\right) \times \left(\frac{1\ L}{0.010\ mol\ Ba(OH)_2}\right) = 0.025\ L = 25\ mL$$

(c) At the beginning of the titration, $H^+(aq)$, $HSO_4^-(aq)$, and $SO_4^{2-}(aq)$ are available to carry the current. As the titration proceeds, $BaSO_4(s)$ precipitates, leaving $H^+(aq)$ and $OH^-(aq)$ to carry the current. The concentrations of H^+ and OH^- are very small ($\sim 10^{-7}$ mol/L) at the stoichiometric point. As more $Ba(OH)_2$ is added, Ba^{2+} and OH^- concentrations increase and these ions carry the current.

16.119 Let NaB = sodium barbiturate, HB = barbituric acid

$$[B^-] = [NaB] = \left(\frac{10.0\ mg}{250.\ mL}\right) \times \left(\frac{1\ mL}{10^{-3}\ L}\right) \times \left(\frac{10^{-3}\ g}{1\ mg}\right) \times \left(\frac{1\ mol}{150.\ g}\right)$$
$$= 2.67 \times 10^{-4}\ mol/L$$

pH = 7.71, pOH = 6.29, 5.1×10^{-7} mol/L OH^-

Concentration (mol/L)	$B^-(aq)$	+	$H_2O(l)$	$\rightleftharpoons$	$OH^-(aq)$	+	$HB(aq)$
initial	2.67×10^{-4}		—		0		0
change	-5.1×10^{-7}		—		$+5.1 \times 10^{-7}$		$+5.1 \times 10^{-7}$
equilibrium	$2.67 \times 10^{-4} - 5.1 \times 10^{-7}$		—		5.1×10^{-7}		5.1×10^{-7}

Assume $[OH^-] = [HB]$

(a) $\%\ \text{protonation} = \dfrac{[HB]}{[B^-]_0} \times 100\% = \dfrac{5.1 \times 10^{-7}}{2.67 \times 10^{-4}} \times 100\% = 0.19\%$

(b) $K_b = \dfrac{K_w}{K_a} = \dfrac{[OH^-][HB]}{[B^-]}$

$$K_a = \frac{K_w[B^-]}{[OH^-][HB]} = \frac{(1.0 \times 10^{-14}) \times (2.67 \times 10^{-4} - 5.1 \times 10^{-7})}{(5.1 \times 10^{-7})^2} = 1.0 \times 10^{-5}$$

16.121 (a) The Case Study in Chapter 15 indicates that two reactions are possible between SO_2 and water. They are

$SO_2(g) + 2\ H_2O(l) \longrightarrow H_3O^+(aq) + HSO_3^-(aq)$ $(H_2SO_3 + H_2O)$

and $2\ SO_2(g) + O_2(g) \longrightarrow 2\ SO_3(g)$

$SO_3(g) + 2\ H_2O(l) \longrightarrow H_3O^+(aq) + HSO_4^-(aq)$

Because particulate matter and aerosols are needed for the second reaction, assume the first reaction occurs.

(b) $NaOH(aq) + HCl(aq) \longrightarrow H_2O(l) + NaCl(aq)$

$$0.0302\ L \times \left(\frac{0.000\ 15\ mol\ HCl}{L}\right) \times \left(\frac{1\ mol\ NaOH}{1\ mol\ HCl}\right)$$

$$= 4.5\bar{3} \times 10^{-6}\ mol\ NaOH\ titrated\ with\ HCl$$

The amount of NaOH that reacted with SO_2 is needed. By subtracting the initial concentration of NaOH from that reacted with HCl, we find the amount of NaOH remaining, and thus reacted, with SO_2.

$$0.000\ 10\ mol/L\ NaOH - \frac{4.5\bar{3} \times 10^{-6}\ mol\ NaOH}{0.050\ 00\ L} =$$

9.4×10^{-6} mol/L NaOH or

$$\frac{9.4 \times 10^{-6}\ mol\ NaOH}{L} \times 0.050\ 00\ L = 4.7 \times 10^{-7}\ mol\ NaOH$$

(c) $2\ NaOH(aq) + H_2SO_3(aq) \longrightarrow Na_2SO_3(aq) + 2\ H_2O(l)$

Thus $4.7 \times 10^{-7}\ mol\ NaOH \times \left(\frac{1\ mol\ H_2SO_3}{2\ mol\ NaOH}\right) = 2.3\bar{5} \times 10^{-7}\ mol\ H_2SO_3$ (or SO_2 gas). This is the amount (moles) of SO_2 in the air. If we find the moles of air, then divide the moles of SO_2 by the moles of air and multiply this ratio by 10^6 (the "million" in ppm), will have the ppm of SO_2 in the air.

$$n = \frac{PV}{RT} = \frac{(753\ Torr)(3.0\ L/h \times 2.5\ h)}{(62.36\ L\cdot Torr/mol\cdot K)(295\ K)} = 0.30\bar{7}\ mol\ air$$

$$\frac{2.3\bar{5} \times 10^{-7}\ mol\ SO_2}{0.30\bar{7}\ mol\ air} \times 10^6 = 0.77\ ppm$$

CHAPTER 17
THE DIRECTION OF CHEMICAL CHANGE

EXERCISES

Entropy

17.1 (a) HBr(g) [Br contains more elementary particles than F does]
(b) NH_3(g) [NH_3 is a more complex structure]
(c) I_2(l) [liquids have more disorder than solids do]
(d) Ar(g) at 1.00 atm [there is more randomness in a gas at low pressure]

17.3 C(s) < H_2O(s) < H_2O(l) < H_2O(g)
Ice has a more complex crystalline structure than carbon (either as diamond or graphite) and so has a higher entropy. H_2O(l) has more disorder than H_2O(s); in turn, H_2O(g) has more disorder than H_2O(l).

17.5 Entropy increases, because liquids are more disordered than solids.

17.7 (a) The entropy decreases. The number of molecules of gas decreases as a result of the reaction, so the disorder of the system decreases.
(b) The gaseous state is more disordered than the solid state; the entropy of the system increases.
(c) As water cools, the amplitude of the random motion of the molecules decreases; the entropy decreases.

17.9 (a) $$\Delta S° = \frac{\Delta H_{\text{freeze}}°}{T_f} = \frac{-\Delta H_{\text{fus}}°}{T_f} = \frac{-6.01\ \text{kJ/mol}}{273.2\ \text{K}} = \frac{-6.01 \times 10^3\ \text{J/mol}}{273.2\ \text{K}}$$
$$= -22.0\ \text{J/K·mol}$$

(b) $$\Delta S_{\text{sys}}° = n\left(\frac{\Delta H_{\text{vap}}°}{T_b}\right)$$

$$n = (50.0\ \text{g}) \times \left(\frac{1\ \text{mol}}{46.07\ \text{g}}\right) = 1.08\overline{5}\ \text{mol}$$

$$\Delta H_{vap}° = 43.5 \text{ kJ/mol}$$

$$\Delta S_{sys}° = (1.08\bar{5} \text{ mol}) \times \left(\frac{4.35 \times 10^4 \text{ J/mol}}{195.3}\right) = 242 \text{ J/K}$$

17.11 (a) $\Delta S_r° = 2S_m°(H_2O, g) - [2S_m°(H_2, g) + S_m°(O_2, g)]$
$= [2 \times 188.83 - (2 \times 130.68 + 205.14)]$ J/K·mol
$= -88.84$ J/K·mol

The entropy change is negative because the number of moles of gas has decreased by one.

(b) $\Delta S_r° = 2S_m°(CO_2, g) - [2S_m°(CO, g) + S_m°(O_2, g)]$
$= [2 \times 213.74 - (2 \times 197.67 + 205.14)]$ J/K·mol
$= (427.48 - 395.34 - 205.14)$ J/K·mol $= -173.00$ J/K·mol

The entropy change is negative because the number of moles of gas has decreased by one.

(c) $\Delta S_r° = S_m°(CaO, s) + S_m°(CO_2, g) - S_m°(CaCO_3, s)$
$= (39.75 + 213.74 - 92.9)$ J/K·mol $= 160.6$ J/K·mol

The entropy change is positive because the number of moles of gas has increased by one.

(d) $\Delta S_r° = 3S_m°(KClO_4, s) + S_m°(KCl, s) - 4S_m°(KClO_3, s)$
$= (3 \times 151.0 + 82.59 - 4 \times 143.1)$ J/K·mol
$= -36.8$ J/K·mol

It is not immediately apparent, but the four moles of solid products are more ordered than the four moles of solid reactants.

Entropy Change in the Surroundings

17.13 (a) ΔH is negative, because heat leaves the coffee as it cools. Therefore, $\Delta S_{surr} = -\Delta H/T$ is positive. Although ΔS of the system is negative, ΔS_{surr} is positive and of a greater magnitude, because the surroundings are at a lower temperature than the system (coffee). Therefore, $\Delta S_{tot} = \Delta S + \Delta S_{surr}$ is positive and the process is spontaneous.

(b) No energy in the form of heat is transferred to or from the surroundings, so $\Delta S_{surr} = 0$. But ΔS of the system is positive as a result of the greater disorder generated by the mixing of the two substances. Thus, $\Delta S_{tot} = \Delta S_{surr} + \Delta S = 0 + \Delta S$ is positive and the process is spontaneous.

(c) When gasoline burns, both ΔS of the system and ΔS_{surr} are positive; thus, ΔS_{tot} is positive. ΔS of the system is positive, because the number of moles of gas increases as CO_2(g) and H_2O(g) are formed by the combustion process. ΔS_{surr} is positive, because the reaction is exothermic and much heat is transferred to the surroundings:

$$\Delta S_{surr} = \frac{-\Delta H}{T} = + \ (\Delta H = -)$$

Because $\Delta S_{tot} = \Delta S + \Delta S_{surr}$ is positive, the process is spontaneous.

(d) The interdiffusion of two gases results in an increase in disorder: the gases are all mixed up; therefore, ΔS of the system is positive. $\Delta S_{surr} = 0$, because there is no energy transferred as heat; thus, $\Delta S_{tot} = \Delta S + \Delta S_{surr} = \Delta S$ is positive and the process is spontaneous.

17.15 A vapor is condensing into a liquid. ΔH is negative, because heat is released to the surroundings. The disorder of the surroundings increases, so ΔS_{surr} is positive, whereas ΔS of the system is negative. The process is spontaneous, ΔS_{tot} is positive. Thus, $\Delta S_{surr} > \Delta S$.

17.17 (a) 100 W = 100 J/s, $\Delta S_{surr} = \dfrac{-\Delta H}{T}$

$$\text{rate of entropy generation} = \frac{\text{rate of heat generation}}{T} = \frac{-(-100 \text{ J/s})}{293 \text{ K}}$$

$$= 0.341 \text{ J/K·d}$$

(b) $\Delta S_{surr} = 0.341 \text{ J/K·s} \times \left(\dfrac{3600 \text{ s}}{1 \text{ h}}\right) \times \left(\dfrac{24 \text{ h}}{1 \text{ d}}\right) = 2.95 \times 10^4 \text{ J/K·d}$

(c) Less, because in the equation $\Delta S_{surr} = -\Delta H/T$, if T is higher, then ΔS_{surr} is smaller.

17.19 (a) $\Delta S_{sys} = \dfrac{+\Delta H}{T} = \dfrac{+5 \text{ J}}{298 \text{ K}} = +0.01\overline{7} \text{ J/K} = +0.02 \text{ J/K}$

(b) $\Delta S_{sys} = \dfrac{+5 \text{ J}}{373 \text{ K}} = +0.01\overline{3} \text{ J/K} = +0.01 \text{ J/K}$

(c) As seen in parts (a) and (b), the magnitude of the entropy change is greater when the block is at 25°C. The same amount of heat has a greater effect on entropy changes at lower temperature. At high temperatures, matter is already more chaotic.

17.21 In each case, $\Delta S_{tot}° = 0$, because these are equilibrium processes. Therefore, because

$$\Delta S_{tot}° = \Delta S_{surr}° + \Delta S_{sys}° = 0$$
$$\Delta S_{sys}° = -\Delta S_{surr}°$$

(a) $\Delta S_{\text{surr}}{}^{\circ} = \dfrac{-1.0 \text{ mol} \times \Delta H_{\text{vap}}{}^{\circ}}{T_{\text{b}}} = \dfrac{-1.0 \text{ mol} \times 8.2 \times 10^{3} \text{ J/mol}}{111.7 \text{ K}}$

$= -73 \text{ J/K}$

$\Delta S_{\text{tot}}{}^{\circ} = 0,\ \Delta S_{\text{sys}}{}^{\circ} = +73 \text{ J/K}$

(b) $\Delta S_{\text{surr}}{}^{\circ} = \dfrac{-1.0 \text{ mol} \times \Delta H_{\text{fus}}{}^{\circ}}{T_{\text{f}}} = \dfrac{-1.0 \text{ mol} \times 4.60 \times 10^{3} \text{ J/mol}}{158.7 \text{ K}}$

$= -29 \text{ J/K}$

$\Delta S_{\text{tot}}{}^{\circ} = 0,\ \Delta S_{\text{sys}}{}^{\circ} = +29 \text{ J/K}$

(c) $\Delta S_{\text{surr}}{}^{\circ} = +29 \text{ J/K}$

$\Delta S_{\text{sys}}{}^{\circ} = -29 \text{ J/K}$

Freezing is the reverse of melting.

17.23 $\Delta S_{\text{surr}}{}^{\circ} = \dfrac{-\Delta H^{\circ}}{T} = \dfrac{-1.96 \times 10^{3} \text{ J}}{298 \text{ K}} = -6.58 \text{ J/K}$

$\Delta S^{\circ} = -109.58 \text{ J/K}$

$\Delta S_{\text{tot}}{}^{\circ} = \Delta S^{\circ} + \Delta S_{\text{surr}}{}^{\circ} = (-109.58 - 6.58) \text{ J/K} = -116.16 \text{ J/K}$

According to the second law, this process is not spontaneous: $\Delta S_{\text{tot}}{}^{\circ}$ is negative; that is, there has been a decrease in disorder.

17.25 Under constant temperature and pressure conditions, a negative value of ΔG_{r} corresponds to spontaneity.

$$\Delta G_{\text{r}} = \Delta H_{\text{r}} - T\Delta S_{\text{r}}$$

Frequently, the magnitude of ΔH_{r} is much greater than the magnitude of $T\Delta S_{\text{r}}$, so no matter what the sign of ΔS_{r}, ΔG_{r} will be negative if ΔH_{r} is negative, and the reaction will be spontaneous.

Free Energy

17.27 (a) $\Delta G = \Delta H - T\Delta S = (-) - (+)(+) = -$

(b) $\Delta G = (+) - (+)(+) = ?$ Sign cannot be predicted. It depends on the magnitudes of ΔH, T, and ΔS.

(c) A temperature change affects the sign in (b) by altering the magnitude of the $-T\Delta S$ term in ΔG.

17.29 $\Delta S_{\text{vap}}{}^{\circ} = \dfrac{\Delta H_{\text{vap}}{}^{\circ}}{T_{\text{b}}}$

$T_{\text{b}} = \dfrac{\Delta H_{\text{vap}}{}^{\circ}}{\Delta S_{\text{vap}}{}^{\circ}} = \dfrac{2.04 \times 10^{4} \text{ J/mol}}{85.4 \text{ J/K}\cdot\text{mol}} = 239 \text{ K or } -34^{\circ}\text{C}$

17.31 $\Delta H_{vap}° = T_b \times \Delta S_{vap}°$

(a) $\Delta H_{vap}° = 353.2\ K \times 85\ J/K\cdot mol = 3.0 \times 10^4\ J/mol = 30.\ kJ/mol$

(b) $$\Delta S_{surr}° = \frac{-n\Delta H_{vap}°}{T_b} = \frac{-(10.\ g) \times \left(\frac{1\ mol}{78.11\ g}\right) \times (3.0 \times 10^4\ J/mol)}{353.2\ K}$$
$$= -11\ J/K$$

Standard Reaction Free Energies

17.33 $\Delta S_r°$ data from Exercise 17.11. Assume $T = 298$ K.

(a) $\Delta H_r° = 2\Delta H_f°(H_2O, g) = 2 \times (-241.82\ kJ/mol) = -483.64\ kJ/mol$

$\Delta G_r° = \Delta H_r° - T\Delta S_r°$

$= -483.64\ kJ/mol - 298\ K \times (-0.088\ 84\ kJ/K\cdot mol)$

$= -457.2\ kJ/mol$

The negative sign indicates that the reaction is spontaneous (has $K > 1$) under standard conditions. The large magnitude of $\Delta G_r°$ implies that a temperature change is not likely to affect the spontaneity, but it is possible at very high temperatures.

(b) $\Delta H_r° = 2 \times (-393.51\ kJ/mol) - 2 \times (-110.53\ kJ/mol)$

$= -565.96\ kJ/mol$

$\Delta G_r° = \Delta H_r° - T\Delta S_r° = -565.96\ kJ/mol - 298\ K \times (-0.173\ 00\ kJ/K\cdot mol)$

$= -514.4\ kJ/mol$

The negative sign indicates the reaction is spontaneous (has $K > 1$) under the standard conditions. The large magnitude of $\Delta G_r°$ implies that a temperature change is not likely to affect the spontaneity, but it is possible at very high temperatures.

(c) $\Delta H_r° = -635.09\ kJ/mol - 393.51\ kJ/mol - (-1206.9\ kJ/mol)$

$= +178.3\ kJ/mol$

$\Delta G_r° = \Delta H_r° - T\Delta S_r° = +178.3\ kJ/mol - 298\ K \times 0.1606\ kJ/K\cdot mol$

$= +130.4\ kJ/mol$

The positive sign of $\Delta G°$ implies that the reaction is not spontaneous (has $K < 1$) under standard conditions, but the magnitude indicates that, at a higher temperature, there could be a change in sign.

(d) $\Delta H_r° = 3 \times (-432.75\ kJ/mol) + 1 \times (-436.75\ kJ/mol)$
$- 4 \times (-397.73\ kJ/mol)$

$= -144.08\ kJ/mol$

$\Delta G_r° = \Delta H_r° - T\Delta S_r° = -144.08\ kJ/mol - (298.2\ K)(-0.0368\ kJ/K\cdot mol)$

$= -133.1\ kJ/mol$

The negative sign implies that the reaction is spontaneous under standard conditions.

17.35 (a) $\frac{1}{2}N_2(g) + \frac{3}{2}H_2(g) \longrightarrow NH_3(g)$

$\Delta S_f^\circ = 1 \times 192.45\ J/K\cdot mol - (\frac{1}{2} \times 191.61\ J/K\cdot mol + \frac{3}{2} \times 130.68\ J/K\cdot mol)$
$= -99.38\ J/K\cdot mol = -0.099\ 38\ kJ/K\cdot mol$

$\Delta G_f^\circ = \Delta H_f^\circ - T\Delta S_f^\circ = -46.11\ kJ/mol - (298\ K)(-0.099\ 38\ kJ/K\cdot mol)$
$= -16.5\ kJ/mol$

(b) $H_2(g) + \frac{1}{2}O_2(g) \longrightarrow H_2O(g)$

$\Delta S_f^\circ = 1 \times 188.83\ J/K\cdot mol - (1 \times 130.68\ J/K\cdot mol + \frac{1}{2} \times 205.14\ J/K\cdot mol)$
$= -44.42\ J/K\cdot mol$

$\Delta G_f^\circ = \Delta H_f^\circ - T\Delta S_f^\circ$
$= -241.82\ kJ/mol - 298\ K \times (-0.044\ 42\ kJ/K\cdot mol) = -228.6\ kJ/mol$

(c) $C(gr) + \frac{1}{2}O_2(g) \longrightarrow CO(g)$

$\Delta S_f^\circ = 1 \times 197.67\ J/K\cdot mol - (1 \times 5.740\ J/K\cdot mol + \frac{1}{2} \times 205.14\ J/K\cdot mol)$
$= 89.36\ J/K\cdot mol = 0.089\ 36\ kJ/K\cdot mol$

$\Delta G_f^\circ = \Delta H_f^\circ - T\Delta S_f^\circ = -110.53\ kJ/mol - (298\ K)(0.089\ 36\ kJ/K\cdot mol)$
$= -137.2\ kJ/mol$

(d) $\frac{1}{2}N_2(g) + O_2(g) \longrightarrow NO_2(g)$

$\Delta S_f^\circ = 1 \times 240.06\ J/K\cdot mol - (\frac{1}{2} \times 191.61\ J/K\cdot mol + 1 \times 205.14\ J/K\cdot mol)$
$= -60.88\ J/K\cdot mol = -0.060\ 88\ kJ/K\cdot mol$

$\Delta G_f^\circ = \Delta H_f^\circ - T\Delta S_f^\circ$
$= 33.18\ kJ/mol - (298\ K)(-0.060\ 88\ kJ/K\cdot mol) = 51.3\ kJ/mol$

17.37 In each case, the decomposition reaction is the reverse of the formation reaction; therefore, in each case, $\Delta G_r^\circ = -\Delta G_f^\circ(\text{cmpd})$. A negative value for ΔG_r° (decomposition) implies instability.

(a) $\Delta G_r^\circ = -\Delta G_f^\circ(PCl_5, g) = +305.0\ kJ/mol$; therefore, stable

(b) $\Delta G_r^\circ = -\Delta G_f^\circ(HCN, g) = -124.7\ kJ/mol$; therefore, unstable and will decompose

(c) $\Delta G_r^\circ = -\Delta G_f^\circ(NO, g) = -86.55\ kJ/mol$; therefore, unstable and will decompose

(d) $\Delta G_r^\circ = -\Delta G_f^\circ(SO_2, g) = +300.19\ kJ/mol$; therefore, stable

17.39 In each case, it is the $-T\Delta S_r^\circ$ term for the decomposition that determines the effect of temperature. $\Delta S_r^\circ(\text{decomposition}) = -\Delta S_r^\circ(\text{formation})$ and $-T\Delta S_r^\circ(\text{decomposition}) = T\Delta S_r^\circ(\text{formation})$. So if $\Delta S_r^\circ(\text{formation})$ is negative, the compound becomes more unstable when the temperature is raised.

(a) $P(s) + \frac{5}{2} Cl_2(g) \rightleftharpoons PCl_5(g)$

$\Delta S_r°(\text{formation}) =$

$$1 \times 364.6 \text{ J/K·mol} - (1 \times 41.09 \text{ J/K·mol} + \tfrac{5}{2} \times 223.07 \text{ J/K·mol})$$

$\Delta S_r°(\text{formation}) = -234.2$ J/K·mol; therefore, more unstable

(b) $\frac{1}{2} H_2(g) + C(s) + \frac{1}{2} N_2(g) \rightleftharpoons HCN(g)$

$\Delta S_r°(\text{formation}) = 1 \times 201.78 \text{ J/K·mol} - (\tfrac{1}{2} \times 130.68 \text{ J/K·mol}$
$+ 1 \times 5.740 \text{ J/K·mol} + \tfrac{1}{2} \times 191.61 \text{ J/K·mol})$

$\Delta S_r°(\text{formation}) = 34.90$ J/K·mol; therefore, more stable

(c) $\frac{1}{2} N_2(g) + \frac{1}{2} O_2(g) \rightleftharpoons NO(g)$

$\Delta S_r°(\text{formation}) =$

$$1 \times 210.76 \text{ J/K·mol} - (\tfrac{1}{2} \times 191.61 \text{ J/K·mol} + \tfrac{1}{2} \times 205.14 \text{ J/K·mol})$$

$\Delta S_r°(\text{formation}) = 12.38$ J/K·mol; therefore, more stable

(d) $S(s) + O_2(g) \rightleftharpoons SO_2(g)$

$\Delta S_r°(\text{formation}) =$

$$1 \times 248.22 \text{ J/K·mol} - (1 \times 31.80 \text{ J/K·mol} + 1 \times 205.14 \text{ J/K·mol})$$

$\Delta S_r°(\text{formation}) = 11.28$ J/K·mol; therefore, more stable

17.41 (a) $\Delta G_r° = 2 \times \Delta G_f°(SO_3, g) - 2 \times \Delta G_f°(SO_2, g)$

$\Delta G_r° = 2 \times (-371.06 \text{ kJ/mol}) - 2 \times (-300.19 \text{ kJ/mol})$

$\Delta G_r° = -141.74$ kJ/mol; spontaneous

(b) $\Delta G_r° = \Delta G_f°(CaO, s) + \Delta G_f°(CO_2, g) - \Delta G_f°(CaCO_3, s)$

$\Delta G_r° = -604.03 \text{ kJ/mol} - 394.36 \text{ kJ/mol} - (-1128.8 \text{ kJ/mol})$

$= +130.4$; kJ/mol; not spontaneous

(c) $\Delta G_r° = 1 \times \Delta G_f°(SbCl_3, g) - 1 \times \Delta G_f°(SbCl_5, g)$

$\Delta G_r° = 1 \times (-301.2 \text{ kJ/mol}) - 1 \times (-334.29 \text{ kJ/mol})$

$\Delta G_r° = 33.1$ kJ/mol; not spontaneous

(d) $\Delta G_r° = 16 \times \Delta G_f°(CO_2, g) + 18 \times \Delta G_f°(H_2O, l) - 2 \times \Delta G_f°(C_8H_{18}, l)$

$\Delta G_r° = 16 \times (-394.36 \text{ kJ/mol}) + 18 \times (-237.13 \text{ kJ/mol}) - 2 \times (6.4 \text{ kJ/mol})$

$\Delta G_r° = -10\,590.9$ kJ/mol; spontaneous

Free Energy and Composition

17.43 Free energy approaches a minimum for the total reacting system as equilibrium is approached. That is, G_{total} decreases until equilibrium is reached.

17.45 In each case, $\ln K_p = \dfrac{-\Delta G_r°}{RT}$

$\Delta G_r°$ data from Exercise 17.33; $RT = 2.479$ kJ/mol at 25°C

(a) $\ln K_p = \dfrac{-(-457.2\ \text{kJ/mol})}{2.479\ \text{kJ/mol}} = 184.4, \quad K_p = 1.2 \times 10^{80}$

(b) $\ln K_p = \dfrac{-(-514.4\ \text{kJ/mol})}{2.479\ \text{kJ/mol}} = 207.5, \quad K_p = 1.3 \times 10^{90}$

(c) $\ln K_p = \dfrac{-(+130.4\ \text{kJ/mol})}{2.479\ \text{kJ/mol}} = -52.60, \quad K_p = 1.4 \times 10^{-23}$

17.47 (a) $\Delta G_r° = -RT \ln K_p = -8.314\ \text{J/K·mol} \times 400.\ \text{K} \times \ln 41$
$= -1.23 \times 10^4\ \text{J/mol} = -12.3\ \text{kJ/mol}$

(b) $\Delta G_r° = -RT \ln K_p = -8.314\ \text{J/K·mol} \times 700.\ \text{K} \times \ln(3.0 \times 10^4)$
$= -6.00 \times 10^4\ \text{J/mol} = -60.0\ \text{kJ/mol}$

17.49 $\Delta G_r° = 2 \times 86.55\ \text{kJ/mol} = +173.1\ \text{kJ/mol}$

$\ln K_p = \dfrac{-\Delta G_r°}{RT} = \dfrac{-(+173.1\ \text{kJ/mol})}{2.479\ \text{kJ/mol}} = -69.83$

$K_p = 4.7 \times 10^{-31}$

$Q_p > K_p$; there is a tendency to form reactants.

17.51 $Q_p = \dfrac{P_{SO_3}{}^2}{P_{SO_2}{}^2 \cdot P_{O_2}} = \dfrac{(5.00)^2}{(0.026)^2 \times 0.026} = 1.4 \times 10^6$

$Q_p > K_p$, so the reaction is spontaneous toward reactants.

$\Delta G_r = \Delta G_r° + RT \ln Q_p$ where $\Delta G_r° = -RT \ln K_p$

$= -RT \ln K_p + RT \ln Q_p = -RT \ln \left(\dfrac{K_p}{Q_p}\right)$

$\Delta G_r = -(8.314\ \text{J/K·mol}) \times (700.\ \text{K}) \times \ln \left(\dfrac{3.0 \times 10^4}{1.4 \times 10^6}\right)$

$= +22$ KJ/mol; spontaneity confirmed

17.53 $Q_p = \dfrac{P_{NH_3}{}^2}{P_{N_2} \cdot P_{H_2}{}^3} = \dfrac{(63)^2}{1.0 \times (4.2)^3} = 53.\overline{6}$

(a) $\Delta G_r = \Delta G_r° + RT \ln Q_p \quad (\Delta G_r° = -RT \ln K_p)$

$= -RT \ln K_p + RT \ln Q_p$

$\Delta G_r = -RT \ln\left(\dfrac{K_p}{Q_p}\right) = -8.314\ \text{J/K·mol} \times 400.\ \text{K} \times \ln\left(\dfrac{41}{53.\overline{6}}\right)$

$= +0.89\ \text{kJ/mol}$

(b) Because $Q_p > K_p$ and $\Delta G_r > 0$, reactants will be formed.

17.55 (a) $\Delta G_r° = \Delta H_r° - T\Delta S_r°$

$= +57.2\ \text{kJ} - 298\ \text{K} \times (+0.175\ 83\ \text{kJ/K·mol})$

$= 4.8\ \text{kJ/mol}$

(b) $\Delta G_r^\circ = +57.2\text{ kJ} - 348\text{ K} \times (+0.175\ 83\text{ kJ/K}\cdot\text{mol}) = -4.0\text{ kJ/mol}$

(c) $0 = \Delta H_r^\circ - T\Delta S_r^\circ$

$$T = \frac{\Delta H_r^\circ}{\Delta S_r^\circ} = \frac{57.2\text{ kJ/mol}}{0.175\ 83\text{ kJ/K}\cdot\text{mol}} = 325\text{ K} = 52°\text{C}$$

$0 = -RT \ln K_p$

$\ln K_p = 0, \quad K_p = 1$

17.57 (a) $2\ Cu^+(aq) \rightleftharpoons Cu^{2+}(aq) + Cu(s)$

$\Delta G_r^\circ = \Delta G_f^\circ(Cu^{2+}, aq) - 2 \times \Delta G_f^\circ(Cu^+, aq)$

$= 65.49\text{ kJ/mol} - 2 \times 49.98\text{ kJ/mol} = -34.47\text{ kJ/mol}$

Clearly, there is a thermodynamic tendency for this to occur at 25°C.

(b) The tendency is greater at lower temperatures, because ΔH_r° and ΔS_r° are both negative.

17.59 (a) $\Delta G_r^\circ = 2 \times (-200.\text{ kJ/mol}) - 1 \times (-762\text{ kJ/mol}) = +362\text{ kJ/mol}$

The positive sign means this reduction is not possible.

(b) $\Delta G_r^\circ = 1 \times (-396\text{ kJ/mol}) - 1 \times (-762\text{ kJ/mol}) = +366\text{ kJ/mol}$

The positive sign means this reduction is not possible.

SUPPLEMENTARY EXERCISES

17.61 Enthalpy is defined in terms of energy as $H = U + PV$. The change in enthalpy, ΔH, is related to the change in U under constant pressure conditions by the relation

$$\Delta H = \Delta U + P\Delta V = q_p \text{ } (q_p \text{ is the heat effect at constant temperature})$$

Changes in enthalpy under constant pressure conditions are thus able to provide the heat, that is ΔH is a measure of the heat that may be obtained from the system when change occurs at constant pressure. Entropy, on the other hand, is a measure of disorder and its total change in system and surroundings determines the spontaneity of a process. And so, both enthalpy and entropy changes are required; the first gives the heat of a process, the second determines whether a process can occur.

17.63 The $Cl_2(g)$ molecule not only has translational energy but also rotational and vibrational energy. The disordering due to this energy contributes to the overall energy disorder. However, matter is more organized in the molecule, so that the molar entropy of the molecule is not twice that of the atom.

17.65 (a) $\Delta S_{surr} = \frac{-\Delta H}{T} = \frac{-(-1 \times 10^{-3}\ J)}{2 \times 10^{-7}\ K} = 5 \times 10^{3}\ J/K$

(b) $\Delta S_{surr} = \frac{-(-1\ J)}{310\ K} = 3 \times 10^{-3}\ J/K$

17.67 (a) $\Delta S_r° = 2 \times S_m°(H_2O, l) + 1 \times S_m°(O_2, g) - 2 \times S_m°(H_2O_2, l)$

$\Delta S_r° = (2 \times 69.91 + 1 \times 205.14 - 2 \times 109.6)\ J/K \cdot mol = +125.8\ J/K \cdot mol$

The positive value is reasonable because there is one more mole of gas in the products.

(b) $\Delta S_r° = 4 \times S_m°(HF, aq) + 1 \times S_m°(O_2, g) - [2 \times S_m°(F_2, g) + 2 \times S_m°(H_2O, l)]$

$\Delta S_r° = [4 \times (-13.8) + 1 \times 205.14 - (2 \times 202.78 + 2 \times 69.91)]\ J/K \cdot mol$
$= -395.4\ kJ/K \cdot mol$

The number of moles of gas decreases, so entropy decreases.

(c) $\Delta S_r° =$
$1 \times S_m°(CO, g) + 3 \times S_m°(H_2, g) - [1 \times S_m°(CH_4, g) + 1 \times S_m°(H_2O, g)]$

$\Delta S_r° = [1 \times 197.67 + 3 \times 130.68 - (1 \times 186.26 + 1 \times 188.83)]\ J/K \cdot mol$
$= +214.62\ J/K \cdot mol$

The positive sign is reasonable because the products contain two more moles of gas than the reactants.

(d) $\Delta S_r° = 1 \times S_m°(N_2O, g) + 2 \times S_m°(H_2O, g) - 1 \times S_m°(NH_4NO_3, s)$

$\Delta S_r° = (1 \times 219.85 + 2 \times 188.83 - 1 \times 151.08)\ J/K \cdot mol = +446.43\ J/K \cdot mol$

The positive sign is reasonable because the products contain three more moles of gas than the reactants.

17.69 (a) $\Delta H_r° = \Delta H_f°(CO, g) - \Delta H_f°(H_2O, g) = [-110.53 - (-241.82)]\ kJ/mol$
$= +131.29\ kJ/mol$

$\Delta S_r° = S_m°(CO, g) + S_m°(H_2, g) - S_m°(C, s, gr) - S_m°(H_2O, g)$
$= (197.67 + 130.68 - 5.740 - 188.83)\ J/K \cdot mol$
$= +133.78\ J/K \cdot mol = 0.133\ 78\ kJ/K \cdot mol$

$\Delta G_r° = \Delta H_r° - T\Delta S_r° = (131.29 - 298 \times 0.133\ 78)\ kJ/mol = +91.4\ kJ/mol$

(b) $T = \frac{\Delta H_r°}{\Delta S_r°} = \frac{+131.29\ kJ/mol}{+0.133\ 78\ kJ/K \cdot mol} = 981.39\ K = 708.24°C$ (5 sfs)

(Note that the sfs retained in the calculated result would not be justified according to the precision that actual experiment could deliver. Estimate 981 K and 708°C.)

$\Delta G_r° = 0 = -RT \ln K_p$ at 981 K

$K_p = 1$

17.71 $RT = 2.4790$ kJ/mol at 25°C

$\Delta G_r° = -RT \ln K_p = -2.4790 \text{ kJ/mol} \times \ln(9.1 \times 10^{-30})$

$= 1.6 \times 10^2$ kJ/mol

17.73 (a) $PCl_3(l) \longrightarrow PCl_3(g)$

$\Delta H_{vap}° = \Delta H_f°(PCl_3, g) - \Delta H_f°(PCl_3, l) = -287.0 \text{ kJ/mol} - (-319.7 \text{ kJ/mol})$

$\Delta H_{vap}° = 32.7 \text{ kJ/mol} = 3.27 \times 10^4$ J/mol

$\Delta S_{vap}° = S_m°(PCl_3, g) - S_m°(PCl_3, l) = (311.78 - 217.18)$ J/K·mol

$\Delta S_{vap}° = 94.60$ J/K·mol

$$T_b = \frac{\Delta H_{vap}°}{\Delta S_{vap}°} = \frac{3.27 \times 10^4 \text{ J/mol}}{94.60 \text{ J/K·mol}} = 346 \text{ K}$$

(b) $\Delta H_{tr}° = \Delta H_f°(S, \text{mono}) - \Delta H_f°(S, \text{rhombic}) = (0.33 - 0)$ kJ/mol

$= 3.3 \times 10^2$ J/mol

$\Delta S_{tr}° = S_m°(S, \text{mono}) - S_m°(S, \text{rhombic}) = (32.6 - 31.80)$ J/K·mol

$= 0.8$ J/K·mol

$$T_{tr} = \frac{\Delta H_{tr}°}{\Delta S_{tr}°} = \frac{3.3 \times 10^2 \text{ J/mol}}{0.8 \text{ J/K·mol}} = 4 \times 10^2 \text{ K}$$

(c) $I_2(s) \longrightarrow I_2(g)$

$\Delta H_{sub}° = \Delta H_f°(I_2, g) - \Delta H_f°(I_2, s) = (62.44 - 0) \text{ kJ/mol} = 62.44$ kJ/mol

$\Delta S_{sub}° = S_m°(I_2, g) - S_m°(I_2, s) = (260.69 - 116.14)$ J/K·mol

$= 144.55$ J/K·mol

$$T_{sub} = \frac{\Delta H_{sub}°}{\Delta S_{sub}°} = \frac{62.44 \times 10^3 \text{ J/mol}}{144.55 \text{ J/K·mol}} = 432 \text{ K}$$

(d) $D_2O(l) \longrightarrow D_2O(g)$

$\Delta H_{vap}° = \Delta H_f°(D_2O, g) - \Delta H_f°(D_2O, l) = [-249.20 - (-294.60)]$ kJ/mol

$= 45.40$ kJ/mol

$\Delta S_{vap}° = S_m°(D_2O, g) - S_m°(D_2O, l) = (198.34 - 75.94)$J/K·mol

$= 122.40$ J/K·mol

$$T_b = \frac{\Delta H_{tr}°}{\Delta S_{tr}°} = \frac{4.540 \times 10^4 \text{ J/mol}}{122.40 \text{ J/K·mol}} = 370.9 \text{ K}$$

17.75 (a) $\Delta S_{vap}° = \dfrac{\Delta H_{vap}°}{T} = +\dfrac{40.7 \text{ kJ/mol}}{373 \text{ K}}$ (see Table 6.2)

$$= \frac{0.109 \text{ kJ}}{\text{mol·K}} \times \left(\frac{1 \text{ mol } H_2O}{18.02 \text{ g}}\right) \times 1.0 \text{ g} = 6.0 \times 10^{-3} \text{ kJ/K or } 6.0 \text{ J/K}$$

(b) $\Delta S_{fus}° = \dfrac{\Delta H_{fus}°}{T} = \dfrac{6.01 \text{ kJ/mol}}{273 \text{ K}}$

$$= \frac{0.0220 \text{ kJ}}{\text{mol·K}} \times \left(\frac{1 \text{ mol } H_2O}{18.02 \text{ g}}\right) \times 1.0 \text{ g} = 1.2 \times 10^{-3} \text{ kJ/K or } 1.2 \text{ J/K}$$

(c) The entropy change for vaporization is larger because gaseous water molecules are more disordered than liquid water molecules.

17.77 (a) The first part of this statement is true but the second part is false. Thermodynamics provides no information about the rates of reaction.

(b) $\Delta G_r = 0$, not $\Delta G_r^\circ = 0$, determines equilibrium.

(c) Pure elements are assigned a free energy of formation equal to zero only for their most stable forms. The most stable form corresponds to only one particular state of matter.

(d) $\Delta G_r^\circ = \Delta H_r^\circ - T\Delta S_r^\circ$

For the case described, ΔH_r° is negative and ΔS_r° is positive, thus

$\Delta G_r^\circ = (-) - (+)(+) = -$

APPLIED EXERCISES

17.79 (a) $C_6H_{12}O_6(s) + 6\ O_2(g) \longrightarrow 6\ CO_2(g) + 6\ H_2O(l)$

$$\begin{aligned}\Delta H_r^\circ &= 6 \times \Delta H_f^\circ(CO_2, g) + 6 \times \Delta H_f^\circ(H_2O, l) - \Delta H_f^\circ(C_6H_{12}O_6, s)\\ &= [6 \times (-393.51) + 6 \times (-285.83) - (-1268)]\ \text{kJ/mol}\\ &= -2808\ \text{kJ/mol}\end{aligned}$$

$$\begin{aligned}\Delta S_r^\circ &= 6 \times S_m^\circ(CO_2, g) + 6 \times S_m^\circ(H_2O, l) - 1 \times S_m^\circ(C_6H_{12}O_6, s)\\ &\qquad - 6 \times S_m^\circ(O_2, g)\\ &= (6 \times 213.74 + 6 \times 69.91 - 212 - 6 \times 205.14)\ \text{J/K·mol}\\ &= 259\ \text{J/K·mol} = 0.259\ \text{kJ/K·mol}\end{aligned}$$

$$\begin{aligned}\Delta G_r^\circ &= \Delta H_r^\circ - T\Delta S_r^\circ = -2808\ \text{kJ/mol} - 298\ \text{K} \times 0.259\ \text{kJ/K·mol}\\ &= -2885\ \text{kJ/mol}\end{aligned}$$

(b) As the temperature is raised, ΔG_r° becomes more negative, and more energy becomes available to do work, $\Delta G_r^\circ = w_{max}$. Because ΔS_r° is positive, the reaction becomes more spontaneous as the temperature is raised.

17.81 ΔG_r° for the photosynthetic synthesis of glucose is the negative of the value for the combustion of glucose calculated in the answer to Exercise 17.79.

$$\Delta G_r^\circ = +2885\ \text{kJ/mol}$$

For ATP $\longrightarrow$ ADP, $\Delta G_r^\circ = -30.5$ kJ/mol; therefore, to provide -2885 kJ, the number of moles required is

$$n = \frac{-2885\ \text{kJ}}{-30.5\ \text{kJ/mol}} = 94.6\ \text{mol ATP}$$

17.83 (a) $\Delta G_r^\circ = -2 \times \Delta G_f^\circ(O_3, g) = -2 \times 163.2\ \text{kJ/mol} = -326.4\ \text{kJ/mol}$

$$\begin{aligned}\Delta S_r^\circ &= 3 \times S_m^\circ(O_2, g) - 2 \times S_m^\circ(O_3, g)\\ &= 3 \times 205.14\ \text{J/K·mol} - 2 \times 238.93\ \text{J/K·mol} = +137.56\ \text{J/K·mol}\end{aligned}$$

(b) $\ln K_p = \dfrac{-\Delta G_r^\circ}{RT} = \dfrac{-(-326.4\ \text{kJ/mol})}{2.479\ \text{kJ/mol}} = 131.\bar{7}$

$K_p = 1.\bar{6} \times 10^{57}$

(c) The conversion of ozone to oxygen is spontaneous; if no matter or energy were added, the ozone would be used up. The rate at which this occurs, however, is another matter; thermodynamics does not give kinetic information.

INTEGRATED EXERCISES

17.85 (a) $\Delta G_{vap}^\circ = 0$; pure liquid water and pure water vapor are in equilibrium at 100.°C and 1 atm pressure.

(b) See Exercise 17.78. For vaporization, the equilibrium constant $K_p = P$. We may use Raoult's law to calculate the vapor pressure of water in the glucose solution. We first calculate the mole fraction of water and then use it in

$$P_{H_2O} = x_{H_2O}P_{H_2O}^\circ = x_{H_2O} \times 1.00 \text{ atm}$$

Assume 100. g of solution, then 90. g H_2O and 10. g $C_6H_{12}O_6$.

For H_2O, $n_{H_2O} = \dfrac{90 \text{ g}}{18.0 \text{ g/mol}} = 5.0 \text{ mol}$

For $C_6H_{12}O_6$, $n_{C_6H_{12}O_6} = \dfrac{10. \text{ g}}{180. \text{ g/mol}} = 5.6 \times 10^{-2} \text{ mol}$

$x_{H_2O} = \dfrac{5.0 \text{ mol}}{5.0 \text{ mol} + 5.6 \times 10^{-2} \text{ mol}} = 0.98\overline{9}$

$P_{H_2O} = 0.98\overline{9} \times 1 \text{ atm} = 0.98\overline{9} \text{ atm}$

$\Delta G_{vap}^\circ = -RT \ln K_p = -RT \ln P$

$= -8.314 \text{ J/K}\cdot\text{mol} \times 373 \text{ K} \times \ln(0.98\overline{9}) = +34 \text{ J/mol}$

Because $\Delta G_{vap}^\circ > 0$, this process is not spontaneous.

(c) The systems are different. The water is not pure in part (b). $\Delta G_{vap}^\circ = 0$, only for pure water at 100.°C and 1 atm pressure. A nonvolatile solute depresses the vapor pressure.

17.87 In each case, $\ln K_{sp} = \dfrac{-\Delta G_r^\circ}{RT} = \dfrac{-\Delta G_r^\circ}{2.479 \text{ kJ/mol}}$, at 25°C

(a) $AgI(s) \rightleftharpoons Ag^+(aq) + I^-(aq)$

$\Delta G_r^\circ = 1 \times (77.11 \text{ kJ/mol}) + 1 \times (-51.57 \text{ kJ/mol}) - 1 \times (-66.19 \text{ kJ/mol})$

$= 91.73 \text{ kJ/mol}$

$\ln K_{sp} = \dfrac{-\Delta G_r^\circ}{RT} = \dfrac{-91.73 \text{ kJ/mol}}{2.479 \text{ kJ/mol}} = -37.00$

$K_{sp} = 8.5 \times 10^{-17}$

$\Delta S_r^\circ = \Delta S_r^\circ (Ag^+, aq) + \Delta S_r^\circ (I^-, aq) - \Delta S_r^\circ (AgI, s)$

$= (72.68 + 111.3 - 115.5) \text{ J/K}\cdot\text{mol} = 68.5 \text{ J/K}\cdot\text{mol}$

ΔS_{surr} must decrease more than ΔS_{sys} increases, so ΔS° must be negative; the dissolution is endothermic.

(b) $CaCO_3(s) \rightleftharpoons Ca^{2+}(aq) + CO_3^{2-}(aq)$

$$\Delta G_r° = 1 \times (-553.58 \text{ kJ/mol}) + 1 \times (-527.9 \text{ kJ/mol}) - 1 \times (-1127.8 \text{ kJ/mol}) = 46.32 \text{ kJ/mol}$$

$$\ln K_{sp} = \frac{-\Delta G_r°}{RT} = \frac{-46.32 \text{ kJ/mol}}{2.479 \text{ kJ/mol}} = -18.68$$

$K_{sp} = 7.7 \times 10^{-9}$

$$\Delta S_r° = \Delta S_r° (Ca^{2+}, aq) + \Delta S_r° (CO_3^{2-}, aq) - \Delta S_r° (CaCO_3, s)$$
$$= (-53.1 + -56.9 - 88.7) \text{ J/K·mol} = -198.7 \text{ J/K·mol}$$

(c) $Ca(OH)_2(s) \rightleftharpoons Ca^{2+}(aq) + 2\ OH^-(aq)$

$$\Delta G_r° = 1 \times (-553.58 \text{ kJ/mol}) + 2 \times (-157.24 \text{ kJ/mol}) - 1 \times (-898.49 \text{ kJ/mol}) = 30.43 \text{ kJ/mol}$$

$$\ln K_{sp} = \frac{-\Delta G_r°}{RT} = \frac{-(30.43 \text{ kJ/mol})}{2.479 \text{ kJ/mol}} = -12.28$$

$K_{sp} = 4.6 \times 10^{-6}$

$$\Delta S_r° = \Delta S_r° (Ca^{2+}, aq) + 2\Delta S_r° (OH^-, aq) - \Delta S_r° [Ca(OH)_2, s]$$
$$= (-53.1 - 2(10.75) - 83.39) \text{ J/K·mol} = -158.0 \text{ J/K·mol}$$

17.89 (a) $K_w = 1.00 \times 10^{-14}$ at 25°C

$$\Delta G_r° = -RT \ln K_w = -2.479 \text{ kJ/mol} \times \ln(1.00 \times 10^{-14})$$
$$= +79.91 \text{ kJ/mol}$$

(b) $\Delta H_r° = (-285.83 - 229.99 - 2(-285.83)) \text{ kJ/mol} = +55.84 \text{ kJ/mol}$

$\Delta S_r° = (69.91 + -10.75 - 2(69.91)) \text{ J/K·mol} = -80.66 \text{ J/K·mol}$

So (+) = (+) − (−) This reaction is never spontaneous; higher temperatures will lead to a larger, positive ΔG and a smaller K_w ($\Delta G = -RT \ln K_w$). A smaller K_w corresponds to a smaller hydronium ion concentration.

17.91 $$Q_c = \frac{[I]^2}{[I_2]} = \frac{(0.0084)^2}{0.026} = 2.7 \times 10^{-3}$$

$Q_c < K_c$; therefore, the reaction is spontaneous in the direction of I(g).

$$\Delta G_r = \Delta G_r° + RT \ln Q_c \quad (\Delta G_r° = -RT \ln K_c)$$
$$= -RT \ln K_c + RT \ln Q_c, \text{ so}$$

$$\Delta G_r = -RT \ln\left(\frac{K_c}{Q_c}\right) = -8.314 \text{ J/K·mol} \times 1200 \text{ K} \times \ln\left(\frac{6.8 \times 10^{-2}}{2.7 \times 10^{-3}}\right)$$
$$= -32 \text{ kJ/mol} \quad \text{(confirming the spontaneity)}$$

CHAPTER 18
ELECTROCHEMISTRY

EXERCISES

Assume a temperature of 25°C (298 K) for the following exercises unless instructed otherwise.

Balancing Redox Equations

18.1 (a) $VO^{2+}(aq) + 2\ H^+(aq) + e^- \longrightarrow V^{3+}(aq) + H_2O(l)$
gain of electron; reduction
(b) $PbSO_4(s) + 2\ H_2O(l) \longrightarrow PbO_2(s) + SO_4^{\ 2-}(aq) + 4\ H^+(aq) + 2\ e^-$
loss of electrons; oxidation
(c) $H_2O_2(aq) \longrightarrow O_2(g) + 2\ H^+(aq) + 2\ e^-$ loss of electrons; oxidation

18.3 (a) $ClO^-(aq) \longrightarrow Cl^-(aq) + H_2O(l)$ (balances O's); then,
$ClO^-(aq) + 2\ H_2O(l) \longrightarrow Cl^-(aq) + H_2O(l) + 2\ OH^-(aq)$ (balances H's); and,
$ClO^-(aq) + H_2O(l) + 2\ e^- \longrightarrow Cl^-(aq) + 2\ OH^-(aq)$ (balances charge)
(b) $IO_3^{\ -}(aq) \longrightarrow IO^-(aq) + 2\ H_2O(l)$ (balances O's); then,
$IO_3^{\ -}(aq) + 4\ H_2O(l) \longrightarrow$
$IO^-(aq) + 2\ H_2O(l) + 4\ OH^-(aq)$ (balances H's); and,
$IO_3^{\ -}(aq) + 2\ H_2O(l) + 4\ e^- \longrightarrow IO^-(aq) + 4\ OH^-(aq)$ (balances charge)
(c) $2\ SO_3^{\ 2-}(aq) \longrightarrow S_2O_4^{\ 2-}(aq) + 2\ H_2O(l)$ (balances O's); then,
$2\ SO_3^{\ 2-}(aq) + 4\ H_2O(l) \longrightarrow$
$S_2O_4^{\ 2-}(aq) + 2\ H_2O(l) + 4\ OH^-(aq)$ (balances H's); and,
$2\ SO_3^{\ 2-}(aq) + 2\ H_2O(l) + 2\ e^- \longrightarrow S_2O_4^{\ 2-}(aq) + 4\ OH^-(aq)$ (balances charge)
In each case, the reactants gain electrons, so all are reduction half-reactions.

18.5 In each case, first obtain the balanced half-reactions by using the methods employed in the solutions to Exercises 18.1 to 18.4. Then multiply the oxidation and reduction half-reactions by appropriate factors that will result in the same number of electrons being present in both half-reactions. Then add the half-reactions, canceling electrons in the process, to obtain the balanced equation for the whole reaction. Check to see that the final equation is balanced.

(a) $4[Cl_2(g) + 2\,e^- \longrightarrow 2\,Cl^-(aq)]$

$1[S_2O_3^{2-}(aq) + 5\,H_2O(l) \longrightarrow 2\,SO_4^{2-}(aq) + 10\,H^+(aq) + 8\,e^-]$

$4\,Cl_2(g) + S_2O_3^{2-}(aq) + 5\,H_2O(l) + 8\,e^- \longrightarrow 8\,Cl^-(aq) + 2\,SO_4^{2-}(aq) + 10\,H^+(aq) + 8\,e^-$

$4\,Cl_2(g) + S_2O_3^{2-}(aq) + 5\,H_2O(l) \longrightarrow 8\,Cl^-(aq) + 2\,SO_4^{2-}(aq) + 10\,H^+(aq)$

Cl_2 is the oxidizing agent and $S_2O_3^{2-}$ is the reducing agent.

(b) $2[MnO_4^-(aq) + 8\,H^+(aq) + 5\,e^- \longrightarrow Mn^{2+}(aq) + 4\,H_2O(l)]$

$5[H_2SO_3(aq) + H_2O(l) \longrightarrow HSO_4^-(aq) + 3\,H^+(aq) + 2\,e^-]$

$2\,MnO_4^-(aq) + 16\,H^+(aq) + 5\,H_2SO_3(aq) + 5\,H_2O(l) + 10\,e^- \longrightarrow 2\,Mn^{2+}(aq) + 8\,H_2O(l) + 5\,HSO_4^-(aq) + 15\,H^+(aq) + 10\,e^-$

$2\,MnO_4^-(aq) + H^+(aq) + 5\,H_2SO_3(aq) \longrightarrow 2\,Mn^{2+}(aq) + 3\,H_2O(l) + 5\,HSO_4^-(aq)$

MnO_4^- is the oxidizing agent and H_2SO_3 is the reducing agent.

(c) $Cl_2(g) + 2\,e^- \longrightarrow 2\,Cl^-(aq)$

$H_2S(aq) \longrightarrow S(s) + 2\,H^+(aq) + 2\,e^-$

$Cl_2(g) + H_2S(aq) + 2\,e^- \longrightarrow 2\,Cl^-(aq) + S(s) + 2\,H^+(aq) + 2\,e^-$

$Cl_2(g) + H_2S(aq) \longrightarrow 2\,Cl^-(aq) + S(s) + 2\,H^+(aq)$

Cl_2 is the oxidizing agent and H_2S is the reducing agent.

(d) $Cl_2(g) + 2\,e^- \longrightarrow 2\,Cl^-(aq)$

$2\,H_2O(l) + Cl_2(g) \longrightarrow 2\,HClO(aq) + 2\,H^+(aq) + 2\,e^-$

$2\,H_2O(l) + 2\,Cl_2(g) + 2\,e^- \longrightarrow 2\,HClO(aq) + 2\,H^+(aq) + 2\,Cl^-(aq) + 2\,e^-$

or $H_2O(l) + Cl_2(g) \longrightarrow HClO(aq) + H^+(aq) + Cl^-(aq)$

Cl_2 is both the oxidizing and the reducing agent.

18.7 (a) $O_3(g) \longrightarrow O_2(g)$

$O_3(g) \longrightarrow O_2(g) + H_2O(l)$ (balances O's)

$2\,H_2O(l) + O_3(g) \longrightarrow O_2(g) + H_2O(l) + 2\,OH^-(aq)$ (balances H's)

$H_2O(l) + O_3(g) \longrightarrow O_2(g) + 2\,OH^-(aq)$ (cancels H_2O)

$H_2O(l) + O_3(g) + 2\,e^- \longrightarrow O_2(g) + 2\,OH^-(aq)$ (balances charge)

$Br^-(aq) \longrightarrow BrO_3^-(aq)$

$3\,H_2O(l) + Br^-(aq) \longrightarrow BrO_3^-(aq)$ (balances O's)

$6\,OH^-(aq) + 3\,H_2O(l) + Br^-(aq) \longrightarrow BrO_3^-(aq) + 6\,H_2O(l)$ (balances H's)

$6\,OH^-(aq) + 3\,H_2O(l) + Br^-(aq) \longrightarrow BrO^-(aq) + 6\,H_2O(l) + 6\,e^-$ (balances charge)

Combining half-reactions yields

$3[H_2O(l) + O_3(g) + 2\,e^- \longrightarrow O_2(g) + 2\,OH^-(aq)]$

$6\,OH^-(aq) + 3\,H_2O(l) + Br^-(aq) \longrightarrow BrO_3^-(aq) + 6\,H_2O(l) + 6\,e^-$

$6\,H_2O(l) + 3\,O_3(g) + 6\,OH^-(aq) + Br^-(aq) + 6\,e^- \longrightarrow 3\,O_2(g) + 6\,OH^-(aq) + BrO_3^-(aq) + 6\,H_2O(l) + 6\,e^-$

and, $3\ O_3(g) + Br^-(aq) \longrightarrow 3\ O_2(g) + BrO_3^-(aq)$

O_3 is the oxidizing agent and Br^- is the reducing agent.

(b) $Br_2(l) + 2\ e^- \longrightarrow 2\ Br^-(aq)$ (balanced reduction half-reaction)

$Br_2(l) + 6\ H_2O(l) \longrightarrow 2\ BrO_3^-(aq)$ (O's balanced); then

$Br_2(l) + 6\ H_2O(l) + 12\ OH^-(aq) \longrightarrow 2\ BrO_3^-(aq) + 12\ H_2O(l)$ (H's balanced)

$Br_2(l) + 12\ OH^-(aq) \longrightarrow 2\ BrO_3^-(aq) + 6\ H_2O(l) + 10\ e^-$

(electrons balanced)

Combining half-reactions yields

$5[Br_2(l) + 2\ e^- \longrightarrow 2\ Br^-(aq)]$

$1[Br_2(l) + 12\ OH^-(aq) \longrightarrow 2\ BrO_3^-(aq) + 6\ H_2O(l) + 10\ e^-]$

$6\ Br_2(l) + 12\ OH^-(aq) + 10\ e^- \longrightarrow$

$10\ Br^-(aq) + 2\ BrO_3^-(aq) + 6\ H_2O(l) + 10\ e^-$

$6\ Br_2(l) + 12\ OH^-(aq) \longrightarrow 10\ Br^-(aq) + 2\ BrO_3^-(aq) + 6\ H_2O(l)$

or $3\ Br_2(l) + 6\ OH^-(aq) \longrightarrow 5\ Br^-(aq) + BrO_3^-(aq) + 3\ H_2O(l)$

Br_2 is both the oxidizing agent and the reducing agent.

(c) $Cr^{3+}(aq) + 4\ H_2O(l) \longrightarrow CrO_4^{2-}(aq)$ (O's balanced); then

$Cr^{3+}(aq) + 4\ H_2O(l) + 8\ OH^-(aq) \longrightarrow CrO_4^{2-}(aq) + 8\ H_2O(l)$ (H's balanced)

$Cr^{3+}(aq) + 8\ OH^-(aq) \longrightarrow CrO_4^{2-}(aq) + 4\ H_2O(l) + 3\ e^-$ (charge balanced)

$MnO_2(s) \longrightarrow Mn^{2+}(aq) + 2\ H_2O(l)$; then

$MnO_2(s) + 4\ H_2O(l) \longrightarrow Mn^{2+}(aq) + 2\ H_2O(l) + 4\ OH^-(aq)$ (H's balanced)

$MnO_2(s) + 2\ H_2O(l) + 2\ e^- \longrightarrow Mn^{2+}(aq) + 4\ OH^-(aq)$ (charge balanced)

Combining half-reactions yields

$2[Cr^{3+}(aq) + 8\ OH^-(aq) \longrightarrow CrO_4^{2-}(aq) + 4\ H_2O(l) + 3\ e^-]$

$3[MnO_2(s) + 2\ H_2O(l) + 2\ e^- \longrightarrow Mn^{2+}(aq) + 4\ OH^-(aq)]$

$2\ Cr^{3+}(aq) + 16\ OH^-(aq) + 3\ MnO_2(s) + 6\ H_2O(l) + 6\ e^- \longrightarrow$

$2\ CrO_4^{2-}(aq) + 8\ H_2O(l) + 3\ Mn^{2+}(aq) + 12\ OH^-(aq) + 6\ e^-$

$2\ Cr^{3+}(aq) + 4\ OH^-(aq) + 3\ MnO_2(s) \longrightarrow$

$2\ CrO_4^{2-}(aq) + 2\ H_2O(l) + 3\ Mn^{2+}(aq)$

Cr^{3+} is the reducing agent and MnO_2 is the oxidizing agent.

(d) $3[P_4(s) + 8\ OH^-(aq) \longrightarrow 4\ H_2PO_2^-(aq) + 4\ e^-]$

$P_4(s) + 12\ H_2O(l) + 12\ e^- \longrightarrow 4\ PH_3(g) + 12\ OH^-(aq)$

$4\ P_4(s) + 12\ H_2O(l) + 24\ OH^-(aq) + 12\ e^- \longrightarrow$

$12\ H_2PO_2^-(aq) + 4\ PH_3(g) + 12\ OH^-(aq) + 12\ e^-$

$4\ P_4(s) + 12\ H_2O(l) + 12\ OH^-(aq) \longrightarrow 12\ H_2PO_2^-(aq) + 4\ PH_3(g)$

or $P_4(s) + 3\ H_2O(l) + 3\ OH^-(aq) \longrightarrow 3\ H_2PO_2^-(aq) + PH_3(g)$

$P_4(s)$ is both the oxidizing and the reducing agent.

18.9 (a) $HSO_3^-(aq) + H_2O(l) \longrightarrow HSO_4^-(aq) + 2\,H^+(aq) + 2\,e^-$; and
$2\,HSO_3^-(aq) \longrightarrow S_2O_6^{2-}(aq) + 2H^+(aq) + 2\,e^-$

(b) $I_2(aq) + 2\,e^- \longrightarrow 2\,I^-(aq)$ is the reduction reaction, thus

$$HSO_3^-(aq) + H_2O(l) \longrightarrow HSO_4^-(aq) + 2\,H^+(aq) + 2\,e^-$$

$$I_2(aq) + HSO_3^-(aq) + H_2O(l) + 2\,e^- \longrightarrow 2\,I^-(aq) + HSO_4^-(aq) + 2\,H^+(aq) + 2\,e^-$$

$$I_2(aq) + HSO_3^-(aq) + H_2O(l) \longrightarrow 2\,I^-(aq) + HSO_4^-(aq) + 2\,H^+(aq)$$

18.11 The half-reactions are

$MnO_4^-(aq) \longrightarrow Mn^{2+}(aq) + 4\,H_2O(l)$ (O's balanced)

$MnO_4^-(aq) + 8\,H^+(aq) \longrightarrow Mn^{2+}(aq) + 4\,H_2O(l)$ (H's balanced)

$MnO_4^-(aq) + 8\,H^+(aq) + 5\,e^- \longrightarrow Mn^{2+}(aq) + 4\,H_2O(l)$ (charge balanced)

$C_6H_{12}O_6(aq) + 6\,H_2O(l) \longrightarrow 6\,CO_2(g)$ (O's balanced)

$C_6H_{12}O_6(aq) + 6\,H_2O(l) \longrightarrow 6\,CO_2(g) + 24\,H^+(aq)$ (H's balanced)

$C_6H_{12}O_6(aq) + 6\,H_2O(l) \longrightarrow 6\,CO_2(g) + 24\,H^+(aq) + 24\,e^-$ (charge balanced)

Adding half-reactions yields

$$24[MnO_4^-(aq) + 8\,H^+(aq) + 5\,e^- \longrightarrow Mn^{2+}(aq) + 4\,H_2O(l)]$$

$$5[C_6H_{12}O_6(aq) + 6\,H_2O(l) \longrightarrow 6\,CO_2(g) + 24\,H^+(aq) + 24\,e^-]$$

$$24\,MnO_4^-(aq) + 192\,H^+(aq) + 5\,C_6H_{12}O_6(aq) + 30\,H_2O(l) + 120\,e^- \longrightarrow 24\,Mn^{2+}(aq) + 96\,H_2O(l) + 30\,CO_2(g) + 120\,H^+(aq) + 120\,e^-$$

$$24\,MnO_4^-(aq) + 72\,H^+(aq) + 5\,C_6H_{12}O_6(aq) \longrightarrow 24\,Mn^{2+}(aq) + 66\,H_2O(l) + 30\,CO_2(g)$$

Galvanic Cells

18.13 See Figs. 18.2 and 18.8
(a) anode (b) positive

18.15 (a) $Zn^{2+}(aq) + 2\,e^- \longrightarrow Zn(s)$
(b) $Fe^{3+}(aq) + e^- \longrightarrow Fe^{2+}(aq)$
(c) $Cl_2(g) + 2\,e^- \longrightarrow 2\,Cl^-(aq)$
(d) $Hg_2Cl_2(s) + 2\,e^- \longrightarrow 2\,Hg(l) + 2\,Cl^-(aq)$

18.17 (a) $Ni^{2+}(aq) + 2\,e^- \longrightarrow Ni(s)$ $E°(\text{cathode}) = -0.23$ V
$Zn^{2+}(aq) + 2\,e^- \longrightarrow Zn(s)$ $E°(\text{anode}) = -0.76$ V
Reversing the anode reaction yields
$Zn(s) \longrightarrow Zn^{2+}(aq) + 2\,e^-$ (at anode), then, upon addition,

$Ni^{2+}(aq) + Zn(s) \longrightarrow Ni(s) + Zn^{2+}(aq)$ (overall cell)

$E° = -0.23\ V - (-0.76\ V) = +0.53\ V$

and $Zn(s)|Zn^{2+}(aq)||Ni^{2+}(aq)|Ni(s)$

(b) $2[Ce^{4+}(aq) + e^- \longrightarrow Ce^{3+}(aq)]$ $E°(\text{cathode}) = +1.61\ V$

$I_2(s) + 2\ e^- \longrightarrow 2\ I^-(aq)$ $E°(\text{anode}) = +0.54\ V$

Reversing the anode reaction yields

$2\ I^-(aq) \longrightarrow 2\ e^- + I_2(s)$ (at anode), then, upon addition,

$2\ I^-(aq) + 2\ Ce^{4+}(aq) \longrightarrow 2\ Ce^{3+}(aq) + I_2(s)$ (overall cell)

$E° = +1.61\ V - 0.54\ V = +1.07\ V$

and $Pt(s)|I^-(aq)|I_2(s)||Ce^{4+}(aq), Ce^{3+}|Pt(s)$

An inert electrode such as Pt is necessary when both oxidized and reduced species are in the same solution.

(c) $Cl_2(g) + 2\ e^- \longrightarrow 2\ Cl^-(aq)$ $E°(\text{cathode}) = +1.36\ V$

$2\ H^+(aq) + 2\ e^- \longrightarrow H_2(g)$ $E°(\text{anode}) = 0.00\ V$

Reversing the anode reaction yields

$H_2(g) \longrightarrow 2\ H^+(aq) + 2\ e^-$ (at anode), then, upon addition,

$H_2(g) + Cl_2(g) \longrightarrow 2\ HCl(aq)$ (overall cell) $E° = +1.36\ V - 0.00\ V$

$= +1.36\ V$

and $Pt(s)|H_2(g)|H^+(aq), Cl_2(g)|Cl^-(aq)|Pt(s)$

An inert electrode such as Pt is necessary for gas/ion electrode reactions.

(d) $3[Au^+(aq) + e^- \longrightarrow Au(s)]$ $E°(\text{cathode}) = +1.69\ V$

$Au^{3+}(aq) + 3\ e^- \longrightarrow Au(s)$ $E°(\text{anode}) = +1.40\ V$

Reversing the anode reaction yields

$Au(s) \longrightarrow Au^{3+}(aq) + 3\ e^-$ then, upon addition, (anode),

$3\ Au^+(aq) \longrightarrow 2\ Au(s) + Au^{3+}(aq)$ (overall cell) $E° = +1.69\ V - 1.40\ V$

$= +0.29\ V$

and $Au(s)|Au^{3+}(aq)||Au^+(aq)|Au(s)$

18.19 (a) $2\ H^+(aq) + 2\ e^- \longrightarrow H_2(g)$ $E°(\text{anode}) = 0.00\ V$

$Cl_2(g) + 2\ e^- \longrightarrow 2\ Cl^-(aq)$ $E°(\text{cathode}) = +1.36\ V$

Therefore, at the anode, after reversal,

$H_2(g) \longrightarrow 2\ H^+(aq) + 2\ e^-$

and, the cell reaction is, upon addition of the half-reactions,

$Cl_2(g) + H_2(g) \longrightarrow 2\ H^+(aq) + 2\ Cl^-(aq)$ $E° = +1.36\ V - 0.00\ V = +1.36\ V$

(b) $2[U^{3+}(aq) + 3\ e^- \longrightarrow U(s)]$ $E°(\text{anode}) = -1.79\ V$

$3[V^{2+}(aq) + 2\ e^- \longrightarrow V(s)]$ $E°(\text{cathode}) = -1.19\ V$

Therefore, at the anode, after reversal,

$2[U(s) \longrightarrow U^{3+}(aq) + 3\ e^-]$

and, the cell reaction is, upon addition of the half-reactions,

$3\ V^{2+}(aq) + 2\ U(s) \longrightarrow 2\ U^{3+}(aq) + 3\ V(s) \quad E° = -1.19\ V - (-1.79\ V)$
$= +0.60\ V$

(c) $O_2(g) + 2\ H_2O(l) + 4\ e^- \longrightarrow 4\ OH^-(aq) \quad E°(\text{cathode}) = 0.40\ V$

$O_2(g) + 4\ H^+(aq) + 4\ e^- \longrightarrow 2\ H_2O(l) \quad E°(\text{anode}) = 1.23\ V$

Reversing the anode half-reaction yields

$2\ H_2O(l) \longrightarrow O_2(g) + 4\ H^+(aq) + 4\ e^-$

and the cell reaction is, upon addition of the half-reactions,

$4\ H_2O(l) \longrightarrow 4\ H^+(aq) + 4\ OH^-(aq) \quad E° = 0.40\ V - 1.23\ V = -0.83\ V$

or, $H_2O(l) \longrightarrow H^+(aq) + OH^-(aq)$

Note: This balanced equation corresponds to the cell notation given. The spontaneous process is the reverse of this reaction.

(d) $Sn^{4+}(aq) + 2\ e^- \longrightarrow Sn^{2+}(aq) \quad E°(\text{anode}) = +0.15\ V$

$Hg_2Cl_2(s) + 2\ e^- \longrightarrow 2\ Hg(l) + 2\ Cl^-(aq) \quad E°(\text{cathode}) = +0.27\ V$

Therefore, at the anode, after reversal,

$Sn^{2+}(aq) \longrightarrow Sn^{4+}(aq) + 2\ e^-$

and, the cell reaction is, upon addition of the half-reactions,

$Sn^{2+}(aq) + Hg_2Cl_2(s) \longrightarrow 2\ Hg(l) + 2\ Cl^-(aq) + Sn^{4+}(aq)$

$E° = 0.27\ V - 0.15\ V = 0.12\ V$

18.21 (a) $MnO_4^-(aq) + 8\ H^+(aq) + 5\ e^- \longrightarrow Mn^{2+}(aq) + 4\ H_2O(l)$

(cathode half-reaction)

$5[Fe^{3+}(aq) + e^- \longrightarrow Fe^{2+}(aq)]$ (anode half-reaction)

(b) Reversing the anode reaction and adding the two equations yields

$MnO_4^-(aq) + 5\ Fe^{2+}(aq) + 8\ H^+(aq) \longrightarrow Mn^{2+}(aq) + 5\ Fe^{3+}(aq) + 4\ H_2O(l)$

(c) The cell diagram is

$Pt(s)\,|\,Fe^{2+}(aq), Fe^{3+}(aq)\,||\,MnO_4^-(aq), Mn^{2+}(aq), H^+(aq)\,|\,Pt(s)$

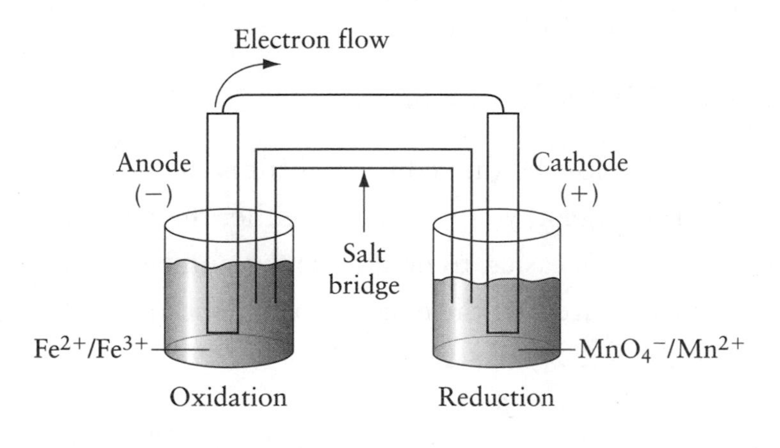

18.23 (a) $Ag^+(aq) + e^- \longrightarrow Ag(s)$ $E°(\text{cathode}) = +0.80\ V$

$AgBr(s) + e^- \longrightarrow Ag(s) + Br^-(aq)$ $E°(\text{anode}) = +0.07\ V$

Reversing the anode reaction yields

$Ag(s) + Br^-(aq) \longrightarrow AgBr(s) + e^-$ then, upon addition,

$Ag^+(aq) + Br^-(aq) \longrightarrow AgBr(s)$ (overall cell) $E° = +0.80\ V - 0.07\ V = +0.73\ V$

This is the direction of the spontaneous cell reaction that could be used to study the given solubility equilibrium. The reverse of this cell reaction corresponds to the reaction as given. It is not spontaneous. Thus,

$AgBr(s) \longrightarrow Ag^+(aq) + Br^-(aq)$ $E° = -0.73\ V$

For this, the cathode and anode reactions are reversed relative to those above. A cell diagram for the nonspontaneous process is

$Ag(s)|Ag^+(aq)||Br^-(aq)|AgBr(s)|Ag(s)$

(b) To conform to the notation of this chapter, the neutralization is rewritten as

$H^+(aq) + OH^- \longrightarrow H_2O(l)$

$O_2(g) + 4\ H^+(aq) + 4\ e^- \longrightarrow 2\ H_2O(l)$ $E°(\text{cathode}) = +1.23\ V$

$O_2(g) + 2\ H_2O(l) + 4\ e^- \longrightarrow 4\ OH^-(aq)$ $E°(\text{anode}) = +0.40\ V$

Reversing the anode reaction yields

$4\ OH^-(aq) \longrightarrow O_2(g) + 2\ H_2O(l) + 4\ e^-$, then, upon addition,

$4\ H^+(aq) + 4\ OH^-(aq) \longrightarrow 4\ H_2O(l)$

or $H^+(aq) + OH^-(aq) \longrightarrow H_2O(l)$ (overall cell) $E° = +1.23\ V - 0.40\ V = +0.83\ V$

and $Pt|O_2(g)|OH^-(aq)||H^+(aq)|O_2(g)|Pt$

(c) $Cd(OH)_2(s) + 2\ e^- \longrightarrow Cd(s) + 2\ OH^-(aq)$ $E°(\text{anode}) = -0.81\ V$

$Ni(OH)_3(s) + e^- \longrightarrow Ni(OH)_2(s) + OH^-(aq)$ $E°(\text{cathode}) = +0.49\ V$

Reversing the anode reaction and multiplying the cathode reaction by 2 yields

$Cd(s) + 2\ OH^-(aq) \longrightarrow Cd(OH)_2(s) + 2\ e^-$

$2\ Ni(OH)_3 + 2\ e^- \longrightarrow 2\ Ni(OH)_2(s) + 2\ OH^-(aq)$ then, upon addition,

$2\ Ni(OH)_2(s) + Cd(s) \longrightarrow Cd(OH)_2(s) + 2\ Ni(OH)_2(s)$

and $Cd(s)|Cd(OH)_2(s)|KOH(aq), Ni(OH)_3(s)|Ni(OH)_2(s)|Ni(s)$

Cell Potential, Free Energy, and the Electrochemical Series

18.25 Because the number of electrons lost by the reducing agent is the same as the number gained by the oxidizing agent, you can choose either the oxidizing agent or the reducing agent, whichever is easier, to determine the number of electrons transferred. But first check to ensure that the equation is balanced.

(a) $6\ O_2(0) \longrightarrow 12\ O(-2)$; $n = 2 \times 12 = 24$

(b) $2\ B(0) \longrightarrow 2\ B(+3)$; $n = 2 \times 3 = 6$

(c) $Si(+4) \longrightarrow Si(0)$; $n = 4$

18.27 A galvanic cell has a positive potential difference; therefore, identify as cathode and anode the electrodes that make $E°(\text{cell})$ positive upon calculating

$$E°(\text{cell}) = E°(\text{cathode}) - E°(\text{anode})$$

(a) $Cu^{2+}(aq) + 2\ e^- \longrightarrow Cu(s)$ $E°(\text{cathode}) = +0.34$ V
$Cr^{3+}(aq) + 1\ e^- \longrightarrow Cr^{2+}(aq)$ $E°(\text{anode}) = -0.41$ V
$E°(\text{cell}) = +0.34\ V - (-0.41\ V) = +0.75\ V$
(b) $AgCl(s) + e^- \longrightarrow Ag(s) + Cl^-(aq)$ $E°(\text{cathode}) = +0.22$ V
$AgI(s) + e^- \longrightarrow Ag(s) + I^-(aq)$ $E°(\text{anode}) = -0.15$ V
$E°(\text{cell}) = +0.22V - (-0.15V) = +0.37\ V$
(c) $Hg_2{}^{2+}(aq) + 2\ e^- \longrightarrow 2\ Hg(l)$ $E°(\text{cathode}) = +0.79$ V
$Hg_2Cl_2(s) + 2\ e^- \longrightarrow 2\ Hg(l) + 2\ Cl^-(aq)$ $E°(\text{anode}) = +0.27$ V
$E°(\text{cell}) = +0.79\ V - (+0.27\ V) = +0.52\ V$

18.29 See Exercise 18.27 solutions for $E°$ (cell) values. In each case, $\Delta G_r° = -nFE°$. $1\ V = 1\ J/C$. n is determined by balancing the equation for the cell reaction constructed from the half-reactions given in Exercise 18.27.
(a) $Cu^{2+}(aq) + 2\ Cr^{2+}(aq) \longrightarrow Cu(s) + 2\ Cr^{3+}(aq)$, $n = 2$
$E° = +0.75\ V$ and $\Delta G_r° = -nFE° = -(2)(9.65 \times 10^4\ C/mol)(0.75\ J/C)$
$= -1.4 \times 10^2\ kJ/mol$
(b) $AgCl(s) + I^-(aq) \longrightarrow AgI(s) + Cl^-(aq)$, $n = 1$
$E° = +0.37\ V$ and $\Delta G_r° = -nFE° = -1 \times 9.65 \times 10^4\ C/mol \times 0.37\ J/C$
$= -36\ kJ/mol$
(c) $Hg_2{}^{2+}(aq) + 2\ Cl^-(aq) \longrightarrow Hg_2Cl_2(s)$, $n = 2$
$E° = +0.52\ V$ and $\Delta G_r° = -nFE° = -(2)(9.65 \times 10^4\ C/mol)(0.52\ J/C)$
$= -1.0 \times 10^2\ kJ/mol$

18.31 Refer to Appendix 2B. The more negative (less positive) the standard reduction potential, the stronger the metal is as a reducing agent. If several reductions are listed, take the most negative, regardless of the charge.
(a) Cu $<$ Fe $<$ Zn $<$ Cr
(b) Mg $<$ Na $<$ K $<$ Li
(c) V $<$ Ti $<$ Al $<$ U
(d) Au $<$ Ag $<$ Sn $<$ Ni

18.33 In each case, identify the couple with the more positive reduction potential. This will be the couple at which reduction occurs and therefore contains the oxidizing agent. The other couple contains the reducing agent.
(a) Co^{2+}/Co $E° = -0.28$ V, Co^{2+} is the oxidizing agent (cathode);
Ti^{2+} is the reducing agent (anode).

$Pt | Ti^{2+}(aq), Ti^{3+}(aq) || Co^{2+}(aq) | Co(s)$

$E° = E°(\text{cathode}) - E°(\text{anode});$

$0.09\text{ V} = -0.28\text{ V} - E°(\text{anode})$ and $E°(\text{anode})$

$= -0.28\text{ V} - 0.09\text{ V} = -0.37\text{ V}$

18.35 (a) Pt^{2+}/Pt $E° = +1.20\text{ V}$, Pt^{2+} is oxidizing agent (cathode)

$AgF/Ag, F^-$ $E° = +0.78\text{ V}$, Ag is reducing agent (anode)

$Ag(s) | AgF(s) | F^-(aq) || Pt^{2+}(aq) | Pt(s)$

$E° = E°(\text{cathode}) - E°(\text{anode}) = +1.20\text{ V} - 0.78\text{ V} = +0.42\text{ V}$

(b) I_3^-/I^- $E° = +0.53\text{ V}$, I_3^- is oxidizing agent (cathode)

Cr^{3+}/Cr^{2+} $E° = -0.41\text{ V}$, Cr^{2+} is reducing agent (anode)

$Pt(s) | Cr^{2+}(aq), Cr^{3+}(aq) || I^-(aq), I_3^-(aq) | Pt(s)$

$E° = +0.53\text{ V} - (-0.41\text{ V}) = +0.94\text{ V}$

18.37 (a) H_2/H^+ $E° = 0.00\text{ V}$; Ti^{2+}/Ti $E° = -1.63\text{ V}$

No, Ti is the better reducing agent; its couple has a more negative $E°$.

(b) Cr^{3+}/Cr $E° = -0.74\text{ V}$; Pb^{2+}/Pb $E° = -0.13\text{ V}$

Yes, Cr is the better reducing agent; its couple has a more negative $E°$.

$2[Cr(s) \longrightarrow Cr^{3+}(aq) + 3\ e^-]$ (anode)

$3[Pb^{2+}(aq) + 2\ e^- \longrightarrow Pb(s)]$ (cathode)

$3\ Pb^{2+}(aq) + 2\ Cr(s) \longrightarrow 3\ Pb(s) + 2\ Cr^{3+}(aq)$

$E° = E°(\text{cathode}) - E°(\text{anode}) = -0.13\text{ V} - (-0.74\text{ V}) = +0.61\text{ V}$

(c) $MnO_4^-, H^+/Mn^{2+}$ $E° = +1.51\text{ V}$; Cu^{2+}/Cu $E° = +0.34\text{ V}$

Yes, MnO_4^- is the better oxidizing agent; its couple has a more positive $E°$.

$5[Cu(s) \longrightarrow Cu^{2+}(aq) + 2\ e^-]$ (anode)

$2[MnO_4^-(aq) + 8\ H^+(aq) + 5\ e^- \longrightarrow Mn^{2+}(aq) + 4\ H_2O(l)]$ (cathode)

$5\ Cu(s) + 2\ MnO_4^-(aq) + 16\ H^+(aq) \longrightarrow 5\ Cu^{2+}(aq) + 2\ Mn^{2+}(aq) + 8\ H_2O(l)$

$E° = 1.51\text{ V} - 0.34\text{ V} = 1.17\text{ V}$

(d) Fe^{3+}/Fe^{2+} $E° = +0.77\text{ V}$; Hg_2^{2+}/Hg $E° = +0.79\text{ V}$

No, Hg_2^{2+} is the better oxidizing agent; its couple has the more positive $E°$.

18.39 (a) $I_2 + H_2 \longrightarrow 2\ I^- + 2\ H^+$

$I_2 + 2\ e^- \longrightarrow 2\ I^-$ reduction (cathode) $E° = +0.54\text{ V}$

$H_2 \longrightarrow 2\ H^+ + 2\ e^-$ oxidation (anode) $E° = 0.00\text{ V}$

$E° = E°(\text{cathode}) - E°(\text{anode}) = +0.54\text{ V} - 0.00\text{ V} = +0.54\text{ V}$

$\Delta G_r° = -nFE° = -(2)(9.65 \times 10^4\ \text{C/mol})(0.54\ \text{J/C}) = -1.0 \times 10^2\ \text{kJ/mol};$

therefore, spontaneous under standard conditions

(b) $Mg^{2+} + Cu \longrightarrow$ no reaction; $E°$ for $Cu^{2+}/Cu > E°$ for Mg^{2+}/Mg

(c) $2\ Al + 3\ Pb^{2+} \longrightarrow 2\ Al^{3+} + 3\ Pb$

$Pb^{2+} + 2\ e^- \longrightarrow Pb$ reduction (cathode) $E° = -0.13$ V

$Al \longrightarrow Al^{3+} + 3\ e^-$ oxidation (anode) $E° = -1.66$ V

$E° = E°(\text{cathode}) - E°(\text{anode}) = -0.13\ \text{V} - (-1.66\ \text{V}) = +1.53\ \text{V}$

$\Delta G_r° = -nFE° = -(6)(9.65 \times 10^4\ \text{C/mol})(+1.53\ \text{J/C}) = -886\ \text{kJ/mol}$;
therefore, spontaneous under standard conditions

18.41 $E°\ (Br_2/Br^-) = +1.09$ V $E°\ (Cl_2, Cl^-) = +1.36$ V

$E°\ (O_2, H^+/H_2O) = +1.23$ V

O_2 could be used because $E°\ (O_2, H^+, H_2O) > E°\ (Br_2, Br^-)$. It is not used because that reaction is so much slower than the one with Cl_2.

Equilibrium Constants

18.43 (a) From the cell diagram,

cathode, reduction $2[AgCl + e^- \longrightarrow Ag + Cl^-]$

anode, oxidation $H_2 \longrightarrow 2\ H^+ - 2\ e^-$

overall $2\ AgCl(s) + H_2(g) \longrightarrow 2\ Ag(s) + 2\ Cl^-(aq) + 2\ H^+(aq)$

$$K_c = \frac{[H^+]^2[Cl^-]^2}{[H_2]}$$

(b) From the cell diagram,

cathode, reduction $NO_3^- + 4\ H^+ + 3\ e^- \longrightarrow NO + 2\ H_2O$

anode, oxidation $Fe^{2+} \longrightarrow Fe^{3+} + e^-$

overall $NO_3^-(aq) + 4\ H^+(aq) + 3\ Fe^{2+}(aq) \longrightarrow$

$NO(g) + 3\ Fe^{3+}(aq) + 2\ H_2O(l)$

$$K_c = \frac{[NO][Fe^{3+}]^3}{[Fe^{2+}]^3[H^+]^4[NO_3^-]}$$

18.45 (a) $Ti^{2+}(aq) + 2\ e^- \longrightarrow Ti(s)$ $E°(\text{cathode}) = -1.63$ V

$Mn^{2+}(aq) + 2\ e^- \longrightarrow Mn(s)$ $E°(\text{anode}) = -1.18$ V

Note: These equations represent the cathode and anode half-reactions for the overall reaction as written. The spontaneous direction of this reaction under standard conditions is the opposite of that given.

$E° = E°(\text{cathode}) - E°(\text{anode}) = -1.63\ \text{V} - (-1.18\ \text{V}) = -0.45$ V, and

$$\ln K = \frac{nFE°}{RT} \text{ at } 25°\text{C} = \frac{nE°}{0.025\ 69\ \text{V}}$$

$$\ln K = \frac{(2)(-0.45\ \text{V})}{0.025\ 69\ \text{V}} = -35.\bar{0} \quad \text{and} \quad K = \bar{6} \times 10^{-16} \approx 10^{-15}$$

(b) $E°$ for $Hg_2^{2+}/Hg = +0.79$ V > $E°$ for $Pb^{2+}/Pb = -0.13$ V

Therefore, Hg_2^{2+} is reduced and Pb is oxidized.

$Hg_2^{2+}(aq) + Pb(s) \longrightarrow Pb^{2+}(aq) + 2\ Hg(l)$

$Hg_2^{2+} + 2\ e^- \longrightarrow 2\ Hg(l) \quad E°(\text{cathode}) = 0.79$ V

$Pb^{2+} + 2\ e^- \longrightarrow Pb(s) \quad E°(\text{anode}) = -0.13$ V

$E° = E°(\text{cathode}) - E°(\text{anode}) = 0.79\ \text{V} - (-0.13\ \text{V}) = 0.92$ V, and

$$\ln K = \frac{nFE°}{RT} \text{ at } 25°\text{C} = \frac{nE°}{0.025\ 69\ \text{V}}$$

$$\ln K = \frac{(2)(0.92\ \text{V})}{0.025\ 69\ \text{V}} = +71.\bar{6} \quad \text{and} \quad K = \bar{1} \times 10^{31} \approx 10^{31}$$

(c) $In^{3+}(aq) + e^- \longrightarrow In^{2+}(aq) \quad E°(\text{cathode}) = -0.49$ V

$U^{4+}(aq) + e^- \longrightarrow U^{3+}(aq) \quad E°(\text{anode}) = -0.61$ V

$E° = E°(\text{cathode}) - E°(\text{anode}) = -0.49\ \text{V} - (-0.61\ \text{V}) = +0.12$ V

$$\ln K = \frac{nFE°}{RT} = \frac{nE°}{0.025\ 69\ \text{V}} \text{ at } 25°\text{C} \quad \ln K = \frac{1 \times (+0.12\ \text{V})}{0.025\ 69\ \text{V}} = 4.7 \quad K = 1 \times 10^2$$

18.47 Consider the half-reactions involved. Construct a cell reaction from them and calculate its standard potential. If positive, the oxidation will occur.

$S_2O_8^{2-}(aq) + 2\ e^- \longrightarrow 2\ SO_4^{2-}(aq) \quad E°(\text{cathode}) = +2.05$ V

$Ag^{2+}(aq) + e^- \longrightarrow Ag^+(aq) \quad E°(\text{anode}) = +1.98$ V

Reverse the anode reaction and multiply by 2.

$S_2O_8^{2-} + 2\ e^- \longrightarrow 2\ SO_4^{2-}$

$2[Ag^+ \longrightarrow Ag^{2+} + e^-]$

$S_2O_8^{2-}(aq) + 2\ Ag^+(aq) \longrightarrow 2\ Ag^{2+}(aq) + 2\ SO_4^{2-}(aq)$

$E° = 2.05\ \text{V} - (+1.98\ \text{V}) = +0.07$ V

Yes, the oxidation will work.

$$\ln K = \frac{nFE°}{RT}, \text{ at } 25°\text{C} = \frac{nE°}{0.025\ 69\ \text{V}} = \frac{(2)(0.07\ \text{V})}{0.025\ 69\ \text{V}} = 5.\overline{45}$$

and $K = \bar{2} \times 10^2 \approx 10^2$

The Nernst Equation

18.49 (a) For this cell, the Nernst equation is

$$E = 1.10\ \text{V} - \left(\frac{0.025\ 693\ \text{V}}{2}\right)\ln\left(\frac{[Zn^{2+}]}{[Cu^{2+}]}\right)$$

if $[Zn^{2+}]$ increases, E decreases.

(b) Adding NaOH to the cathode will precipitate Cu^{2+} as $Cu(OH)_2$. When the Cu^{2+} concentration decreases, E increases.

(c) Diluting the anode compartment decreases $[Zn^{2+}]$, which increases E.

(d) The size of the electrode should not change the cell potential. Note that [Zn](s) does not occur in the Nernst expression, therefore, no change in the cell potential.

18.51 (a) $Pb^{4+}(aq) + 2\,e^- \longrightarrow Pb^{2+}(aq) \quad E°(\text{cathode}) = +1.67\text{ V}$

$Sn^{2+}(aq) \longrightarrow Sn^{4+}(aq) + 2\,e^- \quad E°(\text{anode}) = +0.15\text{ V}$

$Pb^{4+}(aq) + Sn^{2+}(aq) \longrightarrow Pb^{2+}(aq) + Sn^{4+}(aq) \quad E° = 1.67\text{ V} - (0.15\text{ V}) = +1.52\text{ V}$

Then, $E = E° - \left(\frac{0.0257\text{ V}}{n}\right)\ln Q; \quad 1.33\text{ V} = 1.52\text{ V} - \left(\frac{0.0257\text{ V}}{2}\right)\ln Q$

$\ln Q = \frac{1.52\text{ V} - 1.33\text{ V}}{0.0129\text{ V}} = \frac{0.19\text{ V}}{0.0129\text{ V}} = 15 \quad Q = 3 \times 10^6$

(b) $2[Cr_2O_7^{2-}(aq) + 14\,H^+(aq) + 6\,e^- \longrightarrow 2\,Cr^{3+}(aq) + 7\,H_2O(l)]$

$E°(\text{cathode}) = 1.33\text{ V}$

$3[2\,H_2O(l) \longrightarrow O_2(g) + 4\,H^+(aq) + 4\,e^-] \quad E°(\text{anode}) = +1.23\text{ V}$

$2\,Cr_2O_7^{2-}(aq) + 16\,H^+(aq) \longrightarrow 4\,Cr^{3+}(aq) + 8\,H_2O(l) + 3\,O_2(g)$

$E° = 0.10\text{ V}$

Then, $E = E° - \left(\frac{0.0257\text{ V}}{n}\right)\ln Q; \quad 0.10\text{ V} = +0.10\text{ V} - \left(\frac{0.0257\text{ V}}{12}\right)\ln Q$

$\ln Q = 0.00 \quad Q = 1.0$

18.53 (a) $Cu^{2+}(aq, 0.010\text{ M}) + 2\,e^- \longrightarrow Cu(s)$ (cathode)

$Cu^{2+}(aq, 0.0010\text{ M}) + 2\,e^- \longrightarrow Cu(s)$ (anode)

$Cu^{2+}(aq, 0.010\text{ M}) \longrightarrow Cu^{2+}(aq, 0.0010\text{ M}),\ n = 2$

$E° = E°(\text{cathode}) - E°(\text{anode}) = 0\text{ V}$

$E = E° - \left(\frac{RT}{nF}\right)\ln Q = -\left(\frac{0.0257\text{ V}}{2}\right)\ln Q$ at 25°C

$E = -\left(\frac{0.0257\text{ V}}{2}\right)\ln\left(\frac{0.0010\text{ M}}{0.010\text{ M}}\right) = +0.030\text{ V}$

(b) at pH = 3.0, $[H^+] = 1 \times 10^{-3}$ M

at pH = 4.0, $[H^+] = 1 \times 10^{-4}$ M

Cell reaction is $H^+(aq, 1 \times 10^{-3}\text{ M}) \longrightarrow H^+(aq, 1 \times 10^{-4}\text{ M}),\ n = 1$

$E° = 0\text{ V} \quad E = E° - \left(\frac{RT}{nF}\right)\ln Q = -\left(\frac{0.0257\text{ V}}{1}\right)\ln\left(\frac{1 \times 10^{-4}}{1 \times 10^{-3}}\right)$

$= +6 \times 10^{-2}\text{ V}$

18.55 In each case, $E° = E°(\text{cathode}) - E°(\text{anode})$. Recall that the values for $E°$ at the electrodes refer to the electrode potential for the half-reaction written as a reduction reaction. In balancing the cell reaction, the half-reaction at the anode is reversed. However, this does not reverse the sign of electrode potential used at the anode, as the value always refers to the reduction potential.

(a) $Ni^{2+}(aq) + 2\,e^- \longrightarrow Ni(s)$ $E°(\text{cathode}) = -0.23$ V

$Zn(s) \longrightarrow Zn^{2+}(aq) + 2\,e^-$ $E°(\text{anode}) = -0.76$ V

$Ni^{2+}(aq) + Zn(s) \longrightarrow Ni(s) + Zn^{2+}(aq)$ $E° = +0.53$ V

$$\text{Then, } E = E° - \left(\frac{0.0257\text{ V}}{n}\right)\ln\left(\frac{[Zn^{2+}]}{[Ni^{2+}]}\right)$$

$$E = 0.53\text{ V} - \left(\frac{0.0257\text{ V}}{2}\right)\ln\left(\frac{0.10}{0.001}\right) = 0.53\text{ V} - 0.06\text{ V} = 0.47\text{ V}$$

(b) $2\,H^+(aq) + 2\,e^- \longrightarrow H_2(g)$ $E°(\text{cathode}) = 0.00$ V

$2\,Cl^-(aq) \longrightarrow Cl_2(g) + 2\,e^-$ $E°(\text{anode}) = +1.36$ V

$2\,H^+(aq) + 2\,Cl^-(aq) \longrightarrow H_2(g) + Cl_2(g)$ $E° = -1.36$ V

Then,

$$E = E° - \left(\frac{0.0257\text{ V}}{n}\right)\ln\left(\frac{P_{H_2}P_{Cl_2}}{[H^+]^2[Cl^-]^2}\right)$$

$$E = -1.36\text{ V} - \left(\frac{0.0257\text{ V}}{2}\right)\ln\left(\frac{\left(\frac{450.}{760.}\right)\left(\frac{100.}{760.}\right)}{(0.010)^2(1.0)^2}\right)$$

$$E = -1.36\text{ V} - (0.0129\text{ V})\ln\left(\frac{(0.592)(0.132)}{(1.0\times10^{-4})(1)}\right) = -1.36\text{ V} - 0.08\text{ V}$$

$$= -1.44\text{ V}$$

(c) $Sn^{4+}(aq, 0.060\text{ M}) + 2\,e^- \longrightarrow Sn^{2+}(aq, 1.0\text{ M})$ $E°(\text{cathode}) = +0.15$ V

$Sn(s) \longrightarrow Sn^{2+}(aq, 0.020\text{ M}) + 2\,e^-$ $E°(\text{anode}) = -0.14$ V

$Sn^{4+}(aq, 0.060\text{ M}) + Sn(s) \longrightarrow Sn^{2+}(aq, 1.0\text{ M}) + Sn^{2+}(aq, 0.020\text{ M})$

$E° = 0.29$ V

$$E = E° - \left(\frac{0.0257\text{ V}}{n}\right)\ln\left(\frac{[Sn^{2+}, 1.0\text{ M}][Sn^{2+}, 0.020\text{ M}]}{[Sn^{4+}, 0.060\text{ M}]}\right)$$

$$E = 0.29\text{ V} - \left(\frac{0.0257\text{ V}}{2}\right)\ln\left(\frac{(1)(0.020)}{(0.060)}\right) = 0.30\text{ V}$$

18.57 In each case, obtain the balanced equation for the cell reaction from the half-cell reactions at the electrodes by reversing the reduction equation for the half-reaction at the anode, multiplying the half-reaction equations by an appropriate factor to balance the number of electrons, and then adding the half-reactions. Calculate $E° = E°(\text{cathode}) - E°(\text{anode})$. Then write the Nernst equation for the cell reaction and solve for the unknown.

(a) $Hg_2Cl_2(s) + 2\,e^- \longrightarrow 2\,Hg(l) + 2\,Cl^-(aq)$ $E°(\text{cathode}) = +0.27$ V

$H_2(g) \longrightarrow 2\,H^+(aq) + 2\,e^-$ $E°(\text{anode}) = 0.00$ V

$H_2(g) + Hg_2Cl_2(s) \longrightarrow 2\,H^+(aq) + 2\,Hg(l) + 2\,Cl^-(aq)$ $E° = +0.27$ V

$$E = E° - \left(\frac{0.0257\ V}{n}\right) \ln\left(\frac{[H^+]^2[Cl^-]^2}{[H_2]}\right)$$

$$0.33\ V = 0.27\ V - \left(\frac{0.0257\ V}{2}\right) \ln\left(\frac{[H^+]^2(1)^2}{(1)}\right)$$

$$= 0.27\ V - (0.0129\ V)\ \ln[H^+]^2$$

$$0.06\ V = -0.0257\ V\ \ln[H^+] = [H^+] = 0.09\overline{7}$$

$$pH = -\log(0.09\overline{7}) = 1$$

(b) $2[MnO_4^-(aq) + 8\ H^+(aq) + 5\ e^- \longrightarrow Mn^{2+}(aq) + 4\ H_2O(l)]$

$E°(\text{cathode}) = +1.51\ V$

$5[2\ Cl^-(aq) \longrightarrow Cl_2(g) + 2e^-]\quad E°(\text{anode}) = +1.36\ V$

$2\ MnO_4^-(aq) + 16\ H^+(aq) + 10\ Cl^-(aq) \longrightarrow$

$5\ Cl_2(g) + 2\ Mn^{2+}(aq) + 8\ H_2O(l)\quad E° = +0.15\ V$

$$E = E° - \left(\frac{0.0257\ V}{n}\right) \ln\left(\frac{[Cl_2]^5[Mn^{2+}]^2}{[MnO_4]^2[H^+]^{16}[Cl^-]^{10}}\right)$$

$$-0.30\ V = +0.15\ V - \left(\frac{0.0257\ V}{10}\right) \ln\left(\frac{(1)^5 \times (0.10)^2}{(0.010)^2(1 \times 10^{-4})^{16}(Cl^-)^{10}}\right)$$

$$-0.45\ V = -(0.002\ 57\ V)\ \ln\left(\frac{1 \times 10^{-2}}{(1 \times 10^{-4}) \times (1 \times 10^{-64})[Cl^-]^{10}}\right)$$

$$= -0.002\ 57\ V\left[\ln(1 \times 10^{66}) + \ln\left(\frac{1}{[Cl^-]^{10}}\right)\right]$$

$$= -0.3906\ V + (0.002\ 57\ V)\ \ln[Cl^-]^{10}$$

$$-0.0594\ V = 0.002\ 57\ V\ \ln[Cl^-]^{10}$$

$$= (0.0257\ V)\ \ln[Cl^-]$$

$$\ln[Cl^-] = \frac{-0.0594\ V}{0.0257\ V} = -2.31$$

$$[Cl^-] = 9.9 \times 10^{-2}\ \text{mol/L}$$

Practical Cells

18.59 A primary cell is the primary source of the electrical energy produced by its operation. A secondary cell is an energy storage device that stores electrical energy produced elsewhere and releases it upon operation (discharge). It is the secondary source of the electrical energy. Most primary cells produce electricity from chemicals that were sealed into them when they were made. They are not normally rechargeable. A secondary cell is one that must be charged from some other electrical supply before use. It is normally rechargeable.

18.61 See Table 18.2

(a) The electrolyte is KOH(aq)/HgO(s), which will have the consistency of a moist paste.

(b) The oxidizing agent is $HgO(s)$. See the cathode reaction given in Table 18.2.
(c) $HgO(s) + Zn(s) \longrightarrow Hg(l) + ZnO(s)$

18.63 See Table 18.2.
The anode reaction is $Zn(s) \longrightarrow Zn^{2+}(aq) + 2\ e^-$; this reaction supplies the electrons to the external circuit. The cathode reaction is $MnO_2(s) + H_2O(l) + e^- \longrightarrow MnO(OH)_2(s) + OH^-(aq)$. The $OH^-(aq)$ produced reacts with $NH_4^+(aq)$ from the $NH_4Cl(aq)$ present: $NH_4^+(aq) + OH^-(aq) \longrightarrow H_2O(l) + NH_3(g)$. The $NH_3(g)$ produced complexes with the $Zn^{2+}(aq)$ produced in the anode reaction: $Zn^{2+}(aq) + 4\ NH_3(g) \longrightarrow [Zn(NH_3)_4]^{2+}(aq)$. The overall reaction is complicated.

18.65 See Table 18.2. (a) $KOH(aq)$ (b) In the charging process, the cell reaction is the reverse of what occurs in discharge. Therefore, at the anode,
$2\ Ni(OH)_2(s) + 2\ OH^-(aq) \longrightarrow 2\ Ni(OH)_3(s) + 2\ e^-$.

Corrosion

18.67 $Fe^{3+}(aq) + 3\ e^- \longrightarrow Fe(s) \quad E° = -0.04\ V$
$Cr^{3+}(aq) + 3\ e^- \longrightarrow Cr(s) \quad E° = -0.74\ V$
$Fe^{2+}(aq) + 2\ e^- \longrightarrow Fe(s) \quad E° = -0.44\ V$
$Cr^{2+}(aq) + 2\ e^- \longrightarrow Cr(s) \quad E° = -0.91\ V$
Comparison of the reduction potentials shows that Cr is more easily oxidized than Fe, so the presence of Cr retards the rusting of Fe. At the position of the scratch, the gap is filled with oxidation products of Cr, thereby preventing contact of air and water with the iron.

18.69 (a) $Fe_2O_3 \cdot H_2O$ (b) H_2O and O_2 jointly oxidize iron. (c) Water is more highly conducting if it contains dissolved ions, so the rate of rusting is increased.

18.71 (a) aluminum or magnesium; both are below titanium in the electrochemical series.
(b) cost, availability, and toxicity of products in the environment
(c) $Cu^{2+} + 2\ e^- \longrightarrow Cu(s) \quad E° = +0.34\ V$
$Cu^+ + e^- \longrightarrow Cu(s) \quad E° = +0.52\ V$
$Fe^{3+} + 3\ e^- \longrightarrow Fe(s) \quad E° = -0.04\ V$
$Fe^{2+} + 2\ e^- \longrightarrow Fe(s) \quad E° = -0.44\ V$
Fe could act as the anode of an electrochemical cell if Cu^{2+} or Cu^+ are present; thus, it could be oxidized at the point of contact. Water with dissolved ions acts as the electrolyte.

Electrolysis

18.73 See Figs. 18.25 and 18.26. (a) anode (same as galvanic cell) (b) positive (opposite of galvanic cell)

18.75 The strategy is to consider the possible competing cathode and anode reactions. At the cathode, choose the reduction reaction with the most positive (least negative) standard reduction potential ($E°$ value). At the anode, choose the oxidation reaction with the least positive (most negative) standard reduction potential ($E°$ value, as given in the table). Then calculate $E° = E°(\text{cathode}) - E°(\text{anode})$. The negative of this value is the minimum potential that must be supplied.

(a) cathode: $Co^{2+}(aq) + 2\ e^- \longrightarrow Co(s)$ $E° = -0.28$ V

(rather than $2\ H_2O(l) + 2\ e^- \longrightarrow H_2(g) + 2\ OH^-(aq)$ $E° = -0.83$ V)

(b) anode: $2\ H_2O(l) \longrightarrow O_2(g) + 4\ H^+(aq) + 4\ e^-$ $E° = +0.82$ V at pH = 7

(the SO_4^{2-} ion will not oxidize)

(c) $E° = E°(\text{cathode}) - E°(\text{anode}) = -0.28\ \text{V} - (+0.82\ \text{V}) = -1.10$ V

Therefore, E (supplied) must be > +1.10 V; (1.10 V is the minimum).

18.77

(a) $Cu^{2+}(aq) + 2\ e^- \longrightarrow Cu(s)$	reduction, cathode
(b) $Na^+(l) + e^- \longrightarrow Na(l)$	reduction, cathode
(c) $2\ Cl^-(l) \longrightarrow Cl_2(g) + 2\ e^-$	oxidation, anode
(d) $2\ H_2O(l) + 2\ e^- \longrightarrow H_2(g) + 2\ OH^-(aq)$	reduction, cathode

18.79 In each case, compare the reduction potential of the ion to the reduction potential of water ($E° = -0.83$ V) and choose the process with the least negative $E°$ value.

(a) $Mn^{2+}(aq) + 2\ e^- \longrightarrow Mn(s)$ $E° = -1.18$ V

(b) $Al^{3+}(aq) + 3\ e^- \longrightarrow Al(s)$ $E° = -1.66$ V

The reactions in (a) and (b) evolve hydrogen rather than yield a metallic deposit because water is reduced according to $2\ H_2O(l) + 2\ e^- \longrightarrow H_2(g) + 2\ OH^-(aq)$ ($E° = -0.83$ V, at pH = 7)

(c) $Ni^{2+}(aq) + 2\ e^- \longrightarrow Ni(s)$ $E° = -0.23$ V

(d) $Au^{3+}(aq) + 3\ e^- \longrightarrow Au(s)$ $E° = +1.69$ V

In (c) and (d) the metal ion will be reduced.

18.81 (a) $Cu^{2+} + 2\ e^- \longrightarrow Cu$

$$\text{amount (moles) of } e^- = (5.12\ \text{g Cu}) \times \left(\frac{1\ \text{mol Cu}}{63.5\ \text{g Cu}}\right) \times \left(\frac{2\ \text{mol } e^-}{1\ \text{mol Cu}}\right)$$

$$= 0.161\ \text{mol } e^-$$

(b) $Al^{3+} + 3\ e^- \longrightarrow Al$

$$\text{amount (moles) of } e^- = (200.\text{ g Al}) \times \left(\frac{1 \text{ mol Al}}{27.0 \text{ g Al}}\right) \times \left(\frac{3 \text{ mol } e^-}{1 \text{ mol Al}}\right)$$

$$= 22.2 \text{ mol } e^-$$

(c) $2\ H_2O \longrightarrow O_2 + 4\ H^+ + 4\ e^-$

$$\text{amount (moles) of } e^- = (200.\text{ L } O_2) \times \left(\frac{1 \text{ mol } O_2}{22.41 \text{ L } O_2}\right) \times \left(\frac{4 \text{ mol } e^-}{1 \text{ mol } O_2}\right)$$

$$= 35.7 \text{ mol } e^-$$

Note: 22.41 L is the volume of 1 mole of a perfect gas at 273 K and 1 atm.

18.83 (a) $Ag^+(aq) + e^- \longrightarrow Ag(s)$

$$\text{time} = (4.4 \text{ mg Ag}) \times \left(\frac{10^{-3} \text{ g}}{1 \text{ mg}}\right) \times \left(\frac{1 \text{ mol Ag}}{108 \text{ g Ag}}\right) \times \left(\frac{1 \text{ mol } e^-}{1 \text{ mol Ag}}\right)$$

$$\times \left(\frac{9.65 \times 10^4 \text{ C}}{1 \text{ mol } e^-}\right) \times \left(\frac{1 \text{ A}\cdot\text{s}}{1 \text{ C}}\right) \times \left(\frac{1}{0.50 \text{ A}}\right) = 7.9 \text{ s}$$

(b) $Cu^{2+}(aq) + 2\ e^- \longrightarrow Cu(s)$

$$\text{mass Cu} = (7.9 \text{ s}) \times (0.50 \text{ A}) \times \left(\frac{1 \text{ C}}{1 \text{ A}\cdot\text{s}}\right) \times \left(\frac{1 \text{ mol } e^-}{9.65 \times 10^4 \text{ C}}\right)$$

$$\times \left(\frac{0.50 \text{ mol Cu}}{1 \text{ mol } e^-}\right) \times \left(\frac{63.5 \text{ g Cu}}{1 \text{ mol Cu}}\right) = 1.3 \text{ mg Cu}$$

17.85 (a) $Cr(VI) + 6\ e^- \longrightarrow Cr(s)$

$$\text{current} = \frac{\text{charge}}{\text{time}}$$

$$= \frac{4.0 \text{ g Cr} \times \left(\frac{1 \text{ mol Cr}}{52.00 \text{ g Cr}}\right) \times \left(\frac{6 \text{ mol } e^-}{1 \text{ mol Cr}}\right) \times \left(\frac{9.65 \times 10^4 \text{ C}}{1 \text{ mol } e^-}\right)}{24 \text{ h} \times 3600 \text{ s/h}}$$

$$= 0.52 \text{ C/s} = 0.52 \text{ A}$$

(b) $Na^+ + e^- \longrightarrow Na(s)$

$$\text{current} = \frac{4.0 \text{ g Na} \times \left(\frac{1 \text{ mol Na}}{22.99 \text{ g Na}}\right) \times \left(\frac{1 \text{ mol } e^-}{1 \text{ mol Na}}\right) \times \left(\frac{9.65 \times 10^4 \text{ C}}{1 \text{ mol } e^-}\right)}{24 \text{ h} \times 3600 \text{ s/h}}$$

$$= 0.19 \text{ C/s} = 0.19 \text{ A}$$

18.81 $Ti^{n+}(aq) + n\ e^- \longrightarrow Ti(s)$; solve for n

$$\text{moles of Ti} = (0.0150 \text{ g Ti}) \times \left(\frac{1 \text{ mol Ti}}{47.88 \text{ g Ti}}\right) = 3.13 \times 10^{-4}$$

$$\text{total charge} = (500.\text{ s}) \times (120.\text{ mA}) \times \left(\frac{10^{-3}\text{ A}}{1\text{ mA}}\right) \times \left(\frac{1\text{ C/s}}{1\text{ A}}\right) = 60.0\text{ C}$$

$$\text{moles of e}^- = (60.0\text{ C}) \times \left(\frac{1\text{ mol e}^-}{96\,500\text{ C}}\right) = 6.22 \times 10^{-4}\text{ mol e}^-$$

$$n = \frac{6.22 \times 10^{-4}\text{ mol e}^-}{3.13 \times 10^{-4}\text{ mol Ti}} = \frac{2\text{ mol charge}}{1\text{ mol Ti}};$$

therefore, Ti^{2+}, oxidation number of +2

18.89 $Zn^{2+} + 2\,e^- \longrightarrow Zn$ $(2\text{ mol e}^-/1\text{ mol Zn})$

$$\text{charge used} = (1.0\text{ mA}) \times \left(\frac{10^{-3}\text{ A}}{1\text{ mA}}\right) \times (31\text{ d}) \times \left(\frac{24\text{ h}}{1\text{ d}}\right) \times \left(\frac{3600\text{ s}}{1\text{ h}}\right)$$

$$= 2.6\overline{8} \times 10^3\text{ C}$$

$$\text{moles of e}^-\text{ used} = (2.6\overline{8} \times 10^3\text{ C}) \times \left(\frac{1\text{ mol e}^-}{96\,500\text{ C}}\right) = 2.8 \times 10^{-2}\text{ mol e}^-$$

$$\text{moles of Zn} = (2.8 \times 10^{-2}\text{ mol e}^-) \times \left(\frac{1\text{ mol Zn}}{2\text{ mol e}^-}\right) = 1.4 \times 10^{-2}\text{ mol Zn}$$

$$\text{mass of Zn} = (1.4 \times 10^{-2}\text{ mol Zn}) \times \left(\frac{65.4\text{ g Zn}}{1\text{ mol Zn}}\right) = 0.92\text{ g Zn}$$

SUPPLEMENTARY EXERCISES

18.91 (a) $3\,I^-(aq) \longrightarrow I_3^-(aq) + 2\,e^-$; electron loss; oxidation

(b) $SeO_4^{2-}(aq) \longrightarrow SeO_3^{2-}(aq) + H_2O(l)$ (O's balanced)

$SeO_4^{2-}(aq) + 2\,H_2O(l) \longrightarrow SeO_3^{2-}(aq) + H_2O(l) + 2\,OH^-(aq)$ (H's balanced)

$SeO_4^{2-}(aq) + H_2O(l) + 2\,e^- \longrightarrow SeO_3^{2-}(aq) + 2\,OH^-(aq)$ (charge balanced)

electron gain; reduction

18.93 For the standard calomel electrode, $E° = +0.27$ V. If this were set equal to 0, all other potentials would also be decreased by 0.27 V. (a) Therefore, the standard hydrogen electrode's standard reduction potential would be 0.00 V − 0.27 V or −0.27 V. (b) The standard reduction potential for Cu^{2+}/Cu would be 0.34 V − 0.27 V or +0.07 V.

18.95 $Au \longrightarrow Au^{3+} + 3\,e^-$ $E°(\text{anode}) = +1.40$ V

$MnO_4^- + 8\,H^+ + 5\,e^- \longrightarrow Mn^{2+} + 4\,H_2O$ $E°(\text{cathode}) = +1.51$ V

$E° = E°(\text{cathode}) - E°(\text{anode}) = 1.51\text{ V} - (+1.40\text{ V}) = +0.11\text{ V}$

therefore, spontaneous; so gold will be oxidized

however, $Au \longrightarrow Au^{3+} + 3\,e^-$ $\quad E°(\text{anode}) = +1.40\text{ V}$
$Cr_2O_7^{2-} + 14\,H^+ + 6\,e^- \longrightarrow 2\,Cr^{3+} + 7\,H_2O$ $\quad E°(\text{cathode}) = 1.33\text{ V}$
$E° = E°(\text{cathode}) - E°(\text{anode}) = 1.33\text{ V} - (+1.40\text{ V}) = -0.07\text{ V}$
therefore, not spontaneous; so gold will not be oxidized

18.97 In each case, determine the cathode and anode half-reactions corresponding to the reaction *as written*. Look up the standard reduction potentials for these half-reactions and then calculate $E° = E°(\text{cathode}) - E°(\text{anode})$. If $E°$ is negative, the reaction is spontaneous under standard conditions.

(a) $E° = E°(\text{cathode}) - E°(\text{anode}) = +0.96\text{ V} - (+0.79\text{ V}) = +0.17\text{ V}$
therefore, spontaneous galvanic cell
$Hg(l)\,|\,Hg_2^{2+}(aq)\,||\,NO_3^-(aq), H^+(aq)\,|\,NO(g)\,|\,Pt$
$\Delta G_r° = -nFE° = -(6)(9.65 \times 10^4\text{ C/mol})(+0.17\text{ J/C}) = -98\text{ kJ/mol}$

(b) $E° = E°(\text{cathode}) - E°(\text{anode}) = +0.92\text{ V} - (+1.09\text{ V}) = -0.17\text{ V}$
therefore, not spontaneous

(c) $E° = E°(\text{cathode}) - E°(\text{anode}) = +1.33\text{ V} - (+0.97\text{ V}) = +0.36\text{ V}$
therefore, spontaneous galvanic cell
$Pt\,|\,Pu^{3+}(aq), Pu^{4+}(aq)\,||\,Cr_2O_7^{2-}(aq), Cr^{3+}(aq), H^+(aq)\,|\,Pt$
$\Delta G_r° = -nFE° = -(6)(9.65 \times 10^4\text{ C/mol})(0.36\text{ J/C}) = -2.1 \times 10^2\text{ kJ/mol}$

18.99 In each case, break down the solution equilibrium reaction into two half-reactions and identify the cathode and anode half-reactions corresponding to the dissolution reaction. Then calculate $E° = E°(\text{cathode}) - E°(\text{anode})$. From this, calculate K_{sp} and, finally, the solubility.

(a) $Cd(OH)_2(s) + 2\,e^- \longrightarrow Cd(s) + 2\,OH^-(aq)$ $\quad E°(\text{cathode}) = -0.81\text{ V}$
$Cd(s) \longrightarrow Cd^{2+}(aq) + 2\,e^-$ $\quad E°(\text{anode}) = -0.40\text{ V}$
$Cd(OH)_2(s) \longrightarrow Cd^{2+}(aq) + 2\,OH^-(aq)$ $\quad E° = -0.41\text{ V}$

$$\ln K_{sp} = \frac{n(E°)}{0.0257\text{ V}} = \frac{(1)(-0.41)}{0.0257} = -15.\overline{9}5$$

$K_{sp} = \overline{1} \times 10^{-7}$
Because the solubility of $Cd(OH)_2 = \sqrt[3]{K_{sp}/4}$
the solubility of $Cd(OH)_2 = \overline{3} \times 10^{-3}\text{ mol/L} = 10^{-3}\text{ mol/L}$

(b) $Hg_2Cl_2(s) + 2\,e^- \longrightarrow 2\,Hg(l) + 2\,Cl^-(aq)$ $\quad E°(\text{cathode}) = +0.27\text{ V}$
$Hg^0(l) \longrightarrow Hg_2^{2+}(aq) + 2\,e^-$ $\quad E°(\text{anode}) = +0.79\text{ V}$
$Hg_2Cl_2(s) \longrightarrow Hg_2^{2+}(aq) + 2\,Cl^-(aq)$ $\quad E° = -0.52\text{ V}$

$$\ln K_{sp} = \frac{n(E°)}{0.0257\text{ V}} = \frac{(2)(-0.52)}{0.0257} = -40.\overline{5}$$

$K_{sp} = \overline{3} \times 10^{-18} \approx 10^{-18}$

For Hg_2Cl_2, $K_{sp} = \overline{3} \times 10^{-18} = (S)(2S)^2 = 4S^3$

$S = \overline{9} \times 10^{-7}$ mol/L $\approx 10^{-6}$ mol/L

(c) $PbSO_4(s) + 2\,e^- \longrightarrow Pb + SO_4^{2-}(aq) \quad E°(\text{cathode}) = -0.36$ V

$Pb \longrightarrow Pb^{2+}(aq) + 2\,e^- \quad E°(\text{anode}) = -0.13$ V

$PbSO_4(s) \longrightarrow Pb^{2+}(aq) + SO_4^{2-}(aq) \quad E° = -0.23$ V

$$\ln K_{sp} = \frac{n(E°)}{0.0257\text{ V}} = \frac{(2)(-0.23)}{0.0257\text{ V}} = -17.\overline{9}$$

$K_{sp} = \overline{2} \times 10^{-8}$

Solving the K_{sp} expression for $PbSO_4$ yields the solubility of $PbSO_4 = \sqrt{\overline{2} \times 10^{-8}} = \overline{1} \times 10^{-4}$ mol/L $= 10^{-4}$ mol/L

18.101 (a) $Ag^+(aq) + e^- \longrightarrow Ag \quad E°(\text{cathode}) = +0.80$ V

$Fe^{2+}(aq) \longrightarrow Fe^{3+}(aq) + e^- \quad E°(\text{anode}) = +0.77$ V

$Ag^+(aq) + Fe^{2+}(aq) \longrightarrow Fe^{3+}(aq) + Ag(s) \quad E° = +0.03$ V

(b) $$E = E° - \left(\frac{0.0257\text{ V}}{n}\right)\ln\left(\frac{[Fe^{3+}]}{[Ag^+][Fe^{2+}]}\right)$$

$$= 0.03\text{ V} - (0.0257\text{ V})\ln\left(\frac{1}{(0.010)(0.0010)}\right) = 0.03\text{ V} - 0.30\text{ V} = -0.27\text{ V}$$

Note: The cell changes from spontaneous to nonspontaneous as a function of concentration.

18.103 (a) $Pb^{2+}(aq) + 2\,e^- \longrightarrow Pb(s) \quad E°(\text{cathode}) = -0.13$ V

$Zn(s) \longrightarrow Zn^{2+}(aq) + 2\,e^- \quad E°(\text{anode}) = -0.76$ V

$Pb^{2+}(aq) + Zn(s) \longrightarrow Zn^{2+}(aq) + Pb(s) \quad E° = +0.63$ V

(b) $$E = E° - \left(\frac{0.0257\text{ V}}{n}\right)\ln\left(\frac{[Zn^{2+}]}{[Pb^{2+}]}\right)$$

$$0.66\text{ V} = 0.63\text{ V} - \left(\frac{0.0257\text{ V}}{2}\right)\ln\left(\frac{[Zn^{2+}]}{0.10}\right)$$

$$-2.33 = \ln\left(\frac{[Zn^{2+}]}{0.10}\right) = \ln[Zn^{2+}] - \ln(0.10) = \ln[Zn^{2+}] + 2.30;$$

$\ln[Zn^{2+}] = -4.63$

$[Zn^{2+}] = 9.8 \times 10^{-3}$ mol/L

18.105 $Hf^{n+} + n\,e^- \longrightarrow Hf(s)$; solve for n.

charge consumed $= 15.0$ C/s $\times$ 2.00 h $\times$ 3600 s/h $= 1.08 \times 10^5$ C

$$\text{moles of charge consumed} = (1.08 \times 10^5\text{ C}) \times \left(\frac{1\text{ mol e}^-}{9.65 \times 10^4\text{ C}}\right) = 1.12\text{ mol e}^-$$

$$\text{moles of Hf plated} = (50.0\text{ g Hf}) \times \left(\frac{1\text{ mol Hf}}{178.5\text{ g Hf}}\right) = 0.280\text{ mol Hf}$$

Then, $n = \frac{1.12 \text{ mol e}^-}{0.280 \text{ mol Hf}} = 4.0 \text{ mol e}^-/\text{mol Hf}$

Therefore, the oxidation number is 4, that is, Hf^{4+}.

18.107 First, calculate the volume of Ag(s) to be plated on the surface of the copper metal; then determine its mass and the number of moles.

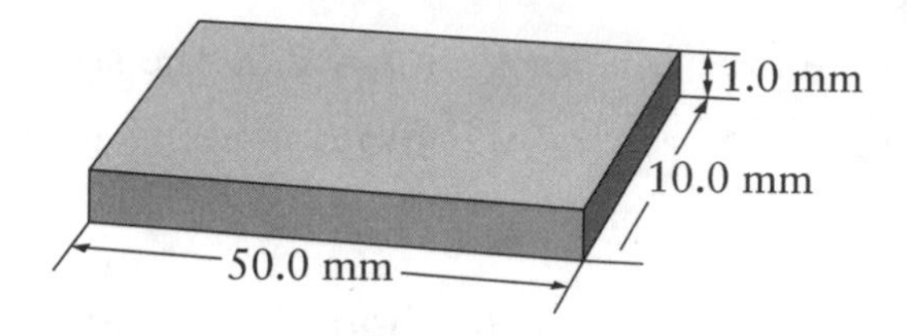

Calculate the surface area:

$$2(50.0 \text{ mm} \times 10.0 \text{ mm}) + 2(10.0 \text{ mm} \times 1.0 \text{ mm}) + 2(50.0 \text{ mm} \times 1.0 \text{ mm})$$
$$= 2(500.\text{ mm}^2) + 2(10.\text{ mm}^2) + 2(50.\text{ mm}^2) = 1120.\text{ mm}^2.$$

Then, $1120.\text{ mm}^2 \times 1\ \mu\text{m} \times \left(\frac{10^{-3}\text{ mm}}{1\ \mu\text{m}}\right) = 1.120\text{ mm}^3 = \text{volume of Ag(s)}$

For $Ag^+ + e^- \longrightarrow Ag$, we require (1 mol e^-/1 mol Ag).

mass of Ag = volume × density

$$= (1.120\text{ mm}^3) \times \left(\frac{1\text{ cm}}{10\text{ mm}}\right)^3 \times \left(\frac{10.5\text{ g}}{1\text{ cm}^3}\right) = 1.17\overline{6} \times 10^{-2}\text{ g}$$

$$\text{moles of e}^- = 1.17\overline{6} \times 10^{-2}\text{ g} \times \left(\frac{1\text{ mol Ag}}{107.87\text{ g Ag}}\right) \times \left(\frac{1\text{ mol e}^-}{1\text{ mol Ag}}\right)$$
$$= 1.09 \times 10^{-4}\text{ mol e}^-$$

$$\text{charge required} = (1.09 \times 10^{-4}\text{ mol e}^-) \times \left(\frac{9.65 \times 10^4\text{ C}}{1\text{ mol e}^-}\right) = 10.5\text{ C}$$

charge = current × time

$$\text{time} = \frac{\text{charge}}{\text{current}} = \frac{10.5\text{ C}}{0.100\text{ C/s}} = 105\text{ s}$$

18.109 (a) $MCl_3 \longrightarrow M^{3+} + 3\ Cl^-$ $\quad M^{3+} + 3\ e^- \longrightarrow M(s)$

First, determine the number of moles of electrons consumed; the number of moles of M^{3+} reduced is three times this number.

$$\text{charge used} = (6.63\text{ h}) \times \left(\frac{3600\text{ s}}{1\text{ h}}\right) \times \left(\frac{0.70\text{ C}}{1\text{ s}}\right) = 1.6\overline{7} \times 10^4\text{ C}$$

$$\text{number of moles of e}^- = (1.6\overline{7} \times 10^4\text{ C}) \times \left(\frac{1\text{ mol e}^-}{9.65 \times 10^4\text{ C}}\right) = 0.17\overline{3}\text{ mol e}^-$$

$$\text{number of moles of } M^{3+} \text{ (and M)} = 0.17\overline{3} \text{ mol } e^- \times \frac{1 \text{ mol } M^{3+}}{3 \text{ mol } e^-}$$

$$= 0.057\overline{7} \text{ mol } M^{3+}$$

$$\text{molar mass M} = \frac{3.00 \text{ g}}{0.057\overline{7} \text{ mol}} = 52 \text{ g/mol}$$

(b) The metal is chromium.

18.111 Set up a cell in which one electrode is the silver-silver chloride electrode and the other is the hydrogen electrode. The E of this cell will be sensitive to $[H^+]$ and therefore can be used to obtain pH.

$2\,H^+(aq) + 2\,e^- \longrightarrow H_2(g) \quad E°(\text{cathode}) = 0.00 \text{ V}$

$2\,AgCl(s) + 2\,e^- \longrightarrow 2\,Ag(s) + 2\,Cl^-(aq) \quad E°(\text{anode}) = +0.22 \text{ V}$

$2\,AgCl(s) + H_2(g, 1 \text{ atm}) \longrightarrow 2\,Ag(s) + 2\,Cl^-(aq) + 2\,H^+(aq) \quad E° = 0.22 \text{ V}$

If $[Cl^-] = 1.0$ mol/L

(a) $E = E° - \left(\frac{0.0257 \text{ V}}{2}\right) \ln([H^+]^2) = 0.22 \text{ V} - (0.0257) \ln[H^+]$

$\ln[H^+] = 2.303 \log[H^+] = -2.303(\text{pH})$, so $E = 0.22 \text{ V} + 0.0592 \text{ V} \times \text{pH}$, and

$$\text{pH} = \frac{E - 0.22 \text{ V}}{0.0592 \text{ V}}$$

So, by measuring E of this cell, pH can be obtained.

(b) pOH = 14.00 − pH

18.113 (a) charge consumed $= 4.00 \text{ A} \times 1800 \text{ s} = 7.20 \times 10^3 \text{ C}$

$$\text{moles of } e^- = (7.20 \times 10^3 \text{ C}) \times \left(\frac{1 \text{ mol } e^-}{9.65 \times 10^4 \text{ C}}\right) = 7.46 \times 10^{-2} \text{ mol } e^-$$

The reaction at the platinum anode is

$$2\,H_2O \longrightarrow O_2 + 4\,H^+ + 4e^-$$

and the ratio of $H^+(H_3O^+)$ to e^- is

$$\frac{1 \text{ mol } H_3O^+}{1 \text{ mol } e^-}$$

Therefore, amount (moles) of $H_3O^+ = 7.46 \times 10^{-2}$ mol H_3O^+

(b) $[H_3O^+] = \dfrac{7.46 \times 10^{-2} \text{ mol } H_3O^+}{0.2000 \text{ L}} = 0.373$ mol/L

pH = −log(0.373) = 0.43 (very acidic!)

APPLIED EXERCISES

18.115 (a) The reaction of hydrogen with nitrogen produces ammonia; water is needed so that ammonia can ultimately serve as an electrolyte.

$2\,H_2O(l) + 3\,H_2(g) + N_2(g) \longrightarrow 2\,NH_4^+(aq) + 2\,OH^-(aq)$

(b) $\Delta G_r° = 2\Delta G°(NH_4^+ aq) + 2\Delta G°(OH^-, aq) - 2\Delta G°(H_2O, l) =$

$2(-79.31\text{ kJ/mol}) + 2(-157.24\text{ kJ/mol}) - 2(-237.13\text{ kJ/mol}) = 1.16\text{ kJ/mol}$

$$\frac{1.16\text{ kJ}}{\text{mol}} \times \left(\frac{1\text{ mol}}{28.02\text{ g}}\right) \times \left(\frac{10^3\text{ g}}{\text{kg}}\right) \times 28.0\text{ kg} = 1.16 \times 10^3\text{ kJ or } 1.16\text{ MJ}$$

(c) This fuel cell is not thermodynamically feasible.

18.117 In this reaction, $O_2(0) \longrightarrow 2\,O(-2)$. Thus, there is a transfer of 4 e^- for each O_2, or $6 \times 4\,e^- = 24\,e^-$ for each glucose molecule oxidized. For the reaction:

$\Delta G_r° = (6)(-394.36) + (6)(-237.13) - (-910) = -2879\text{ kJ/mol}$

$$\text{Then, current} = \left(\frac{1.0 \times 10^7\text{ J}}{1\text{ day}}\right) \times \left(\frac{1\text{ mol glucose}}{2.879 \times 10^6\text{ J}}\right) \times \left(\frac{24\text{ mol e}^-}{1\text{ mol glucose}}\right) \times \left(\frac{9.65 \times 10^4\text{ C}}{1\text{ mol e}^-}\right) \times \left(\frac{1\text{ day}}{86\,400\text{ s}}\right) = 93\text{ C/s } (= 93\text{ A})$$

18.119 The wording of this exercise suggests that K^+ ions participate in an electrolyte concentration cell reaction. Therefore, $E° = 0.00$ V, because the two half cells would be identical under standard conditions.

Then,

$$E = E° - \left(\frac{0.0257\text{ V}}{n}\right) \ln\left(\frac{[K^+_{out}]}{[K^+_{in}]}\right) = 0.00\text{ V} - \left(\frac{0.0257\text{ V}}{1}\right) \ln\left(\frac{1}{30}\right)$$

$$= +0.09\text{ V} \quad \text{and} \quad E = 0.00\text{ V} - \left(\frac{0.0257\text{ V}}{1}\right) \ln\left(\frac{1}{20}\right) = +0.08\text{ V}$$

The range of potentials is 0.08 V to 0.09 V.

18.121 (a) Closed: energy is exchanged, but not matter.

(b) Open: matter, in the form of fuel, continuously enters the system, and reaction products are continuously expelled.

(c) Closed: energy, but not matter, is exchanged with the surroundings.

INTEGRATED EXERCISES

18.123 $F_2(g) + 2\,e^- \longrightarrow 2\,F^-(aq)\quad E°(\text{cathode}) = +2.87\ V$

$2\,HF(aq) \longrightarrow F_2(g) + 2\,H^+(aq) + 2\,e^-\quad E°(\text{anode}) = +3.03\ V$

$2\,HF(aq) \longrightarrow 2\,H^+(aq) + 2\,F^-(aq)\quad E° = -0.16\ V$

For the above reaction, $K = \dfrac{[H^+]^2[F^-]^2}{[HF]^2}$

and $\ln K = \dfrac{nFE°}{RT}$ at 25°C $= \dfrac{nE°}{0.025\ 69\ V} = \dfrac{(2)(-0.16\ V)}{0.025\ 69\ V} = -12.4\overline{6}$

$K = 4 \times 10^{-6}$

18.125 $2\,Cl^-(aq) \longrightarrow Cl_2(g) + 2\,e^-$

$P_{Cl_2} = P_{total} - P_{H_2O} = 770.\ \text{Torr} - 17.54\ \text{Torr} = 752\ \text{Torr}$

$PV = nRT,\ n = \dfrac{PV}{RT}$

$$n_{Cl_2} = \frac{\left(\frac{752}{760}\right)\text{atm} \times 20.0\ L}{0.082\ 06\ L\cdot atm/K\cdot mol \times 293\ K} = 0.823\ \text{mol}\ Cl_2$$

$$\text{time} = 0.823\ \text{mol}\ Cl_2 \times \left(\frac{2\ \text{mol}\ e^-}{1\ \text{mol}\ Cl_2}\right) \times \left(\frac{9.65 \times 10^4\ C}{1\ \text{mol}\ e^-}\right) \times \left(\frac{1\ A}{1\ C/s}\right) \times \left(\frac{1}{2.0\ A}\right)$$

$= 7.94 \times 10^4\ s$

18.127 $Mg^{2+} + 2\,e^- \longrightarrow Mg$

$2\,Cl^- \longrightarrow Cl_2 + 2\,e^-$

$MgCl_2(l) \longrightarrow Mg(s) + Cl_2(g)$

(a) For 24 hours $= 8.64 \times 10^4$ s, the charge in coulombs is

$10. \times 10^3\ C/s \times 8.64 \times 10^4\ s = 8.6\overline{4} \times 10^8\ C$

The number of moles of e^- supplied is

$$(8.6\overline{4} \times 10^8\ C) \times \left(\frac{1\ \text{mol}\ e^-}{9.65 \times 10^4\ C}\right) = 8.9\overline{5} \times 10^3\ \text{mol}\ e^-$$

Because $Mg^{2+} + 2\,e^- \longrightarrow Mg$, $(8.9\overline{5} \times 10^3\ \text{mol}\ e^-) \times \left(\dfrac{1\ \text{mol Mg}}{2\ \text{mol}\ e^-}\right) =$

$4.4\overline{8} \times 10^3$ mol Mg and $(4.4\overline{8} \times 10^3\ \text{mol Mg}) \times \left(\dfrac{24.31\ \text{g Mg}}{1\ \text{mol Mg}}\right) \times \left(\dfrac{1\ \text{kg}}{10^3\ \text{g}}\right)$

$= 10\overline{9}$ kg Mg or 1.1×10^2 kg Mg (estimated)

(b) Use the fact that 1 mol $Cl_2(g)$ occupies 22.41 L at STP. Then,

$$(4.4\overline{8} \times 10^3\ \text{mol}\ Cl_2) \times \left(\frac{22.41\ L\ Cl_2}{1\ \text{mol}\ Cl_2}\right) = 1.0 \times 10^5\ L\ Cl_2\quad \text{(estimated)}$$

CHAPTER 19
THE ELEMENTS:
THE FIRST FOUR MAIN GROUPS

EXERCISES

Periodic Trends

19.1 (a) saline (b) molecular (c) molecular (d) metallic

19.3 (a) $H_2(g) + Cl_2(g) \xrightarrow{\text{light}} 2\ HCl(g)$
(b) $H_2(g) + 2\ Na(l) \xrightarrow{\Delta} 2\ NaH(s)$
(c) $P_4(s) + 6\ H_2(g) \longrightarrow 4\ PH_3(g)$

19.5 (a) $Li^+\ [H:]^-$ (b) $H—\overset{H}{\underset{H}{Si}}—H$ (c) $H—\underset{H}{\ddot{Sb}}—H$

19.7 (a) Binary hydrogen compounds increase in acidity from left to right across Period 2. As the element attached to hydrogen becomes more electronegative, the electrons are pulled more toward that element. The hydrogen can then be transferred to another molecule as H^+ (e.g., $H^+ + H_2O \rightarrow H_3O^+$), and thus the compound serves as an acid. (b) Oxides also become more acidic (and less basic) from left to right across Period 2.

Hydrogen

19.9 In the majority of its reactions, hydrogen acts as a reducing agent. Examples are $2\ H_2(g) + O_2(g) \longrightarrow 2\ H_2O(l)$ and various ore reduction processes, such as $NiO(s) + H_2(g) \xrightarrow{\Delta} Ni(s) + H_2O(g)$. With highly electropositive elements, such as the alkali metals, $H_2(g)$ acts as an oxidizing agent and forms metal hydrides, for example, $2\ K(s) + H_2(g) \longrightarrow 2\ KH(s)$.

19.11 (a) $CO(g) + H_2O(g) \xrightarrow{400\ °C,\ Fe/Cu} CO_2(g) + H_2(g)$
(b) $2\ Li(s) + 2\ H_2O(l) \longrightarrow 2\ LiOH(aq) + H_2(g)$

(c) $Mg(s) + 2\ H_2O(l) \longrightarrow Mg(OH)_2(aq) + H_2(g)$
(d) $2\ K(s) + H_2(g) \longrightarrow 2\ KH(s)$

19.13 (a) $CH_4(g) + H_2O(g) \longrightarrow CO(g) + 3\ H_2(g)$

$$\Delta H_r^\circ = 1 \times \Delta H_f^\circ(CO, g) - [1 \times \Delta H_f^\circ(CH_4, g) + 1 \times \Delta H_f^\circ(H_2O, g)]$$
$$= 1 \times (-110.53\ \text{kJ/mol}) - [1 \times (-74.81\ \text{kJ/mol}) + 1 \times (-241.82\ \text{kJ/mol})] = +206.10\ \text{kJ/mol}$$

(b) $CO(g) + H_2O(g) \longrightarrow CO_2(g) + H_2(g)$

$$\Delta H_r^\circ = 1 \times \Delta H_f^\circ(CO_2, g) - [1 \times \Delta H_f^\circ(CO, g) + 1 \times \Delta H_f^\circ(H_2O, g)]$$
$$= 1 \times (-393.51\ \text{kJ/mol}) - [1 \times (-110.53\ \text{kJ/mol}) + 1 \times (-241.82\ \text{kJ/mol})] = -41.16\ \text{kJ/mol}$$

(c) $CH_4(g) + 2\ H_2O(g) \rightleftharpoons CO_2(g) + 4\ H_2(g)$

$$\Delta H_r^\circ(c) = \Delta H_r^\circ(a) + \Delta H_r^\circ(b) = (206.10 - 41.16)\ \text{kJ/mol} = +164.94\ \text{kJ/mol}$$

19.15 $N_2(g) + 3\ H_2(g) \longrightarrow 2\ NH_3(g)$

$$\text{amount (moles) of } H_2 \text{ used} = (1.5 \times 10^9\ \text{kg } NH_3) \times \left(\frac{1\ \text{mol } NH_3}{0.017\ \text{kg } NH_3}\right) \times \left(\frac{3\ \text{mol } H_2}{2\ \text{mol } NH_3}\right) = 1.3 \times 10^{11}\ \text{mol } H_2$$

$$\text{amount (moles) of } H_2 \text{ produced} = (3 \times 10^8\ \text{kg } H_2) \times \left(\frac{1\ \text{mol } H_2}{0.002\ \text{kg } H_2}\right) = 1.\bar{5} \times 10^{11}\ \text{mol } H_2$$

$$\text{fraction used} = \frac{1.3 \times 10^{11}\ \text{mol } H_2}{1.\bar{5} \times 10^{11}\ \text{mol } H_2} = 0.9$$

Group 1: The Alkali Metals

19.17 (a) red (b) violet (c) yellow (d) violet

19.19 (a) $4\ Li(s) + O_2(g) \longrightarrow 2\ Li_2O(s)$
(b) $6\ Li(s) + N_2(g) \xrightarrow{\Delta} 2\ Li_3N(s)$
(c) $2\ Na(s) + 2\ H_2O(l) \longrightarrow 2\ NaOH(aq) + H_2(g)$
(d) $4\ KO_2(s) + 2\ H_2O(g) \longrightarrow 4\ KOH(s) + 3\ O_2(g)$

19.21 (a) $Ca(s) + H_2(g) \xrightarrow{\Delta} CaH_2(s)$
(b) $2\ NaHCO_3(s) \longrightarrow Na_2O(s) + 2\ CO_2(g) + H_2O(g)$

19.23 (a) sodium chloride, NaCl
(b) potassium chloride, KCl

19.25 1 mol $Na_2CO_3 \cdot 10H_2O$ yields 1 mol Na_2CO_3 in water.

mass of $Na_2CO_3 \cdot 10H_2O = 0.250 \text{ L} \times 0.100 \text{ mol/L}$
$\times 286.15 \text{ g } Na_2CO_3 \cdot 10H_2O\text{/mol} = 7.15 \text{ g } Na_2CO_3 \cdot 10H_2O$

Group 2: The Alkaline Earth Metals

19.27 Be is the weakest reducing agent. Mg is stronger, but weaker than the remaining members of the group, all of which have approximately the same reducing strength. This effect is related to the very small radius of the Be^{2+} ion, 27 pm; its strong polarizing power introduces much covalent character into its compounds. Thus, Be attracts electrons more strongly and does not release them as readily as other members of the group. Mg^{2+} is also a small ion, 58 pm, so the same reasoning applies to it also, but to a lesser extent. The remaining ions of the group are considerably larger, release electrons more readily, and are better reducing agents.

19.29 (a) beryl (b) limestone or dolomite (c) dolomite

19.31 (a) magnesium sulfate heptahydrate, $MgSO_4 \cdot 7H_2O$
(b) calcium carbonate, $CaCO_3$
(c) magnesium hydroxide, $Mg(OH)_2$

19.33 (a) $2\text{ Al(s)} + 2\text{ OH}^-\text{(aq)} + 6\text{ H}_2\text{O(l)} \longrightarrow 2[\text{Al(OH)}_4]^-\text{(aq)} + 3\text{ H}_2\text{(g)}$
(b) $\text{Be(s)} + 2\text{ OH}^-\text{(aq)} + 2\text{ H}_2\text{O(l)} \longrightarrow [\text{Be(OH)}_4]^{2-}\text{(aq)} + \text{H}_2\text{(g)}$
Sodium ions are spectator ions in both (a) and (b) and can be omitted. The similarity of Be and Al in chemical reactions is an example of the diagonal relationship in the periodic table; namely, the similar chemical behavior of elements that are diagonal neighbors of each other, such as Be and Al.

19.35 (a) $\text{Mg(OH)}_2\text{(s)} + 2\text{ HCl(aq)} \longrightarrow \text{MgCl}_2\text{(aq)} + 2\text{ H}_2\text{O(l)}$
(b) $\text{Ca(s)} + 2\text{ H}_2\text{O(l)} \longrightarrow \text{Ca(OH)}_2\text{(aq)} + \text{H}_2\text{(g)}$
(c) $\text{BaCO}_3\text{(s)} \xrightarrow{\Delta} \text{BaO(s)} + \text{CO}_2\text{(g)}$

19.37 (a) $:\ddot{\underset{..}{\text{Cl}}}\text{—Be—}\ddot{\underset{..}{\text{Cl}}}:$ (b) 180° (c) *sp*

19.39 Epsom salts = $MgSO_4 \cdot 7H_2O$; molar mass = 246.48 g/mol

$$\text{mass \% } H_2O = \frac{7 \times 18.02 \text{ g/mol}}{246.48 \text{ g/mol}} \times 100\% = 51.18\%$$

Group 13: The Boron Family

19.41 The overall equation for the electrolytic reduction in the Hall process is
$4\ Al^{3+}(melt) + 6\ O^{2-}(melt) + 3\ C(s, gr) \longrightarrow 4\ Al(l) + 3\ CO_2(g)$

19.43 (a) boric acid, $B(OH)_3$
(b) alumina, Al_2O_3
(c) borax, $Na_2B_4O_7 \cdot 10H_2O$

19.45 (a) $B_2O_3(s) + 3\ Mg(l) \xrightarrow{\Delta} 2\ B(s) + 3\ MgO(s)$
(b) $2\ Al(s) + 3\ Cl_2(g) \longrightarrow 2\ AlCl_3(s)$
(c) $4\ Al(s) + 3\ O_2(g) \longrightarrow 2\ Al_2O_3(s)$

19.47 (a) The hydrate of $AlCl_3$, that is, $AlCl_3 \cdot 6H_2O$, functions as a deodorant and antiperspirant.
(b) α-Alumina is corundum. It is used as an abrasive in sandpaper.
(c) $B(OH)_3$ is an antiseptic and insecticide.

19.49 (a) $B_2H_6(g) + 6\ H_2O(l) \longrightarrow 2\ B(OH)_3(aq) + 6\ H_2\ (g)$
(b) $B_2H_6(g) + 3\ O_2(g) \longrightarrow B_2O_3(g) + 3\ H_2O(l)$

Group 14: The Carbon Family

19.51 Silicon occurs widely in the Earth's crust in the form of silicates in rocks and as silicon dioxide in sand. It is obtained from quartzite, a form of quartz (SiO_2), by the following processes:
(1) reduction in an electric arc furnace
$SiO_2(s) + 2\ C(s) \longrightarrow Si(s, crude) + 2\ CO(g)$
(2) purification of the crude product in two steps
$Si(s, crude) + 2\ Cl_2(g) \longrightarrow SiCl_4(l)$
followed by reduction with hydrogen to the pure element
$SiCl_4(l) + 2\ H_2(g) \longrightarrow Si(s, pure) + 4\ HCl(g)$

19.53 In diamond, carbon is sp^3 hybridized and forms a tetrahedral, three-dimensional network structure, which is extremely rigid. Graphite carbon is sp^2 hybridized and planar, and its application as a lubricant results from the fact that the two-dimensional sheets can "slide" across one another, thereby reducing friction. In graphite, the unhybridized *p*-electrons are free to move from one carbon atom to another, which results in its high electrical conductivity. In diamond, all electrons are localized in sp^3 hybridized C—C σ-bonds, so diamond is a poor conductor of electricity.

19.55 (a) carborundum, SiC
(b) silica, SiO_2
(c) zircon, $ZrSiO_4$

19.57 (a) $SiCl_4(l) + 2\ H_2(g) \longrightarrow Si(s) + 4\ HCl(g)$
(b) $SiO_2(s) + 3\ C(s) \xrightarrow{2000°C} SiC(s) + 2\ CO(g)$
(c) $Ge(s) + 2\ F_2(g) \longrightarrow GeF_4(s)$
(d) $CaC_2(s) + 2\ H_2O(l) \longrightarrow Ca(OH)_2(s) + C_2H_2(g)$

19.59

$$\left[\begin{array}{ccc} & :\ddot{O}: & \\ & | & \\ :\ddot{\underset{..}{O}}- & Si & -\ddot{\underset{..}{O}}: \\ & | & \\ & :\underset{..}{O}: & \end{array}\right]^{4-}$$

Formal charges: Si = 0, O = −1
oxidation numbers: Si = +4, O = −2
This is an AX_4 VSEPR structure; therefore the shape is tetrahedral.

19.61 Silica is SiO_2.

$$\text{mass percentage Si} = \frac{\text{molar mass Si}}{\text{molar mass SiO}_2} \times 100\% = \frac{28.09\ \text{g/mol}}{60.09\ \text{g/mol}} \times 100\% = 46.75\%$$

19.63

$$\text{mass of HF required} = 2.00 \times 10^{-3}\ \text{g} \times \left(\frac{1\ \text{mol SiO}_2}{60.09\ \text{g SiO}_2}\right) \times \left(\frac{6\ \text{mol HF}}{1\ \text{mol SiO}_2}\right) \times \left(\frac{20.01\ \text{g HF}}{1\ \text{mol HF}}\right) = 4.00 \times 10^{-3}\ \text{g HF} = 4.00\ \text{mg HF}$$

19.65

$$\text{surface area} = (1.00\ \text{mol C}) \times \left(\frac{12.01\ \text{g C}}{1\ \text{mol C}}\right) \times \left(\frac{2.0 \times 10^3\ \text{m}^2}{1.0\ \text{g}}\right) = 2.4 \times 10^4\ \text{m}^2$$

19.67 (a) The $Si_2O_7^{6-}$ ion is built from two SiO_4^{4-} tetrahedral ions in which the silicate tetrahedra share one O atom. See Figs. 19.43 and 19.44a. This is the only case in which one O is shared.
(b) The pyroxenes, for example, jade, $NaAl(SiO_3)_2$, consist of chains of SiO_4 units in which two O atoms are shared by neighboring units. The repeating unit has the formula SiO_3^{2-}. See Fig. 19.46.

SUPPLEMENTARY EXERCISES

19.69 (a)

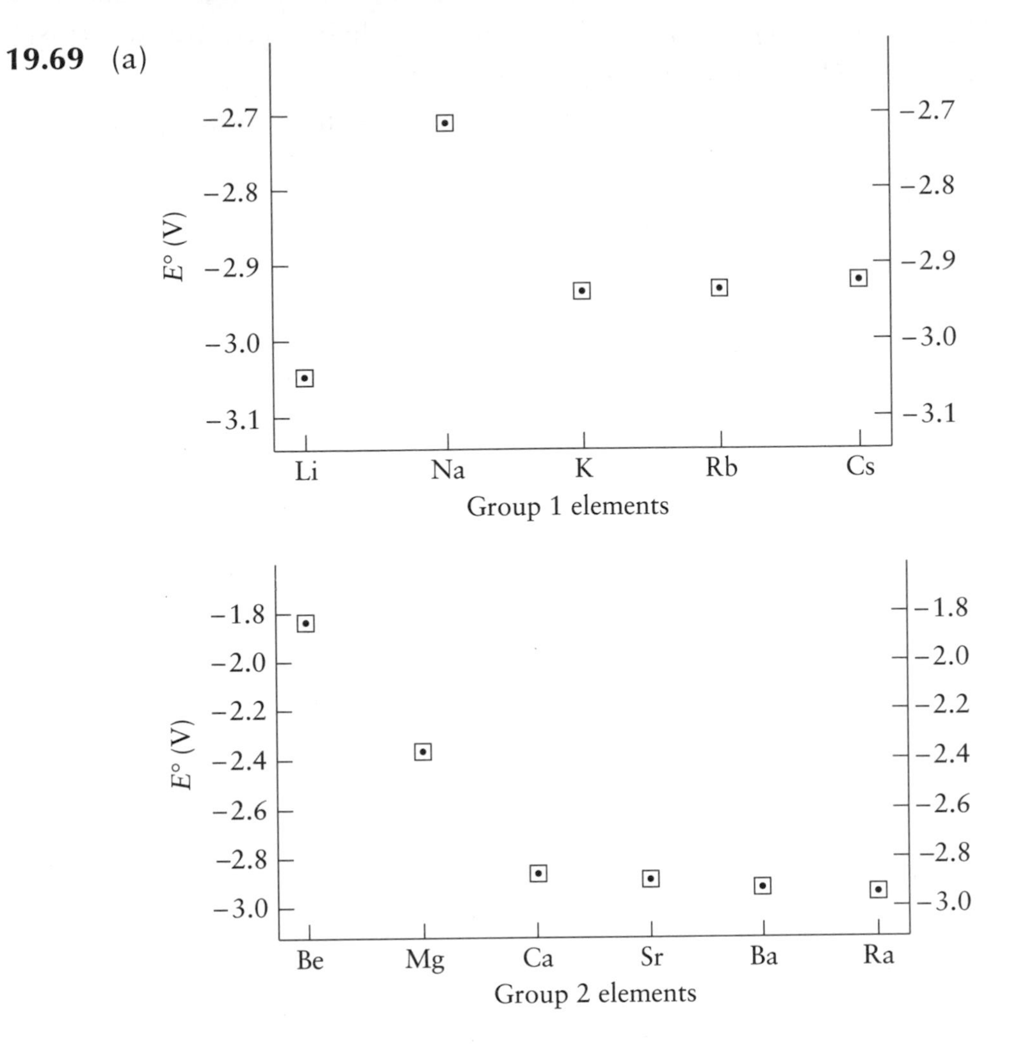

(b) For both groups, the trend in standard potentials with increasing atomic number is downward overall (they become more negative), but lithium is anomalous. This overall downward trend makes sense, because we expect that it is easier to remove electrons that are farther away from the nuclei. However, because there are several factors that influence ease of removal, the trend is not smooth. The potentials are a net composite of the free energies of sublimation of solids, dissociation of gaseous molecules, ionization enthalpies, and enthalpies of hydration of gaseous ions. The origin of the anomalously strong reducing power of Li is the strongly exothermic energy of hydration of the very small Li^+ ion, which favors the ionization of the element in aqueous solution.

19.71 (a, b)

$H_2(g) + F_2(g) \longrightarrow 2\ HF(g)$, explosive
$H_2(g) + Cl_2(g) \longrightarrow 2\ HCl(g)$, explosive
$H_2(g) + Br_2(g) \longrightarrow 2\ HBr(g)$, vigorous
$H_2(g) + I_2(g) \longrightarrow 2\ HI(g)$, less vigorous

The word *vigor* as used here has two components, a thermodynamic one and a kinetic one. $\Delta G_f°$ is most negative for HF and becomes slightly positive for HI. So the formation of HF, HCl, and HBr are all thermodynamically spontaneous. In this case, the kinetics parallel the thermodynamic spontaneity, although this parallel behavior is not necessarily true in other systems.

(c) hydrofluoric acid, hydrochloric acid, hydrobromic acid, hydroiodic acid

19.73 (a) Ba^{2+} $[:\ddot{O}—\ddot{O}:]^{2-}$ (b) H—Be—H

(c) Na^+Na^+ $[:\ddot{O}—\ddot{O}:]^{2-}$ (d) $[Be(OH)_4]^{2-}$: four H—Ö— groups bonded to Be

19.75 (a)

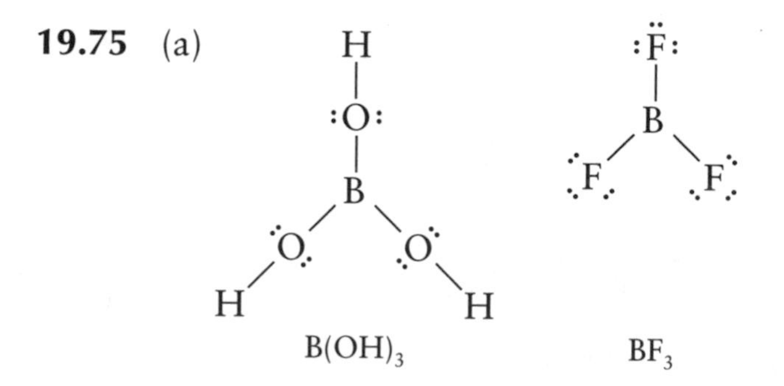

(b) VSEPR predicts a trigonal planar structure with 120° bond angles for each.

(c) sp^2 hybridization correlates with a trigonal planar arrangement

19.77 (a)

Element		Ionization energy, kJ/mol		Atomic radius, pm	
Group 13	B	799	Gen. decreasing ↓	88	Increasing ↓
	Al	577		143	
	Ga	577		153	
	In	556		167	
	Tl	590		171	
Group 14	C	1090	Gen. decreasing ↓	77	Increasing ↓
	Si	786		118	
	Ge	784		122	
	Sn	707		158	
	Pb	716		175	

(b) The ionization energies generally decrease down a group. As the atomic number of an element increases, atomic shells and subshells that are farther from the nucleus are filled. The outermost valence electrons are consequently easier to remove. The radii increase down a group for the same reason. The radii are primarily determined by the outer shell electrons, which are farther from the nucleus in the heavy elements.
(c) The trends correlate well with elemental properties, for example, the greater ease of outermost electron removal correlates with increased metallic character, that is, ability to form positive ions by losing one or more electrons.

19.79 (a) $2\ AlCl_3(s) + 3\ H_2O(l) \longrightarrow 6\ HCl(g) + Al_2O_3(s)$
(b) $B_2H_6(g) \xrightarrow{\text{high temp.}} 2\ B(s) + 3\ H_2(g)$
(c) $4\ BF_3 + 3\ BH_4^- \xrightarrow{\text{organic solvent}} 3\ BF_4^- + 2\ B_2H_6$

19.81 In the majority of its reactions, hydrogen acts as a reducing agent, that is, $H_2(g) \longrightarrow 2\ H^+(aq) + 2\ e^-$, $E° = 0$ V. In these reactions, hydrogen resembles Group 1 elements, such as Na and K. However, as described in the text and in the answer to Exercise 19.9, it may also act as an oxidizing agent; that is, $H_2(g) + 2\ e^- \longrightarrow 2\ H^-(aq)$, $E° = -2.25$ V. In these reactions, hydrogen resembles Group 17 elements, such as Cl and Br. Consequently, H_2 will oxidize elements with standard reduction potentials more negative than -2.25 V, such as the alkali and alkaline earth metals (except Be). The compounds formed are hydrides and contain the H^- ion; the singly charged negative ion is reminiscent of the halide ions. Hydrogen also forms diatomic molecules and covalent bonds like the halogens.
The atomic radius of H is 78 pm, which compares rather well to that of F (64 pm) but not as well to that of Li (157 pm). The ionization energy of H is 1310 kJ/mol which is similar to that of F (1680 kJ/mol) but not similar to that of Li (519 kJ/mol). The electron affinity of H is +73 kJ/mol, that of F is +328 kJ/mol, and that of Li is 60 kJ/mol. So in its atomic radius and ionization energy, H more closely resembles the Period 2 halogen, fluorine, in Group 17, than the Period 2 alkali metal, lithium, in Group 1; whereas in electron affinity, it more closely resembles lithium, Group 1. In electronegativity, H does not resemble elements in either Group 1 or Group 17, although its electronegativity is somewhat closer to those of Group 1. Consequently, hydrogen could be placed in either Group 1 or Group 17. But it is probably best to think of hydrogen as a unique element that has properties in common with both metals and nonmetals;

therefore, it should probably be centered in the periodic table, as it is shown in the table in the text.

19.83 (a)

Ion	Radius, pm	Polarizing ability (×1000)	Ion	Radius, pm	Polarizing ability (×1000)
Li^+	58	17	Be^{2+}	27	74
Na^+	102	9.8	Mg^{2+}	72	28
K^+	138	7.2	Ca^{2+}	100	20
Rb^+	149	6.7	Sr^{2+}	116	17
Cs^+	170	5.9	Ba^{2+}	136	15

(b) These data roughly support the diagonal relationship. Li^+ is more like Mg^{2+} than Be^{2+}, and Na^{2+} is more like Ca^{2+} than Mg^{2+}; but further down the group, the correlation fails. Charge divided by r^3 would be a better measure of polarizing ability.

APPLIED EXERCISES

19.85 Several reactions occur, depending on the silica to alkali ratio:

(1) $SiO_2(s) + OH^-(aq) \longrightarrow HSiO_3^-(aq)$

(2) $SiO_2(s) + 2\ OH^-(aq) \longrightarrow H_2O(l) + SiO_3^{2-}(aq)$ (metasilicate ion)

(3) $SiO_2(s) + 4\ OH^-(aq) \longrightarrow 2\ H_2O(l) + SiO_4^{4-}(aq)$ (orthosilicate ion)

(4) $2\ SiO_2(s) + 6\ OH^-(aq) \longrightarrow 3\ H_2O(l) + Si_2O_7^{6-}(aq)$ (pyrosilicate ion)

19.87 Glasses form from materials in which inhibited recrystallization occurs. Such materials are likely to be covalently bonded materials involving extensive network structures and materials with large complex molecules. On the basis of these principles, we predict that, of the substances listed, the following would probably solidify as a glass: (a) tar; (c) molten granite; (e) low-density polyethylene; and (f) a highly branched polymer. Any substance can form a glass if cooled rapidly enough, but (b) sodium chloride and (d) water would normally form crystalline solids.

19.89 (a) The "hardness" of water is due to the presence of calcium and magnesium salts (particularly their hydrogen carbonates). In laundering and bathing, Ca^{2+} and Mg^{2+} cations convert soluble Na^+ soaps to insoluble Ca^{2+} and Mg^{2+} soaps (thereby reducing the detergent efficiency).

(b) Softening can be achieved by removing the Ca^{2+} and Mg^{2+} cations before they form these insoluble compounds with the use of lime.

$$Ca(HCO_3)_2(aq) + Ca(OH)_2(aq) \longrightarrow 2\ CaCO_3(s) + 2\ H_2O(l)$$
$$Mg(HCO_3)_2(aq) + Ca(OH_2)(aq) \longrightarrow Mg(OH)_2(s) + Ca(HCO_3)_2(aq)$$

INTEGRATED EXERCISES

19.91 The overall reaction is

$$MgCl_2(l) \longrightarrow Mg(l) + Cl_2(g)$$

The half-reaction involving Mg is

$$Mg^{2+}(l) + 2\ e^- \longrightarrow Mg(l)$$

(a) $$\text{mass of Mg} = (100.\ \text{A}) \times (1.5\ \text{h}) \times \left(\frac{3600\ \text{s}}{1\ \text{h}}\right) \times \left(\frac{1\ \text{mol } e^-}{96\ 500\ \text{C}}\right) \times \left(\frac{1\ \text{mol Mg}}{2\ \text{mol } e^-}\right) \times \left(\frac{24.31\ \text{g Mg}}{1\ \text{mol Mg}}\right) = 68\ \text{g Mg}$$

(b) At 273 K and 1.00 atm, 1 mol of gas occupies 22.41 L.

$$\text{volume of } Cl_2 = (1.000 \times 10^6\ \text{g Mg}) \times \left(\frac{1\ \text{mol Mg}}{24.31\ \text{g Mg}}\right) \times \left(\frac{1\ \text{mol } Cl_2}{1\ \text{mol Mg}}\right) \times \left(\frac{22.41\ \text{L } Cl_2}{1\ \text{mol } Cl_2}\right) = 9.21 \times 10^5\ \text{L } Cl_2$$

19.93 $$\text{volume of } H_2(g) = (20.0\ \text{g } WO_3) \times \left(\frac{1\ \text{mol } WO_3}{231.85\ \text{g } WO_3}\right) \times \left(\frac{3\ \text{mol } H_2}{1\ \text{mol } WO_3}\right) \times \left(\frac{22.41\ \text{L } H_2}{1\ \text{mol } H_2}\right) = 5.80\ \text{L } H_2$$

19.95

```
    H           H     H     H
    |            \   / \   /
    B              B     B
   / \            / \   / \
  H   H          H   H     H
```

(1) $B(s) + \frac{3}{2} H_2(g) \longrightarrow BH_3(g)$ $\quad \Delta H_f^\circ = +100$ kJ/mol

(2) $B(s) \longrightarrow B(g)$ $\quad \Delta H_f^\circ = +563$ kJ/mol

(3) $\frac{1}{2} H_2(g) \longrightarrow H(g)$ $\quad \Delta H_f^\circ = +218$ kJ/mol

(4) $2\ B(s) + 3\ H_2(g) \longrightarrow B_2H_6(g)$ $\quad \Delta H_f^\circ = 36$ kJ/mol

(a) Reverse equation 1, multiply equation 3 by 3 and add to equation 2:

$$BH_3(g) \longrightarrow B(s) + \tfrac{3}{2} H_2(g) \qquad \Delta H^\circ = -100\ \text{kJ/mol}$$
$$B(s) \longrightarrow B(g) \qquad \Delta H^\circ = +563\ \text{kJ/mol}$$
$$\tfrac{3}{2} H_2(g) \longrightarrow 3\ H(g) \qquad \Delta H^\circ = 3 \times (218)\ \text{kJ/mol}$$
$$BH_3(g) \longrightarrow B(g) + 3\ H(g) \qquad \Delta H^\circ = +1117\ \text{kJ/mol} = 3 \times \Delta H(B - H)$$

$$\Delta H(B - H) = +372\ \text{kJ/mol}$$

Assume terminal B—H bonds have the same bond enthalpy as B—H bonds in BH_3, that is, 372 kJ/mol.

Reverse equation 4, multiply equation 2 by 2 and equation 3 by 6. Then add.

$B_2H_6(g) \longrightarrow 2\ B(s) + 3\ H_2(g) \quad \Delta H° = -36$ kJ/mol

$2\ B(s) \longrightarrow 2\ B(g) \quad \Delta H° = 2 \times 563$ kJ/mol

$3\ H_2(g) \longrightarrow 6\ H\ (g) \quad \Delta H° = 6 \times 218$ kJ/mol

$B_2H_6(g) \longrightarrow 2\ B(g) + 6\ H(g) \quad \Delta H° = 2398$ kJ/mol

$\Delta H° = 2398 \text{ kJ/mol} = 4 \times \Delta H(\text{B—H}) + 4 \times \Delta H(\text{B—H—B})$

$$\Delta H(\text{B—H—B}) = \frac{2398 \text{ kJ/mol} - 4 \times 372 \text{ kJ/mol}}{4}$$

$$= 228 \text{ kJ/mol}$$

As bond length and bond enthalpy are (very roughly) inversely related, the stronger terminal B—H bonds are expected to be shorter.

19.97 $2\ CO(g) + O_2(q) \longrightarrow 2\ CO_2(g)$

$\Delta H° = \Delta H°(\text{products}) - \Delta H°(\text{reactants})$

$\Delta H° = 2(-393.51 \text{ kJ/mol}) - [(2)(-110.53 \text{ kJ/mol}) + 0]$

$= -565.96 \text{ kJ/mol} = -565.96 \times 10^3 \text{ J/mol}$

and $\Delta S° = S°(\text{products}) - S°(\text{reactants})$

$\Delta S° = (2)(213.74 \text{ J/K·mol}) - [(2)(+197.67 \text{ J/K·mol}) + 205.14 \text{ J/K·mol}]$

$= -173.00 \text{ J/K·mol}$

$\Delta G° = \Delta H° - T\Delta S° = (-565.96 \times 10^3 \text{ J/mol}) - (298 \text{ K})(-173.00 \text{ J/K·mol})$

$= -514.41 \text{ kJ/mol}$

The reaction is spontaneous below

$$T = \frac{-565.96 \times 10^3 \text{ J/mol}}{-173.00 \text{ J/K·mol}} = 3271 \text{ K}$$

19.99 (a) $SnO_2(s) + C(s) \longrightarrow Sn(s) + CO_2(g)$

$SnO_2(s) + 2\ CO(g) \longrightarrow Sn(s) + 2\ CO_2(g)$

(b) $\Delta G°$ can be determined in two ways: (1) from values of $\Delta G°$ of the substance involved in the reactions, or (2) from $\Delta G° = \Delta H° - T\Delta S°$. Because part (c) requests a discussion of temperature and spontaneity, we choose here the second method for the reaction: $SnO_2(s) + C(s) \longrightarrow Sn(s) + CO_2(g)$

$\Delta H° = \Delta H_f°(\text{products}) - \Delta H_f°(\text{reactants})$

$\Delta H° = [0 + (-393.51 \text{ kJ/mol})] - [(-580.7 \text{ kJ/mol}) + 0]$

$\Delta H° = +187.2 \text{ kJ/mol} = 187.2 \times 10^3 \text{ J/mol}$

and

$\Delta S° = S°(\text{products}) - S°(\text{reactants})$

$\Delta S° = [(51.55 + 213.74) - (52.3 + 5.740)] \text{ J/K·mol} = 207.3 \text{ J/K·mol}$

and

$\Delta G° = \Delta H° - T\Delta S°$

$\Delta G° = 187.2 \times 10^3$ J/mol $-$ (298 K) $\times$ (207.3 J/K·mol)

$\Delta G° = 125.4 \times 10^3$ J/mol = 125.4 kJ/mol

For the reaction at 980 K:

$SnO_2(s) + 2\ CO(g) \longrightarrow Sn(l) + 2\ CO_2(g)$ (m.p. Sn = 505 k)

Under standard conditions, the reaction is

$SnO_2(s) + 2\ CO(g) \longrightarrow Sn(s) + 2\ CO_2(g)$ (25°C)

$\Delta H° = \Delta H_f°(\text{products}) - \Delta H_f°(\text{reactants})$

$\Delta H° = [0 + (2)(-393.51\text{ kJ/mol})] - [(-580.7\text{ kJ/mol}) + (2)(-110.53\text{ kJ/mol})]$

$\Delta H° = +14.8$ kJ/mol $= 14.8 \times 10^3$ J/mol

and

$\Delta S° = S°(\text{products}) - S°(\text{reactants})$

$\Delta S° = \{[51.55 + (2)(213.74)] - [52.3 + (2)(197.67)]\}$ J/K·mol = 31.4 J/K·mol

and

$\Delta G° = \Delta H_r° - T\Delta S_r°$

$\Delta G° = 14.8 \times 10^3$ J/mol $-$ (298 K) $\times$ (31.4 J/K·mol)

$\Delta G° = 5.44 \times 10^3$ J/mol = 5.44 kJ/mol

(c) In the first case, the reaction is spontaneous at high temperatures. In the second case, the reaction is also spontaneous at high temperatures.

19.101 The cathode reaction is $Al^{3+}(\text{melt}) + 3\ e^- \longrightarrow Al(l)$

$$\text{charge consumed} = (8.0\text{ h}) \times \left(\frac{3600\text{ s}}{1\text{ h}}\right) \times (1.0 \times 10^5\text{ C/s}) = 2.88 \times 10^9\text{ C}$$

$$\text{mass of Al produced} = (2.88 \times 10^9\text{ C}) \times \left(\frac{1\text{ mol e}^-}{9.65 \times 10^4\text{ C}}\right) \times \left(\frac{1\text{ mol Al}}{3\text{ mol e}^-}\right) \times \left(\frac{26.98\text{ g Al}}{1\text{ mol Al}}\right) = 2.7 \times 10^5\text{ g Al}$$

19.103 (a) $\Delta H° = 1 \times (-986.09\text{ kJ/mol}) - [1 \times (-635.09\text{ kJ/mol}) + 1 \times (-285.83\text{ kJ/mol})] = -65.17\text{ kJ/mol} = -6.517 \times 10^4\text{ J/mol}$

(b) The heat absorbed by the water is

$+6.517 \times 10^4\text{ J} = \text{mass} \times c \times (t_{\text{final}} - t_{\text{initial}})$

$= 250.\text{ g} \times 4.184\text{ J/g·(°C)} \times \Delta t$

Solving for Δt, we get

$$\Delta t = \frac{6.517 \times 10^4\text{ J}}{250.\text{ g} \times 4.184\text{ J/g·(°C)}} = 62.3°\text{C}$$

CHAPTER 20
THE ELEMENTS:
THE LAST FOUR MAIN GROUPS

EXERCISES

Groups 15–18

20.1 (a) He, $1s^2$

(b) O, [He] $2s^2\ 2p^4$

(c) F, [He] $2s^2\ 2p^5$

(d) As, [Ar] $3d^{10}\ 4s^2\ 4p^3$

20.3 The S in SO_3 has an oxidation number of +6 and cannot lose more electrons. S in SO_2 has an oxidation number of +4 and can be oxidized further.

20.5 Refer to Appendix 2D.

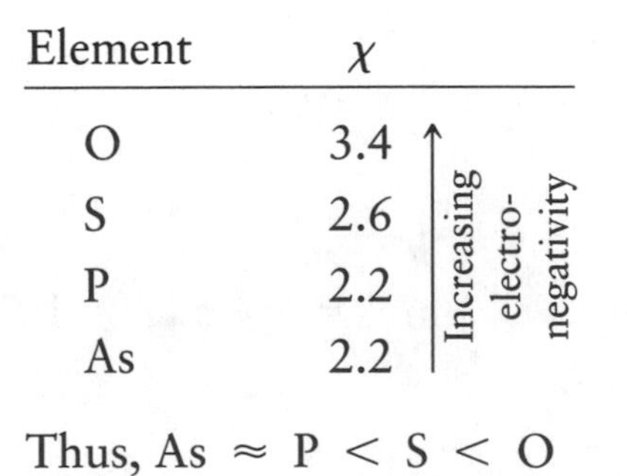

Element	χ
O	3.4
S	2.6
P	2.2
As	2.2

Increasing electro-negativity ↑

Thus, As ≈ P < S < O

Group 15

20.7 The first step is the liquefaction of air, which is 76% by mass nitrogen. Air is cooled to below its boiling point by a series of expansion and compression steps in a kind of refrigerator. Nitrogen gas is then obtained by distillation of liquid air. The nitrogen boils off at $-196°C$, but gases with higher boiling points, principally O_2, remain as a liquid. The pure nitrogen gas is then liquefied by repeating the process.

20.9 (a) HNO_2, nitrous acid
(b) NO, nitrogen monoxide or nitric oxide
(c) H_3PO_4, phosphoric acid
(d) N_2O_3, dinitrogen trioxide

20.11 (a) ammonium nitrate, NH_4NO_3
(b) magnesium nitride, Mg_3N_2
(c) calcium phosphide, Ca_3P_2
(d) hydrazine, H_2NNH_2

20.13 Let N_{ox} = oxidation number. The sum of the oxidation numbers must equal the charge on the species.
(a) $1 \times N_{ox}(N) + 1 \times N_{ox}(O) = 0$
$1 \times N_{ox}(N) + 1 \times (-2) = 0$
Therefore, $N_{ox}(N) = +2$
(b) $2 \times N_{ox}(N) + 1 \times N_{ox}(O) = 0$
$2 \times N_{ox}(N) + 1 \times (-2) = 0$
Therefore, $N_{ox}(N) = +1$
(c) $1 \times N_{ox}(H) + 1 \times N_{ox}(N) + 2 \times N_{ox}(O) = 0$
$1 \times (+1) + 1 \times N_{ox}(N) + 2 \times (-2) = 0$
Therefore, $N_{ox}(N) = -1 - (-4) = +3$
(d) $N_{ox}(N) = \frac{1}{3}(-1) = -\frac{1}{3}$

20.15 $CO(NH_2)_2 + 2\ H_2O \longrightarrow (NH_4)_2CO_3$

$$\text{mass of } (NH_4)_2CO_3 = (5.0\text{ kg urea}) \times \left(\frac{10^3\text{ g urea}}{1\text{ kg urea}}\right) \times \left(\frac{1\text{ mol urea}}{60.06\text{ g urea}}\right)$$
$$\times \left(\frac{1\text{ mol }(NH_4)_2CO_3}{1\text{ mol urea}}\right) \times \left(\frac{96.09\text{ g }(NH_4)_2CO_3}{1\text{ mol }(NH_4)_2CO_3}\right)$$
$$= 8.0 \times 10^3\text{ g (or 8.0 kg)}(NH_4)_2CO_3$$

20.17 (a) N_2O_3: HNO_2; N_2O_5: HNO_3
(b) $N_2O_3(g) + H_2O(l) \longrightarrow 2\ HNO_2(aq)$
$N_2O_5(g) + H_2O(l) \longrightarrow 2\ HNO_3(aq)$

20.19

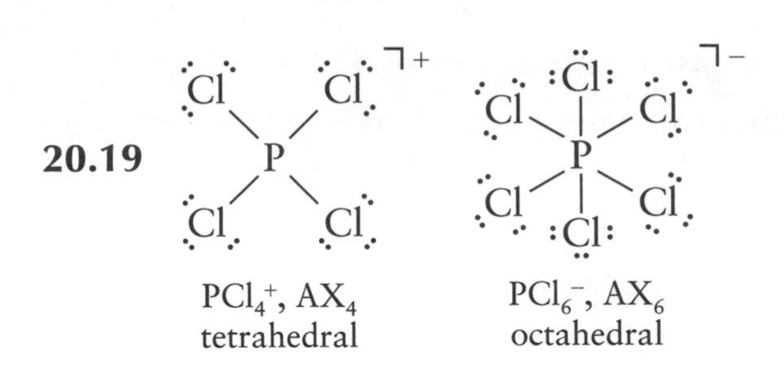

20.21 (a) superphosphate, $2\ CaSO_4 + Ca(H_2PO_4)_2$

$$\%P = \frac{2 \times \text{molar mass P}}{\left(\begin{array}{c}3 \times \text{molar mass Ca} + 2 \times \text{molar mass S} + 16 \times \text{molar mass O} \\ + 4 \times \text{molar mass H} + 2 \times \text{molar mass P}\end{array}\right)} \times 100\%$$

$$\%P = \frac{[2(30.97)]\ \text{g/mol}}{[3(40.08) + 2(32.06) + 16(16.00) + 4(1.008) + 2(30.97)]\ \text{g/mol}} \times 100\%$$

$$\%P = \frac{61.94}{(120.24 + 64.12 + 256.00 + 4.032 + 61.94)} \times 100\% = 12.23\%$$

(b) triple superphosphate, $Ca(H_2PO_4)_2$

$$\%P = \frac{2 \times \text{molar mass P}}{\left(\begin{array}{c}1 \times \text{molar mass Ca} + 4 \times \text{molar mass H} + 2 \times \text{molar mass P} \\ + 8 \times \text{molar mass O}\end{array}\right)} \times 100\%$$

$$\%P = \frac{[2(30.97)]\ \text{g/mol}}{[1(40.08) + 4(1.008) + 2(30.97) + 8(16.00)]\ \text{g/mol}} \times 100\% = 26.46\%$$

Group 16

20.23 (a) H_2SO_4 (b) $CaSO_3$ (c) O_3 (d) BaO_2

20.25 (a) $4\ Li(s) + O_2(g) \xrightarrow{\Delta} 2\ Li_2O(s)$
(b) $2\ Na(s) + 2\ H_2O(l) \longrightarrow 2\ NaOH(aq) + H_2(g)$
(c) $2\ F_2(g) + 2\ H_2O(l) \longrightarrow 4\ HF(aq) + O_2(g)$
(d) $2\ H_2O(l) \longrightarrow O_2(g) + 4\ H^+(aq) + 4\ e^-$

20.27 (a) $2\ H_2S(g) + 3\ O_2(g) \longrightarrow 2\ SO_2(g) + 2\ H_2O(g)$
(b) $PCl_5(s) + 4\ H_2O(l) \longrightarrow H_3PO_4(aq) + 5\ HCl(g)$

20.29 H
$\ddot{O}-\ddot{O}$
H

Each O in H_2O_2 is an AX_2E_2 structure; the bond angle is predicted to be $< 109.5°$. In actuality, it is 97°.

20.31 (a) When we consider formal charges and allow the possibility of expanded octets in S atoms, we can visualize two types of resonance structures that might contribute to the overall structure of SO_2:

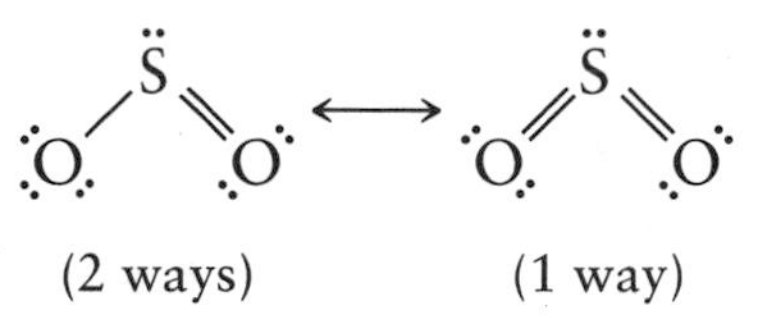

(2 ways) (1 way)

The completely double-bonded structure has zero formal charge on all atoms and, as a result, may predominate in the resonance hybrid. Both structures have angular geometry.

(b)

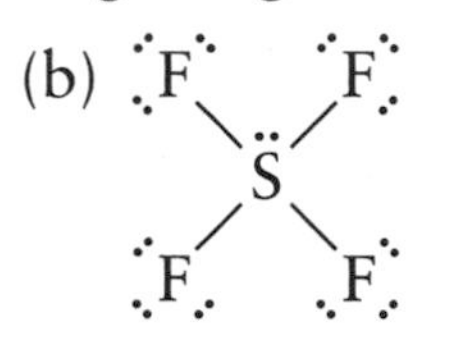

SF_4 is an AX_4E VSEPR structure. Therefore, its shape is seesaw.

(c) When we consider formal charges and the possibility of expanded octets in S atoms, we can visualize three types of resonance structures that might contribute to the overall structure of SO_4^{2-}:

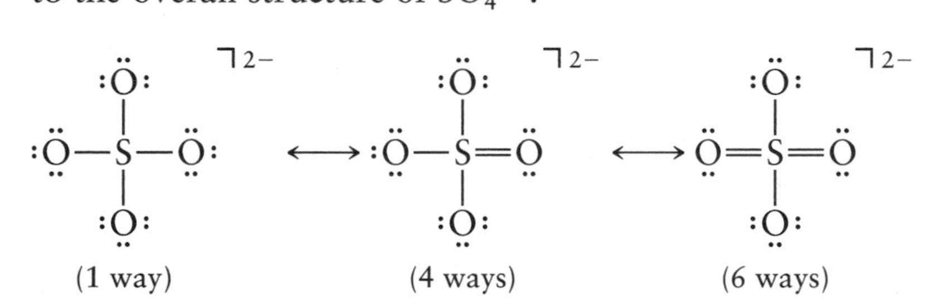

(1 way) (4 ways) (6 ways)

The third type of structure, which has smaller average differences in formal charge between the S and O atoms, may be the predominant structure. All three structures have tetrahedral geometry.

Group 17

20.33 Fluorine comes from the minerals fluorspar, CaF_2; cryolite, Na_3AlF_6; and the fluorapatites, $Ca_5F(PO_4)_3$. The free element is prepared from HF and KF by electrolysis, but the HF and the KF needed for the electrolysis are prepared in the laboratory. Chlorine primarily comes from the mineral rock salt, NaCl. The pure element is obtained by electrolysis of liquid NaCl.

20.35 Fluorine: KF acts as an electrolyte for the electrolytic process; the net reaction is

$$2\,H^+ + 2\,F^- \xrightarrow{\text{current}} H_2(g) + F_2(g)$$

Chlorine:

$$2\ NaCl(l) \xrightarrow{\text{current}} 2\ Na(l) + Cl_2(g)$$

20.37 (a) $HBr(aq)$, hydrobromic acid
(b) IBr, iodine bromide
(c) ClO_2, chlorine dioxide
(d) $NaIO_3$, sodium iodate

20.39 (a) perchloric acid, $HClO_4(aq)$
(b) sodium chlorate, $NaClO_3$
(c) hydroiodic acid, $HI(aq)$
(d) sodium triiodide, NaI_3

20.41 (a) $HIO(aq)$ $H = +1, O = -2$; therefore, $I = +1$
(b) ClO_2 $O = -2$; therefore, $Cl = +4$
(c) Cl_2O_7 $O = -2$; therefore, $Cl = +14/2 = +7$
(d) $NaIO_3$ $Na = +1, O = -2$; therefore, $I = +5$

20.43 (a) $[O{=}Cl({-}O)({=}O){=}O]^-$ (with lone pairs on each O)

AX_4, tetrahedral electronic arrangement and shape

Note: This structure is the preferred structure based on formal charge considerations; alternative structures with 0, 1, 2, and 4 double bonds could be drawn. All structures have the same geometry, however.

(b) $[O{=}I({-}O){=}O]^-$ (with lone pairs on each O and one lone pair on I)

AX_3E, tetrahedral electronic arrangement, trigonal pyramidal shape

(c) 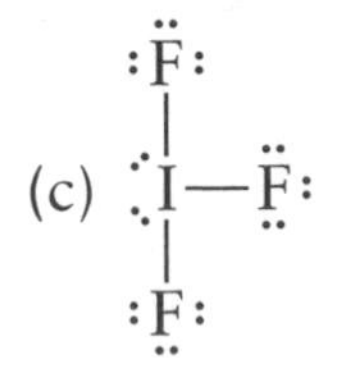

AX_3E_2, trigonal bipyramidal electronic arrangement, T-shaped

20.45 (a) $4\ KClO_3(l) \xrightarrow{\Delta} 3\ KClO_4(s) + KCl(s)$

(b) $Br_2(l) + H_2O(l) \longrightarrow HBrO(aq) + HBr(aq)$

(c) $NaCl(s) + H_2SO_4(aq) \longrightarrow NaHSO_4(aq) + HCl(g)$

(d) (a) and (b) are redox reactions. In (a), Cl is both oxidized and reduced. In (b), Br is both oxidized and reduced. (c) is a Brønsted acid-base reaction; H_2SO_4 is the acid, and Cl^- the base.

20.47 (a) $HClO < HClO_2 < HClO_3 < HClO_4$ ($HClO_4$ is strongest; HClO, weakest)

(b) The oxidation number of Cl increases from HClO to $HClO_4$. In $HClO_4$, chlorine has its highest oxidation number of +7, so $HClO_4$ will be the strongest oxidizing agent.

20.49

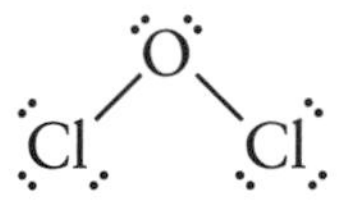

AX_2E_2, angular, slightly less than 109°

20.51

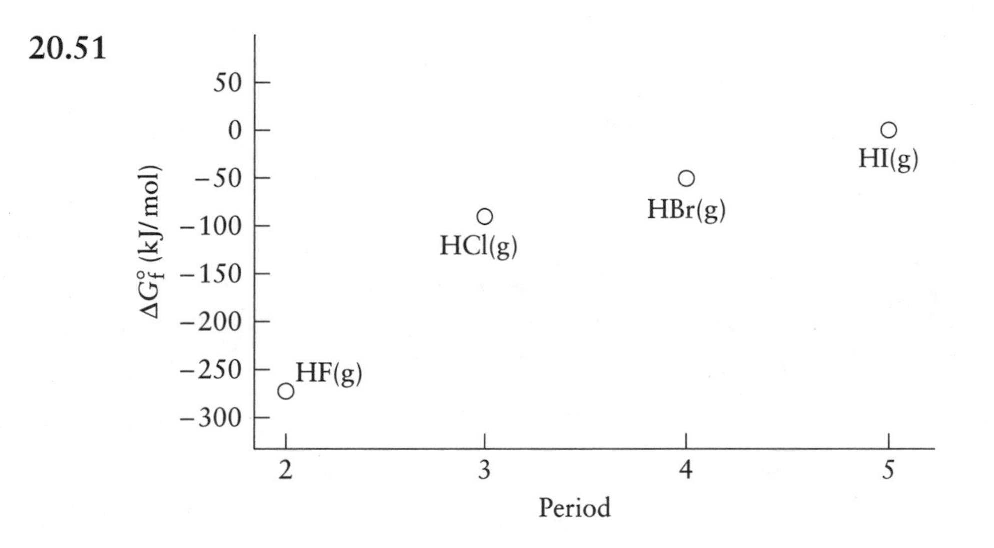

The thermodynamic stability of the hydrogen halides decreases down the group. The $\Delta G_f°$ values of HCl, HBr, and HI fit nicely on a straight line; whereas HF is anomalous. In other properties, HF is also the anomalous member of the group, in particular, its acidity.

Group 18

20.53 Helium occurs as a component of natural gases found under rock formations in certain locations, especially some in Texas. Argon is obtained by distillation of liquid air.

20.55 (a) KrF_2: F = −1; therefore, Kr = +2

(b) XeF_6: F = −1; therefore, Xe = +6

(c) KrF_4: F = −1; therefore, Kr = +4

(d) XeO_4^{2-}: O = −2, $N_{ox}(Xe) + 4 \times (-2) = -2$;

therefore, $N_{ox}(Xe) = -2 - (-8) = +6$

20.57 $XeF_4 + 4\ H^+ + 4\ e^- \longrightarrow Xe + 4\ HF$

20.59 Because H_4XeO_6 has a greater number of highly electronegative O atoms bonded to Xe, we predict that H_4XeO_6 is more acidic than H_2XeO_4.

20.61

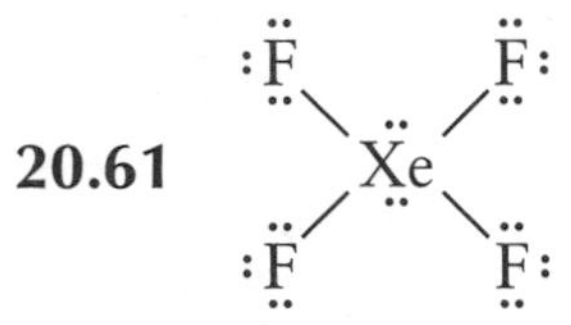

AX_4E_2, square planar, 90°

SUPPLEMENTARY EXERCISES

20.63 Ionization energies increase; electron affinities also increase (in magnitude). Large ionization energies and electron affinities are characteristic of nonmetals. Electronegativities and standard reduction potentials increase as well; large values of these properties are also characteristic of nonmetals.

20.65 oxidation: $As_2S_3(s) + 8\ H_2O(l) \longrightarrow 2\ AsO_4^{3-}(aq) + 3\ S^{2-}(aq) + 16\ H^+(aq) + 4\ e^-$

reduction: $H_2O_2(aq) + 2\ H^+(aq) + 2\ e^- \longrightarrow 2\ H_2O(l)$

Multiply the reduction reaction by 2, cancel electrons, and add.

overall: $As_2S_3(s) + 2\ H_2O_2(aq) + 4\ H_2O(l) \longrightarrow$

$$2\ AsO_4^{3-}(aq) + 3\ S^{2-}(aq) + 12\ H^+(aq)$$

20.67 $Ca_3(PO_4)_2(s) + 3\ H_2SO_4(l) \longrightarrow 2\ H_3PO_4(l) + 3\ CaSO_4(s)$

$$\text{volume} = (1000.\ \text{kg } H_3PO_4) \times \left(\frac{10^3\ \text{g } H_3PO_4}{1\ \text{kg } H_3PO_4}\right) \times \left(\frac{1\ \text{mol } H_3PO_4}{98.0\ \text{g } H_3PO_4}\right)$$

$$\times \left(\frac{3\ \text{mol } H_2SO_4}{2\ \text{mol } H_3PO_4}\right) \times \left(\frac{98.1\ \text{g } H_2SO_4}{1\ \text{mol } H_2SO_4}\right) \times \left(\frac{1\ \text{mL}}{1.84\ \text{g}}\right) \times \left(\frac{10^{-3}\ \text{L}}{1\ \text{mL}}\right)$$

$$= 8.16 \times 10^2\ \text{L conc. } H_2SO_4$$

20.69 (a) $SO_2(g) + H_2O(l) \longrightarrow H_2SO_3(aq)$ This is a Lewis acid-base reaction. SO_2 is the acid and H_2O is the base.

(b) $2\,F_2(g) + 2\,NaOH(aq) \longrightarrow OF_2(g) + 2\,NaF(aq) + H_2O(l)$ This is a redox reaction illustrating the oxidizing ability of F_2 in basic solution and is used for the preparation of $OF_2(g)$. O is oxidized and F is reduced.

(c) $S_2O_3^{2-}(aq) + 4\,Cl_2(g) + 13\,H_2O(l) \longrightarrow$

$$2\,HSO_4^-(aq) + 8\,H_3O^+(aq) + 8\,Cl^-(aq)$$

This is a redox reaction illustrating the oxidizing power of $Cl_2(g)$ in acidic solution. S is oxidized and Cl is reduced.

(d) $2\,XeF_6(s) + 16\,OH^-(aq) \longrightarrow$

$$XeO_6^{4-}(aq) + Xe(g) + 12\,F^-(aq) + 8\,H_2O(l) + O_2(g)$$

This is a redox reaction that is also a disproportionation reaction, in that Xe goes from oxidation number +6 to +8 and to 0. Xe is both oxidized and reduced.

20.71 (a) $I_2(s) + 3\,F_2(g) \longrightarrow 2\,IF_3(s)$

(b) $I_2(aq) + I^-(aq) \longrightarrow I_3^-(aq)$

(c) $Cl_2(g) + H_2O(l) \longrightarrow HCl(aq) + HOCl(aq)$

But there are competing reactions, such as

$2\,Cl_2(g) + 2\,H_2O(l) \longrightarrow 4\,HCl(aq) + O_2(g)$.

The predominant reaction is determined by the temperature and pH.

(d) $2\,F_2(g) + 2\,H_2O(l) \longrightarrow 4\,HF(aq) + O_2(g)$

20.73 (a) $CaCl_2(s) + 2\,H_2SO_4(aq, conc) \longrightarrow Ca(HSO_4)_2(aq) + 2\,HCl(g)$

(b) $KBr(s) + H_3PO_4(aq) \xrightarrow{\Delta} KH_2PO_4(aq) + HBr(g)$

(c) $KI(s) + H_3PO_4(aq) \xrightarrow{\Delta} KH_2PO_4(aq) + HI(g)$

20.75 (a) Some of the unusual properties of fluorine relative to the other halogens are

(1) its much higher electronegativity
(2) its restriction to a negative oxidation number (−1) in all of the compounds
(3) the weak acidity of its hydrogen compound, HF
(4) the high lattice enthalpies of its ionic fluorides
(5) the lower solubility of the ionic fluorides
(6) the weakness of the F—F bond
(7) the high rate of its reactions
(8) the high volatility of most of its covalent compounds
(9) the low volatility of HF

Properties (1), (2), (4), (5), and (6) are related to the smaller size of the fluorine atom; property (3) is due to the greater strength of the H—F bond; property (7) is a result of property (6), the weakness of the F—F bond; property (8) is a result of weaker London forces, which in turn is due to the smaller size of the fluorine atom; and property (9) is related to stronger hydrogen bonding in HF, which in turn is due to its smaller size and greater electronegativity.

(b) This statement relates to its great strength as an oxidizing agent.

20.77 (a) $Xe(g) + 2\ F_2(g) \xrightarrow{\Delta} XeF_4(s)$

(b) $Pt(s) + XeF_4(s) \longrightarrow Xe(g) + PtF_4(s)$

20.79 Physical: melting points and boiling points increase down the group in a regular fashion consistent with the increase in the molar masses of the elements. Ionization energies decrease down the group in a regular manner consistent with the fact that further electron shells are being successively filled.

Chemical: chemical reactivity increases down the group. This is related to the decrease in ionization energy; it is easier for the heavier elements to share their electrons. He, Ne, and Ar form no stable compounds, but Kr, Xe, and Rn do, especially with fluorine.

20.81 Because this chapter deals with elements from groups 15–18, only Period 2 and Period 3 elements from these four groups are included in the answer. Other examples are possible.

(a) N: NH_3, used in the production of fertilizers

O: H_2O, found in all life forms, covers much of earth's surface.

F: Na_3AlF_6, used in the electrolytic refining of aluminum

Ne: no known compounds, but the element is used in "neon" lights.

(b) P: found in phosphate groups that help hold DNA together (see Chapter 11 for the structure of this compound). Without it, the reproduction of higher life forms on this planet would not be possible. Phosphorus is also important in the conversion of ATP to ADP, the energy storage mechanism in living cells.

S: H_2SO_4, sulfuric acid, used in the production of fertilizers and detergents One of the most heavily produced chemicals in the world

Cl: $Ca(ClO)_2$, calcium hypochlorite, used as dry bleach and in swimming pools

Ar: no known compounds, but the element is used in welding and as an inert gas in light bulbs.

20.83 (a) The method of organization is determined by whether one wishes to emphasize similarities or differences. It seems easier and more logical to emphasize similarities rather than differences. Trends within groups from top to bottom tend to be quantitative (small numerical differences), rather than qualitative (large numerical differences leading to distinctly different behavior), as are observed from left to right within a period. Trends from metallic to nonmetallic behavior across a period would be more apparent in the organization by period, as would the related trends in ionization energy and electron affinity. One could also more readily see changes in valence as, say, represented by the formulas of common compounds, such as the oxides and chlorides. Trends in melting points and boiling points of the elements and their compounds would also be more apparent, as well as many other physical properties. There are advantages and disadvantages to both methods of organization, but the organization by group still seems preferable, because organization by period would not permit the generalizations and summaries that are useful features of the present arrangement.

(b) In the case of the transition elements, similarities within a period can be stronger than within groups; here, organization by period allows a logical structuring of properties, and this is the usual choice among authors.

20.85 The ionic radii of the halide ions increase down the group. The smaller the ion, the greater the lattice enthalpy of the compounds of the ion, due to the greater concentration of the charge in the ion. Increased lattice enthalpy results in higher melting and boiling points; the ions cannot as easily break free from each other. Thus, melting and boiling points decrease from fluoride to iodide for ionic halides. The predominant forces between covalent halogen compounds are London forces, which are greater for larger atoms. Thus, the melting and boiling points increase from fluoride to iodide for molecular halides.

APPLIED EXERCISES

20.87 Chlorine is reduced: $Cl(+7) \longrightarrow Cl(-1)$

Aluminium is oxidized: $Al(0) \longrightarrow Al(+3)$

Nitrogen is oxidized: $N(-3) \longrightarrow N(+2)$

The half-reactions are

$3\ Al + 3\ H_2O \longrightarrow Al^{3+} + Al_2O_3 + 6\ H^+ + 9\ e^-$

$NH_4^+ + H_2O \longrightarrow NO + 6\ H^+ + 5\ e^-$

$ClO_4^- + 8\ H^+ + 8\ e^- \longrightarrow Cl^- + 4\ H_2^-$

Multiply the last two equations by 3 and add all the half-reactions together to get the overall equation above.

20.89 (a) $4\ CH_3NHNH_2(l) + 5\ N_2O_4(l) \longrightarrow 9\ N_2(g) + 12\ H_2O(g) + 4\ CO_2(g)$

$$\Delta H_r^\circ = 12 \times \Delta H_f^\circ(H_2O, g) + 4 \times \Delta H_f^\circ(CO_2, g) - 4 \times \Delta H_f^\circ(CH_3NHNH_2, l) - 5 \times \Delta H_f^\circ(N_2O_4, l)$$
$$= [12 \times (-241.82) + 4 \times (-393.51) - 4 \times (+54) - 5 \times (+9.16)] \text{ kJ/mol}$$
$$= -4738 \text{ kJ/mol}$$

(b) $\text{specific enthalpy} = \dfrac{4738 \text{ kJ/mol}}{4 \times 46.08 \text{ g/mol}} = 25.70 \text{ kJ/g}$

20.91 (a) $2\ NaCl(aq) + 2\ H_2O(l) \xrightarrow{\text{current}} 2\ NaOH(aq) + H_2(g) + Cl_2(g) \qquad n = 2$

(b) $\text{charge} = (1.2 \times 10^{10} \text{ kg } Cl_2) \times \left(\dfrac{1 \text{ mol } Cl_2}{0.070\ 90 \text{ kg } Cl_2}\right) \times \left(\dfrac{2 \text{ mol } e^-}{1 \text{ mol } Cl_2}\right) \times \left(\dfrac{96\ 485 \text{ C}}{1 \text{ mol } e^-}\right) = 2.9\bar{9} \times 10^{16} \text{ C}$

(c) $\text{current} = (2.9\bar{9} \times 10^{16} \text{ C}) \times \left(\dfrac{1}{365 \text{ day}}\right) \times \left(\dfrac{1 \text{ day}}{24 \text{ h}}\right) \times \left(\dfrac{1 \text{ h}}{3600 \text{ s}}\right)$

$= 9.5 \times 10^8 \text{ C/s} = 9.5 \times 10^8 \text{ A}$

INTEGRATED EXERCISES

20.93 (a) 1 mol of $N_2(g)$ occupies 22.41 L at STP. For the reaction $Pb(N_3)_2 \longrightarrow Pb + 3\ N_2$, the volume of $N_2(g)$ produced is

$$(1.0 \text{ g } Pb(N_3)_2) \times \left(\frac{1 \text{ mol } Pb(N_3)_2}{291 \text{ g } Pb(N_3)_2}\right) \times \left(\frac{3 \text{ mol } N_2}{1 \text{ mol } Pb(N_3)_2}\right) \times \left(\frac{22.41 \text{ L } N_2}{1 \text{ mol } N_2}\right) = 0.23 \text{ L } N_2(g)$$

(b) $Hg(N_3)_2$ would produce a larger volume, because its molar mass is less. Note that molar mass occurs in the denominator in this calculation.

20.95 $2\ H_2O_2(l) \longrightarrow 2\ H_2O(l) + O_2(g)$

Assume 3% by mass. At 273 K and 1.00 atm, 1 mol O_2 has a volume of 22.41 L.

$$\text{volume of } O_2 = (500. \text{ mL soln}) \times \left(\frac{1.0 \text{ g soln}}{1.0 \text{ mL soln}}\right) \times \left(\frac{0.03 \text{ g } H_2O_2}{1.0 \text{ g soln}}\right) \times \left(\frac{1 \text{ mol } H_2O_2}{34.0 \text{ g } H_2O_2}\right) \times \left(\frac{1 \text{ mol } O_2}{2 \text{ mol } H_2O_2}\right) \times \left(\frac{22.41 \text{ L } O_2}{1 \text{ mol } O_2}\right) = 4.9 \text{ L } O_2$$

20.97 The system is $HS_{(aq)}^- + H_2O(l) \rightleftharpoons H_2S_{(aq)} + OH^-(aq)$

Concentration (mol/L)	HS^-	H_2O	H_2S	OH^-
initial	0.050	—	0	0
change	$-x$	—	$+x$	$+x$
equilibrium	$0.050 - x$	—	x	x

$$K_b = \frac{K_w}{K_a} = \frac{1.0 \times 10^{-14}}{1.3 \times 10^{-7}} = 7.7 \times 10^{-8}$$

$$K_b = \frac{x^2}{0.050 - x} \approx \frac{x^2}{0.050} = 7.7 \times 10^{-8}$$

$x = 6.2 \times 10^{-5}$ mol/L OH^-

pOH = 4.21

pH = 14.00 − 4.21 = 9.79

20.99 Calculate $\Delta G_r°$ for the reaction $PbSO_4(s) \longrightarrow PbO_2(s) + SO_2(g)$

$$\Delta G_r° = \Delta G_f°(PbO_2, s) + \Delta G_f°(SO_2, g) - \Delta G_f°(PbSO_4, s)$$
$$= [-217.33 + (-300.19) - (-813.14)] \text{ kJ/mol} = +295.62 \text{ kJ/mol}$$

The positive $\Delta G_r°$ is unfavorable to the formation of PbO_2. Thus, $PbSO_4$ is favored as a product over $PbO_2 + SO_2$.

20.101 $\frac{1}{2} H_2(g) + \frac{1}{2} Cl_2(g) \rightleftharpoons HCl(g) \quad \Delta G_r° = \Delta G_f° = -95.30$ kJ/mol

$$\Delta G_r° = -RT \ln Kp$$

$$\ln Kp = \frac{-\Delta G_r°}{RT} = \frac{-(-9.530 \times 10^4 \text{ J/mol}}{8.314 \text{ J/K·mol} \times 298 \text{ K}} = +38.46$$

$Kp = 2.92 \times 10^{38}$ (estimate 3×10^{38})

20.103 concentration of I_2 $= 0.028\,45 \text{ L} \times \left(\frac{0.025 \text{ mol } S_2O_3^{2-}}{1 \text{ L}}\right) \times \left(\frac{1 \text{ mol } I_2}{2 \text{ mol } S_2O_3^{2-}}\right) \times \left(\frac{1}{0.025\,00 \text{ L}}\right) = 0.014 \text{ mol/L}$

20.105 (a) mass of sulfur $= (4 \times 10^{10} \text{ kg } H_2SO_4) \times \left(\frac{32.06 \text{ kg S}}{98.08 \text{ kg } H_2SO_4}\right) = 1 \times 10^{10} \text{ kg S}$

(b) $SO_3(g) + H_2O(l) \longrightarrow H_2SO_4(aq)$

moles of SO_3 $= (4 \times 10^{13} \text{ g } H_2SO_4) \times \left(\frac{1 \text{ mol } H_2SO_4}{98.08 \text{ g } H_2SO_4}\right) \times \left(\frac{1 \text{ mol } SO_3}{1 \text{ mol } H_2SO_4}\right)$

$= 4 \times 10^{11} \text{ mol } SO_3$

Molar volume (25°C, 1 atm) = 24.47 L/mol; at 5 atm, this is $\frac{1}{5} \times 24.47$ L/mol

thus $\frac{1}{5} \times 24.47 \text{ L/mol} \times 4 \times 10^{11} \text{ mol} = 2 \times 10^{12} \text{ L } SO_3(g)$

20.107 (a) $4\ Zn(s) + NO_3^-(aq) + 7\ OH^-(aq) + 6\ H_2O(l) \longrightarrow$

$NH_3(aq) + 4\ Zn(OH)_4^{2-}(aq)$

$NH_3(g) + HCl(aq) \longrightarrow NH_4Cl(aq)$

$HCl(aq) + NaOH(aq) \longrightarrow H_2O(l) + NaCl(aq)$

(b) number of moles of unreacted HCl $= 0.028\ 22\ L \times 0.150\ mol/L$
$= 0.004\ 23\ mol$

total number of moles of HCl $= 0.050\ 00\ L \times 0.250\ mol/L = 0.0125\ mol$

moles of HCl reacted = moles of $NH_3 = (0.0125 - 0.004\ 23)\ mol$
$= 0.0083\ mol\ NH_3$

$$[NO_3^-] = \frac{(0.0083\ mol\ NH_3) \times \left(\frac{1\ mol\ NO_3^-}{1\ mol\ NH_3}\right)}{0.025\ 00\ L} = 0.33\ mol/L$$

CHAPTER 21
THE *d* BLOCK: METALS IN TRANSITION

EXERCISES

The d-Block Elements and Their Electron Configurations

21.1 (a) Mn, [Ar] $3d^5 4s^2$
(b) Cd, [Kr] $4d^{10} 5s^2$
(c) Zn, [Ar] $3d^{10} 4s^2$
(d) Zr, [Kr] $4d^2 5s^2$

21.3 (a) Sc, [Ar] $3d^1 4s^2$: one unpaired $3d$-electron
(b) V, [Ar] $3d^3 4s^2$: three unpaired $3d$-electrons
(c) Cu, [Ar] $3d^{10} 4s^1$: one unpaired $4s$-electron

21.5 iron, Fe; cobalt, Co; nickel, Ni

Trends in Properties

21.7 See Figs. 21.2 and 21.4
(a) Sc (b) Au (c) Nb

21.9 See Fig. 7.39 and Appendix 2D.
(a) Ti (b) Cu (c) Zn

21.11 The lanthanide contraction accounts for the failure of the third-row (Period 6) metallic radii to increase as expected, relative to the radii of the second row (Period 5). It results from the presence of the *f*-block orbitals. The *f*-electrons present in the lanthanides are even poorer as nuclear shields than *d*-electrons, and a marked decrease in metallic radius occurs along the *f*-block elements as a result of the increased effective nuclear charge, which pulls the electrons inward. When the *d* block resumes (at lutetium), the metallic radius has "contracted" from 188 to 157 pm. Therefore, all the elements following lutetium have smaller-than-expected radii. Examples of this effect include the high density of the Period 6 elements and lack of reactivity of gold and platinum.

21.13 Hg is much more dense than Cd because the shrinkage in atomic radius that occurs between $Z = 58$ and $Z = 71$ (the lanthanide contraction) causes the atoms following the rare earths to be smaller than might have been expected for their atomic masses and atomic numbers. Zn and Cd have densities that are not too dissimilar because the radius of Cd is subject to only a smaller *d*-block contraction.

21.15 Proceeding down a group in the *d* block (for example, from Cr to Mo to W), there is an increasing probability of finding the elements in higher oxidation states. That is, higher oxidation states become more stable on going down a group.

21.17 In MO_3, M has an oxidation number of +6. Of these three elements, the +6 oxidation state is most stable for Cr. See Fig. 21.7.

Scandium Through Nickel

21.19 (a) $TiO_2(s) + 2\ C(s) + Cl_2(g) \xrightarrow{1000°C} TiCl_4(g) + 2\ CO(g)$,
followed by $TiCl_4(g) + 2\ Mg(l) \xrightarrow{700°C} Ti(s) + 2\ MgCl_2(s)$
(b) $V_2O_5(g) + 5\ Ca(l) \longrightarrow 5\ CaO(s) + 2\ V(s)$
or $VCl_2(s) + Mg(l) \xrightarrow{\Delta} V(s) + MgCl_2(s)$

21.21 (a) $Ti(s)$, $MgCl_2(s)$
$TiCl_4(g) + 2\ Mg(l) \longrightarrow Ti(s) + 2\ MgCl_2(s)$
(b) $Co^{2+}(aq)$, $HCO_3^-(aq)$, $NO_3^-(aq)$
$CoCO_3(s) + HNO_3\ (aq) \longrightarrow Co^{2+}(aq) + HCO_3^-(aq) + NO_3^-(aq)$
(c) $V(s)$, $CaO(s)$
$V_2O_5(s) + 5\ Ca(l) \xrightarrow{\Delta} 2\ V(s) + 5\ CaO(s)$

21.23 (a) titanium(IV) oxide, TiO_2
(b) iron(III) oxide, Fe_2O_3
(c) manganese(IV) oxide, MnO_2

21.25 (a) $V^{2+} + 2\ e^- \longrightarrow V(s)\quad E° = -1.19\ V$
$V^{3+} + e^- \longrightarrow V^{2+}\quad E° = -0.26\ V$
$2\ H^+ + 2\ e^- \longrightarrow H_2(g)\quad E° = 0.00\ V$
Therefore, V(s) will be oxidized to V^{3+}. The products are V^{3+}, H_2, and Cl^-.

(b) $Hg_2^{2+} + 2\ e^- \longrightarrow 2\ Hg \quad E° = +0.79\ V$

$Hg^{2+} + 2\ e^- \longrightarrow Hg \quad E° = +0.85\ V$

$2\ H^+ + 2\ e^- \longrightarrow H_2(g) \quad E° = 0.00\ V$

Therefore, no reaction.

(c) $Co^{2+} + 2\ e^- \longrightarrow Co(s) \quad E° = -0.28\ V$

$Co^{3+} + e^- \longrightarrow Co^{2+} \quad E° = +1.81\ V$

$2\ H^+ + 2\ e^- \longrightarrow H_2(g) \quad E° = 0.00\ V$

Therefore, Co(s) will be oxidized to Co^{2+}. The products are Co^{2+}, H_2, and Cl^-. The further oxidation to Co^{3+} is not favorable.

Groups 11 and 12

21.27 All three elements in this group—Cu, Ag, and Au—are chemically rather inert, Ag more so than Cu, and Au more so than Ag. The standard potentials of their ions are all positive, in the order Au > Ag > Cu, so they are not readily oxidized. They have a common electron configuration, $(n-1)d^{10}ns^1$.

21.29 (a) chalcopyrite, $CuFeS_2$, copper iron sulfide

(b) sphalerite, ZnS, zinc sulfide

(c) cinnabar, HgS, mercury(II) sulfide

21.31 (a) $2\ ZnS(s) + 3\ O_2(g) \xrightarrow{\Delta} 2\ ZnO(s) + 2\ SO_2(g)$,

followed by $ZnO(s) + C(s) \xrightarrow{\Delta} Zn(l) + CO(g)$

(b) $HgS(s) + O_2(g) \xrightarrow{\Delta} Hg(g) + SO_2(g)$

d-Metal Complexes

21.33 Let x = the oxidation number to be determined

(a) $x(Fe) + 6 \times (-1) = -4$

$x(Fe) = -4 - (-6) = +2$

(b) $(Co) + 6 \times (0) = +3$

$x(Co) = +3$

(c) $x(Co) + 5 \times (-1) + 1 \times (0) = -2$

$x(Co) = -2 - (-5) = +3$

(d) $x(Co) + 1 \times (-2) + 5 \times (0) = +1$

$x(Co) = +1 - (-2) = +3$

21.35 (a) 4 (b) 2 (c) 6 (en is bidentate) (d) 6 (EDTA is hexadentate)

21.37 (a) hexacyanoferrate(II) ion
(b) hexaamminecobalt(III) ion
(c) aquapentacyanocobaltate(III) ion
(d) pentaamminesulfatocobalt(III) ion

21.39 (a) $K_3[Cr(CN)_6]$
(b) $[Co(NH_3)_5(SO_4)]Cl$
(c) $[Co(NH_3)_4(H_2O)_2]Br_3$
(d) $Na[Fe(H_2O)_2(C_2O_4)_2]$

Isomerism

21.41 (a) structural isomers, linkage isomers
(b) structural isomers, ionization isomers
(c) structural isomers, linkage isomers
(d) structural isomers, ionization isomers

21.43 $[Co(H_2O)_6]Cl_3$, $[CoCl(H_2O)_5]Cl_2 \cdot H_2O$, $[CoCl_2(H_2O)_4]Cl \cdot 2H_2O$, and $[CoCl_3(H_2O)_3] \cdot 3H_2O$

21.45 $[CoCl(NO_2)(en)_2]Cl$ and $[CoCl(ONO)(en)_2]Cl$

21.47 (a) yes

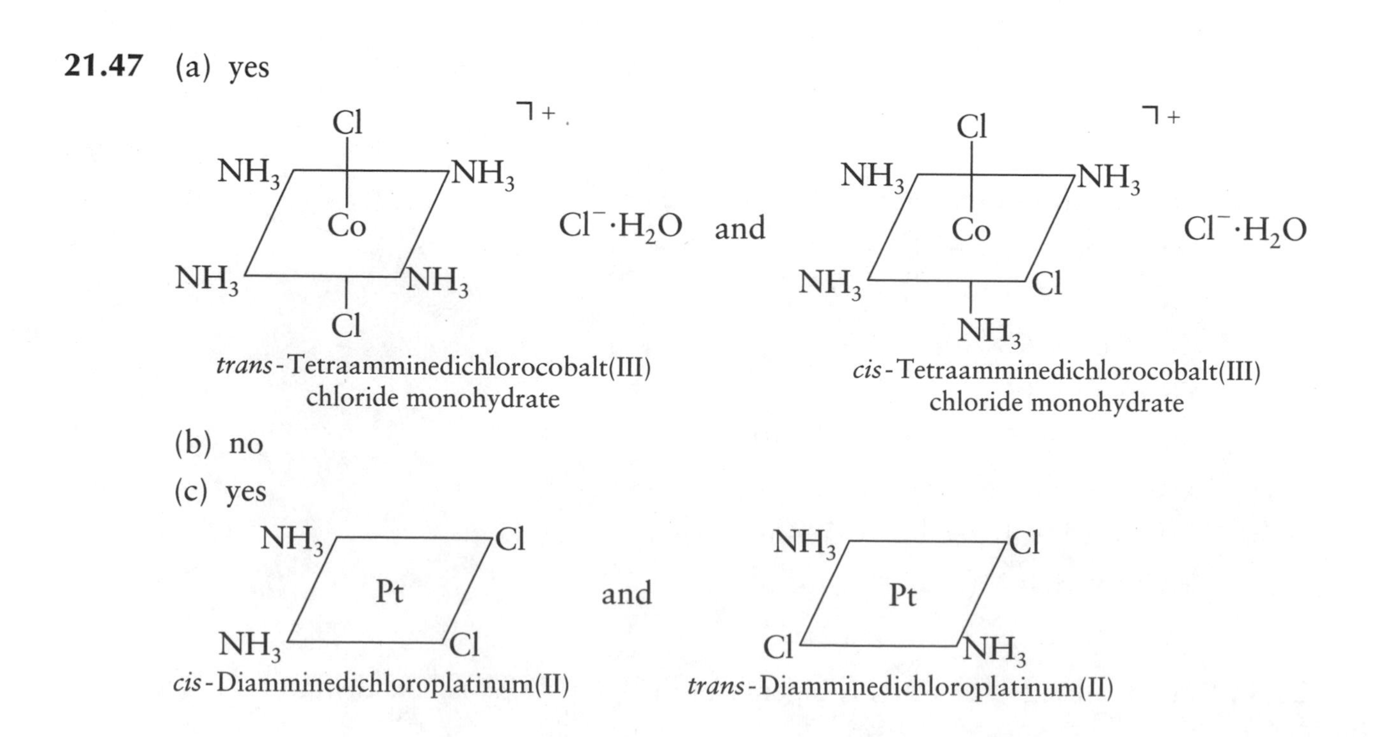

trans-Tetraamminedichlorocobalt(III) chloride monohydrate

cis-Tetraamminedichlorocobalt(III) chloride monohydrate

(b) no

(c) yes

cis-Diamminedichloroplatinum(II)

trans-Diamminedichloroplatinum(II)

21.49 (a) yes, in the form of optical isomerism; see part (c)

(b) no, only 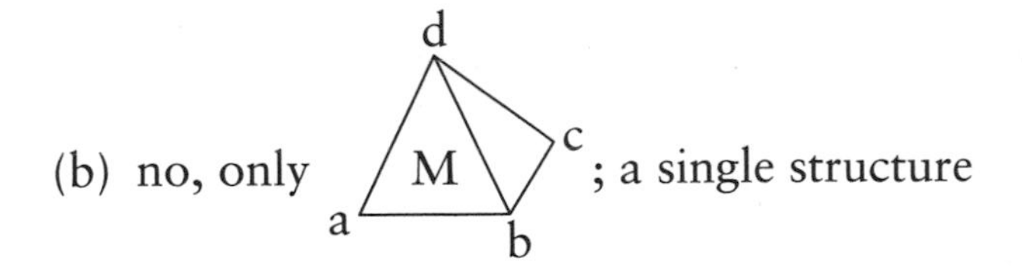

; a single structure

(c) Yes; if four different ligand groups are bonded to the central atom, then the central atom is chiral and exhibits optical activity.

21.51

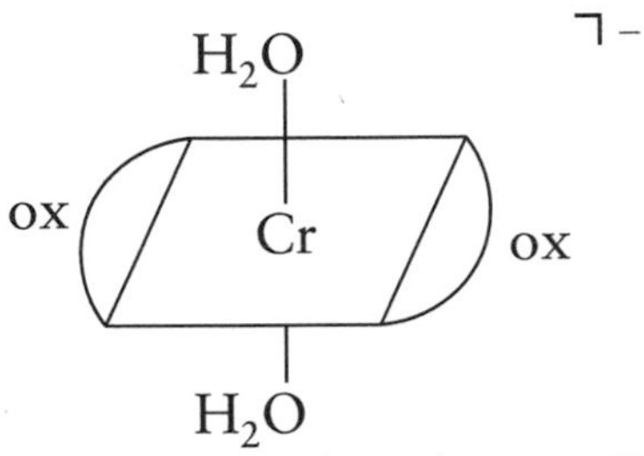

trans-Diaquabis(oxalato)chromate(III) ion

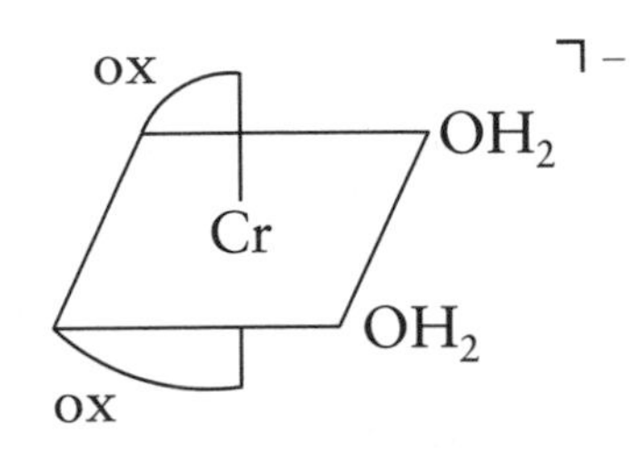

cis-Diaquabis(oxalato)chromate(III) ion

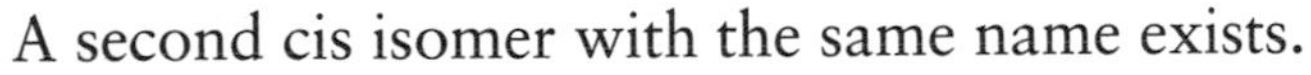
A second cis isomer with the same name exists.

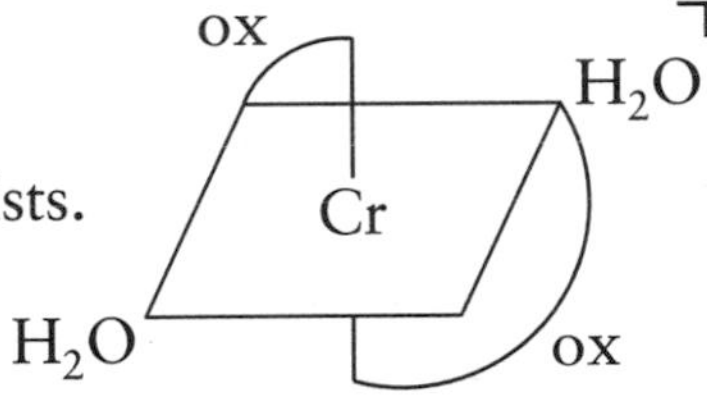

21.53 (a) first complex:

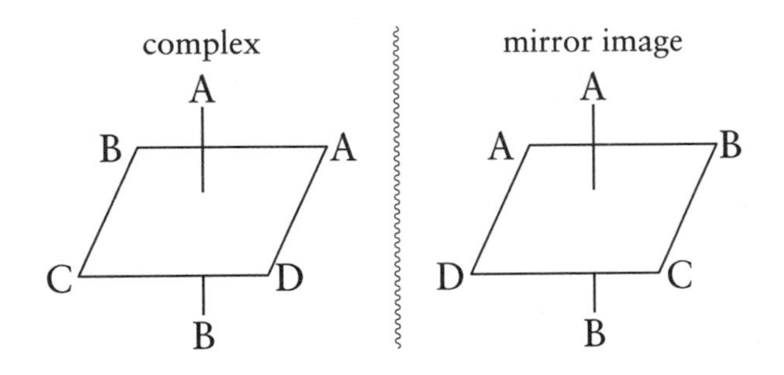

No rotation will make the complex and its mirror image match; therefore it is chiral.

(b) second complex:

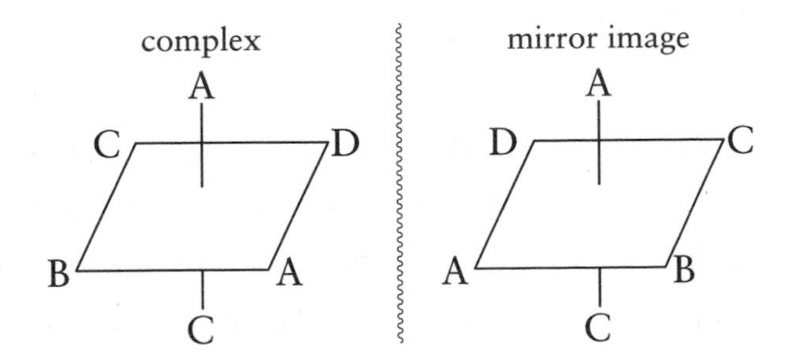

A double rotation shows that the complex and its mirror image are superimposable; hence it is not chiral.

The two complexes are not enantiomers; they are not even isomers.

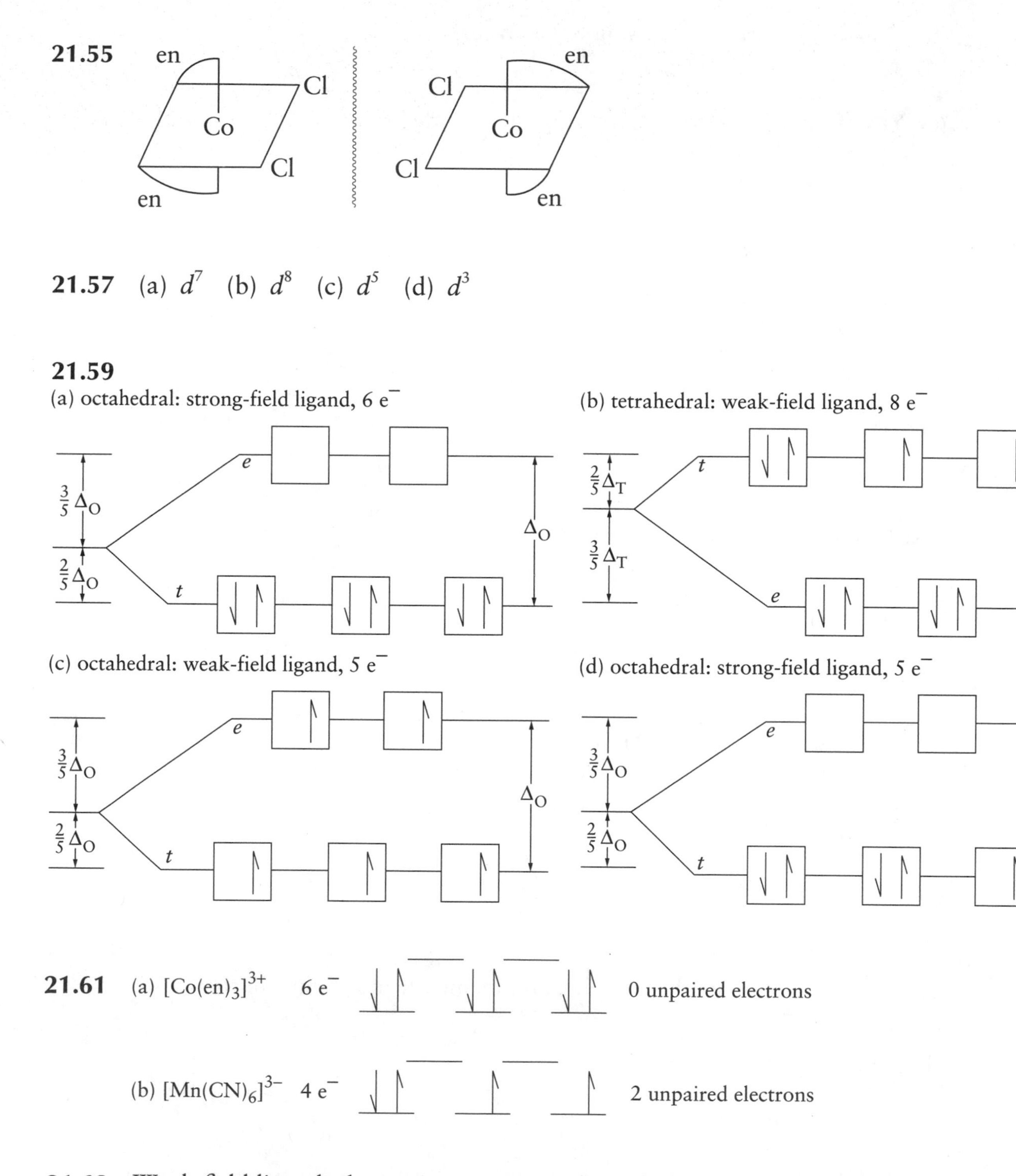

21.63 Weak-field ligands do not interact strongly with the *d*-electrons in the metal ion, so they produce only a small crystal field splitting of the *d*-electron energy states. The opposite is true of strong-field ligands. With weak-field ligands, unpaired electrons remain unpaired if there are unfilled orbitals; therefore, a weak-field ligand is likely to lead to a high-spin complex. Strong-field ligands cause electrons in excess of three to pair up with electrons in lower energy orbitals. A strong-field ligand is likely to lead to a low-spin complex. The arrangement of ligands in the spectrochemical series helps to distinguish strong-field and weak-field ligands. Measurement of magnetic susceptibility (paramagnetism) can be used to determine the number of unpaired electrons, which in turn establishes whether the associated ligand is weak-field or strong-field in nature.

21.65 (a) $[CoF_6]^{3-}$ 6 e^- F^- is a weak-field ligand; therefore,

$$\underline{\uparrow}\quad\underline{\uparrow}$$

$$\underline{\uparrow\downarrow}\quad\underline{\uparrow}\quad\underline{\uparrow}$$

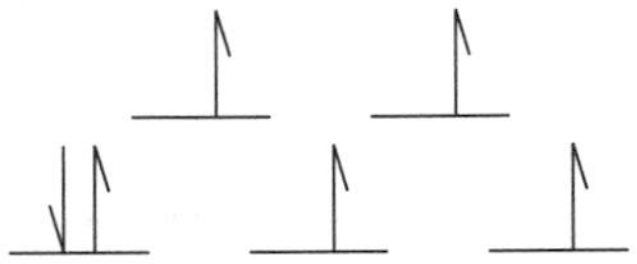

4 unpaired electrons

(b) $[Co(en)_3]^{3+}$ 6 e^- en is a strong-field ligand; therefore,

$$\underline{\quad}\quad\underline{\quad}$$

$$\underline{\uparrow\downarrow}\quad\underline{\uparrow\downarrow}\quad\underline{\uparrow\downarrow}$$

0 unpaired electrons

Because F^- is a weak-field ligand and en a strong-field ligand, the splitting between levels is less in (a) than in (b). Therefore, (a) will absorb light of longer wavelength than will (b) and consequently will display a shorter wavelength color. Blue light is shorter in wavelength than yellow light, so (a) $[CoF_6]^{3-}$ is blue and (b) $[Co(en)_3]^{3+}$ is yellow.

21.67 See Fig. 21.41.

(a) 410. nm, yellow

(b) 650. nm, green

(c) 480. nm, orange

(d) 590. nm, blue

21.69 In Zn^{2+}, the 3*d*-orbitals are filled (d^{10}). Therefore, there can be no electronic transitions between the *t* and *e* levels; so no visible light is absorbed and the aqueous ion is colorless. The d^{10} configuration has no unpaired electrons, so Zn compounds would not be paramagnetic.

21.71 (a) $$\Delta_O = \frac{hc}{\lambda} = \frac{(6.63 \times 10^{-34}\ \text{J/s})(3.00 \times 10^{8}\ \text{m/s})}{740. \times 10^{-9}\ \text{m}} = 2.69 \times 10^{-19}\ \text{J}$$

(b) $$\Delta_O = \frac{hc}{\lambda} = \frac{(6.63 \times 10^{-34}\ \text{J/s})(3.00 \times 10^{8}\ \text{m/s})}{460. \times 10^{-9}\ \text{m}} = 4.32 \times 10^{-19}\ \text{J}$$

(c) $$\Delta_O = \frac{hc}{\lambda} = \frac{(6.63 \times 10^{-34}\ \text{J/s})(3.00 \times 10^{8}\ \text{m/s})}{575 \times 10^{-9}\ \text{m}} = 3.46 \times 10^{-19}\ \text{J}$$

These numbers can be multiplied by 6.02×10^{23} to obtain kJ/mol.
(a) 2.69×10^{-19} J $\times$ 6.022×10^{23}/mol = 162 kJ/mol
(b) 4.32×10^{-19} J $\times$ 6.022×10^{23}/mol = 260. kJ/mol
(c) 3.46×10^{-19} J $\times$ 6.022×10^{23}/mol = 208 kJ/mol
$Cl < H_2O < NH_3$ (spectrochemical series)

SUPPLEMENTARY EXERCISES

21.73 (a) A compound with unpaired electrons is paramagnetic and is pulled into a magnetic field. A diamagnetic substance has no unpaired electrons and is weakly pushed out of a magnetic field.
(b) Paramagnetism is a property of any substance with unpaired electrons, whereas ferromagnetism is a property of certain substances that can become permanently magnetized. Ferromagnetism results when a large number of electrons in the metal have parallel spins. This parallel alignment can be retained even in the absence of a magnetic field. In a paramagnetic substance, the alignment is lost when the magnetic field is removed.

21.75 $[Sc(H_2O)_6]^{3+}(aq) + H_2O(l) \longrightarrow [Sc(H_2O)_5OH]^{2+}(aq) + H_3O^+(aq)$ Here the hydrated ion acts as a proton donor, and therefore it is a Brønsted acid.

21.77 (a) $CuSO_4(s) + 5\,H_2O(l) \longrightarrow CuSO_4 \cdot 5H_2O(s)$
The pentahydrate has a complicated structure. It consists of the $Cu(H_2O)_4{}^{2+}$ ion, which is linked to the sulfate ion through the fifth water molecule. The four water molecules of the tetrahydrate ion form an approximate square planar structure around the copper atom:

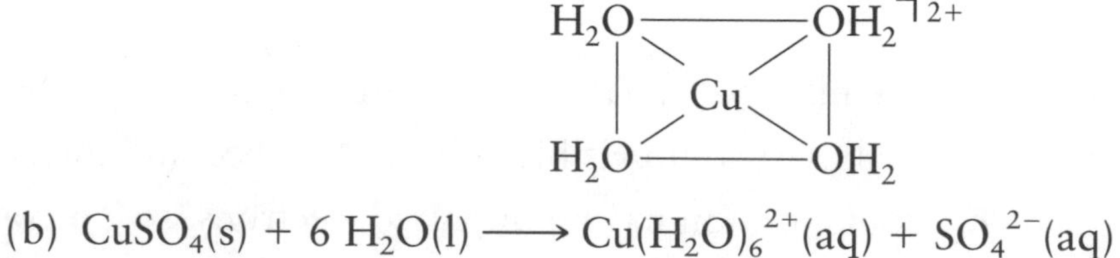

(b) $CuSO_4(s) + 6\,H_2O(l) \longrightarrow Cu(H_2O)_6{}^{2+}(aq) + SO_4{}^{2-}(aq)$
(c) $Cu(H_2O)_6{}^{2+}(aq) + 4\,NH_3(aq) \longrightarrow Cu(NH_3)_4{}^{2+}(aq) + 6\,H_2O(l)$

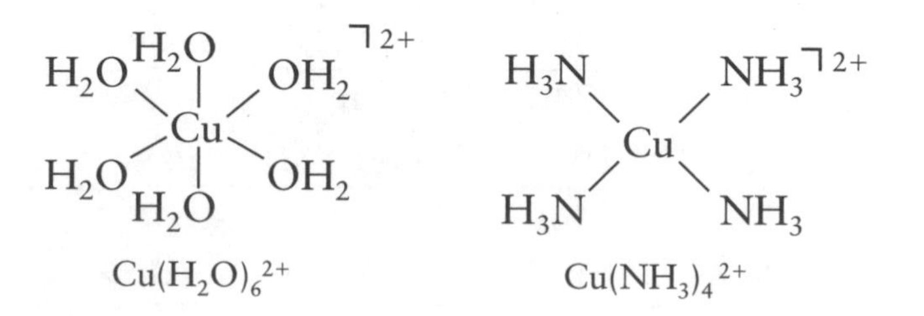

21.79 (a) Cr^{3+} ions in water form the complex $[Cr(H_2O)_6]^{3+}(aq)$, which behaves as a Brønsted acid:

$$[Cr(H_2O)_6]^{3+}(aq) + H_2O(l) \rightleftharpoons [Cr(H_2O)_5OH]^{2+}(aq) + H_3O^+(aq)$$

(b) The gelatinous precipitate is the hydroxide $Cr(OH)_3$. The precipitate dissolves as the $Cr(OH)_4^-$ complex ion is formed:

$$Cr^{3+}(aq) + 3\ OH^-(aq) \longrightarrow Cr(OH)_3(s)$$

$$Cr(OH)_3(s) + OH^-(aq) \longrightarrow Cr(OH)_4^-(aq)$$

21.81 Fe^{2+} $[Ar]3d^6$ Fe^{3+} $[Ar]3d^5$

Co^{2+} $[Ar]3d^7$ Co^{3+} $[Ar]3d^6$

Ni^{2+} $[Ar]3d^8$ Ni^{3+} $[Ar]3d^7$

Oxidation of Fe^{2+} to Fe^{3+} readily occurs because a half-filled *d*-subshell is obtained, which is not the case with either Co^{2+} or Ni^{2+}. Half-filled subshells are low-energy electron arrangements; they have a strong tendency to form.

21.83 If the isomer were tetrahedral, there would be only one compound, not two.

21.85 Molar mass of $CrNH_3Cl_3 \cdot 2H_2O = 211.42$ g/mol. Therefore 2.11 g of the compound is 0.009 98 mole.

$$\text{moles of AgCl} = 2.87\text{ g} \times \frac{1\text{ mole}}{143.32\text{ g}} = 0.0200\text{ mole AgCl}$$

$$\frac{0.0200\text{ mol Cl}}{0.009\ 98\text{ mol cmpd}} = 2\text{ mol Cl/1 mol cmpd}$$

Therefore, the true formula of the compound is $[CrNH_3Cl(H_2O)_2]Cl_2$.
It contains the complex ion $[CrNH_3Cl(H_2O)_2]^{2+}$, which probably has a square planar arrangement of the ligands about Cr. There are cis and trans isomers of this complex ion.

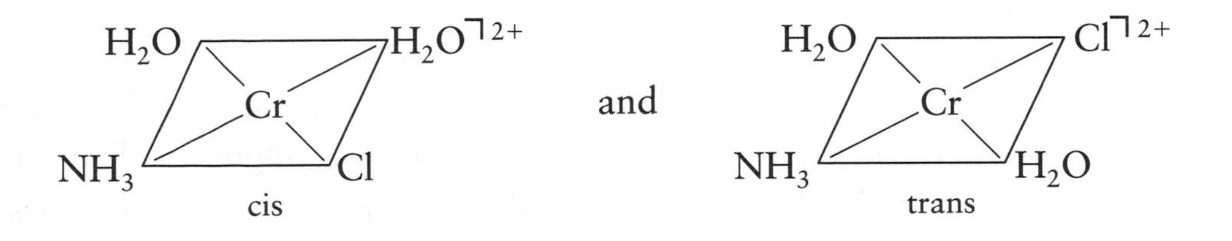

21.87 NH_3 is a strong-field ligand relative to H_2O; as such, it will yield a low-spin complex. Low-spin complexes exhibit diamagnetic properties.

21.89 The correct structure for $[Co(NH_3)_6]Cl_3$ consists of four ions, $Co(NH_3)_6^{3+}$ and 3 Cl^- in aqueous solution. The chloride ions can be easily precipitated as AgCl.

This would not be possible if they were bonded to the other (NH_3) ligands. If the structure were $Co(NH_3—NH_3—Cl)_3$, VSEPR theory would predict that the Co^{3+} ion would have a trigonal planar ligand arrangement. The splitting of the *d*-orbital energies would not be the same as the octahedral arrangement and would lead to different spectroscopic and magnetic properties inconsistent with the experimental evidence. In addition, neither optical nor geometrical isomers would be observed.

APPLIED EXERCISES

21.91 (a) bronze, an alloy of copper and tin

(b) green "patina," a compound, basic copper carbonate, $Cu_2(OH)_2CO_3$

(c) 24-carat gold, "native" or pure gold

(d) densest element, osmium

21.93 "Fixing," that is, removing undeveloped AgI, was not a part of the process of producing a daguerreotype. Consequently, on further exposure to light, reduction of some of the remaining Ag^+ ions continued and caused a darkening or fading of the image.

The images on photographic film are permanent because further reduction of Ag^+ is prevented by "fixing," that is, removing undeveloped AgBr with sodium hyposulfite ($Na_2S_2O_3 \cdot 5H_2O$):

$AgBr(s) + 2\ S_2O_3^{2-}(aq) \longrightarrow Ag(S_2O_3)_2^{3-}(aq) + Br^-(aq)$

The water-soluble ions are then washed away.

In photochromic sunglasses, the darkening of the glass is a result of the reversible redox reaction: $Ag^+(s) + Cu^+(s) \xrightarrow{\text{light}} Ag(s) + Cu^{2+}(s)$

This reaction is driven to the right in the presence of light, and the formation of Ag(s) causes a darkening of the lens. When the light is removed, Ag(s) is oxidized by $Cu^{2+}(s)$ back to Ag^+.

21.95 The standard reduction potential for $Ag^+ + e^- \longrightarrow Ag(s)$ is +0.80 V. Metal ion/metal couples with similar positive reduction potentials might be likely candidates. Referring to Appendix 2B, we see that mercury, gold, platinum, and copper seem to be theoretically possible. These elements also share a common ability to form complex ions, which would aid in the fixing of the film. The solubility of the metal halides of mercury, gold, platinum, and copper should also be considered; more soluble salts would be washed away in the developing process. However, mercury compounds are highly toxic, and gold and platinum are very expensive. Copper is the best choice.

21.97 (a) CO

(b) $Fe_2O_3(s) + 3\ CO(g) \longrightarrow 2\ Fe(s) + 3\ CO_2(g)$ (Zone C)

$Fe_2O_3(s) + CO(g) \longrightarrow 2\ FeO(s) + CO_2(g)$ (Zone D)

followed by $FeO(s) + CO(g) \longrightarrow Fe(s) + CO_2(g)$ (Zone C)

(c) carbon

INTEGRATED EXERCISES

21.99 (a) $E°\ (Cr_2O_7^{2-}/Cr^{3+}) = +1.33\ V$

$E°\ (Br_2/Br^-) = +1.09\ V$

Yes, $Cr_2O_7^{2-}$ is a stronger oxidizing agent than Br_2.

(b) $E°(Ag^{2+}/Ag^+) = +1.98\ V$

No, Ag^{2+} is a stronger oxidizing agent than $Cr_2O_7^{2-}$.

21.101

$$
\begin{array}{ll}
Cu^+ + e^- \longrightarrow Cu & E° = +0.52\ V \\
Cu^+ \longrightarrow Cu^{2+} + e^- & E° = +0.15\ V \\
\hline
2\ Cu^+ \longrightarrow Cu^{2+} + Cu & E° = +0.37\ V
\end{array}
$$

$\Delta G_r° = -nFE° = -RT \ln K \quad (n = 1)$

$nFE° = RT \ln K$

$$\ln K = \frac{(1)(9.65 \times 10^4\ C/mol)(0.37\ V)}{(8.31\ J/K \cdot mol)(298\ K)} = 14.\bar{4}$$

$K = \bar{2} \times 10^6$ or 10^6

21.103 $4\ FeCr_2O_4(s) + 8\ Na_2CO_3(s) + 7\ O_2(g) \longrightarrow$

$8\ Na_2CrO_4(s) + 2\ Fe_2O_3(s) + 8\ CO_2(g)$

$$\text{mass of } FeCr_2O_4 = (1.00\ kg\ Na_2CrO_4) \times \left(\frac{10^3\ g}{1\ kg}\right) \times \left(\frac{1\ mol\ Na_2CrO_4}{161.98\ g\ Na_2CrO_4}\right)$$

$$\times \left(\frac{4\ mol\ FeCr_2O_4}{8\ mol\ Na_2CrO_4}\right) \times \left(\frac{223.85\ g\ FeCr_2O_4}{1\ mol\ FeCr_2O_4}\right) \times \left(\frac{1\ kg}{10^3\ g}\right) = 0.691\ kg\ FeCr_2O_4$$

CHAPTER 22
NUCLEAR CHEMISTRY

EXERCISES

Nuclear Structure and Radiation

22.1

Type of radiation	Composition	Mass	Charge	Relative penetrating power
α	${}^{4}_{2}He$ nuclei	~4.0 u	2+	low
β	electrons	0.000 55 u	1−	moderate
γ	photons	0 u	0	high

22.3 In each case, the number of protons is the atomic number of the element, the number of nucleons is the mass number (not the molar atomic mass), and the number of neutrons is the difference between the mass number and atomic number.

Nuclide	Protons	Neutrons	Nucleons
(a) ${}^{2}H$	1	1	2
(b) ${}^{24}Mg$	12	12	24
(c) ${}^{263}Rf$	104	159	263
(d) ${}^{60}Co$	27	33	60
(e) ${}^{238}Pu$	94	144	238
(f) ${}^{258}Md$	101	157	258

22.5 In each case, the nuclear symbol is obtained from the chemical symbol for the element, with the mass number placed as a superscript and the atomic number as a subscript, both on the left of the symbol. The atomic number is not always specifically included in the symbol, because the chemical symbol itself identifies the atomic number.

Nuclear symbol	Protons	Neutrons	Nucleons
(a) ${}^{81}_{35}Br$	35	46	81
(b) ${}^{90}_{36}Kr$	36	54	90
(c) ${}^{244}_{96}Cm$	96	148	244
(d) ${}^{128}_{53}I$	53	75	128
(e) ${}^{32}_{16}S$	16	16	32
(f) ${}^{241}_{95}Am$	95	146	241

Radioactive Decay

22.7 (a) ${}^{3}_{1}T \longrightarrow {}^{0}_{-1}e + {}^{A}_{Z}E$ $A = 3 - 0 = 3, Z = 1 - (-1) = 2$, E = He
so ${}^{3}_{1}T \longrightarrow {}^{0}_{-1}e + {}^{3}_{2}He$
(b) ${}^{83}_{39}Y \longrightarrow {}^{0}_{+1}e + {}^{A}_{Z}E$ $A = 83 - 0 = 83, Z = 39 - 1 = 38$, E = Sr
so ${}^{83}_{39}Y \longrightarrow {}^{0}_{+1}e + {}^{83}_{38}Sr$
(c) ${}^{87}_{36}Kr \longrightarrow {}^{0}_{-1}e + {}^{A}_{Z}E$ $A = 87 - 0 = 87, Z = 36 - (-1) = 37$, E = Rb
so ${}^{87}_{36}Kr \longrightarrow {}^{0}_{-1}e + {}^{87}_{37}Rb$
(d) ${}^{225}_{91}Pa \longrightarrow {}^{4}_{2}\alpha + {}^{A}_{Z}E$ $A = 225 - 4 = 221, Z = 91 - 2 = 89$, E = Ac
so ${}^{225}_{91}Pa \longrightarrow {}^{4}_{2}\alpha + {}^{221}_{89}Ac$

22.9 (a) ${}^{8}_{5}B \longrightarrow {}^{0}_{+1}e + {}^{A}_{Z}E$ $A = 8 - 0 = 8, Z = 5 - 1 = 4$, E = Be
so ${}^{8}_{5}B \longrightarrow {}^{0}_{+1}e + {}^{8}_{4}Be$
(b) ${}^{63}_{28}Ni \longrightarrow {}^{0}_{-1}e + {}^{A}_{Z}E$ $A = 63 - 0 = 63, Z = 28 - (-1) = 29$, E = Cu
so ${}^{63}_{28}Ni \longrightarrow {}^{0}_{-1}e + {}^{63}_{29}Cu$
(c) ${}^{185}_{79}Au \longrightarrow {}^{4}_{2}\alpha + {}^{A}_{Z}E$ $A = 185 - 4 = 181, Z = 79 - 2 = 77$, E = Ir
so ${}^{185}_{79}Au \longrightarrow {}^{4}_{2}\alpha + {}^{181}_{77}Ir$
(d) ${}^{7}_{4}Be + {}^{0}_{-1}e \longrightarrow {}^{A}_{Z}E$ $A = 7 + 0 = 7, Z = 4 - 1 = 3$, E = Li
so ${}^{7}_{4}Be + {}^{0}_{-1}e \longrightarrow {}^{7}_{3}Li$

22.11 (a) ${}^{24}_{11}Na \longrightarrow {}^{24}_{12}Mg + {}^{0}_{-1}e$; a β particle is emitted.
(b) ${}^{128}_{50}Sn \longrightarrow {}^{128}_{51}Sb + {}^{0}_{-1}e$; a β particle is emitted.
(c) ${}^{140}_{57}La \longrightarrow {}^{140}_{56}Ba + {}^{0}_{+1}e$; a positron ($\beta^+$) is emitted.
(d) ${}^{228}_{90}Th \longrightarrow {}^{224}_{88}Ra + {}^{4}_{2}\alpha$; an α particle is emitted.

The Pattern of Nuclear Stability

22.13 (a) ${}^{40}Ca$ (b) ${}^{208}Pb$, both because nuclei with an even number of protons and neutrons are more stable than nuclei with other combinations.

22.15 Let N = number of neutrons, Z = number of protons, and $A = N + Z$. Then $N = A - Z$. The N/Z ratio is close to 1 for light nuclei, but approaches 1.6 for heavy stable nuclei. Note that Fig. 22.12 shows A as a function of Z, not N as a function of Z. $A/Z = N/Z + 1$ or $N/Z = A/Z - 1$.

(a) α particles are emitted in the decay of nuclei with $Z > 83$. A few other nuclei that have N/Z ratios lying near the band of stability and with $60 < Z < 70$ also emit α particles.

(b) β particles are emitted in the decay of nuclei with N/Z ratios greater than the N/Z ratio of the band of stability. The emission of a β particle is the equivalent of converting a neutron into a proton; thus, the N/Z ratio becomes smaller to fit the values required for stability.

22.17 (a) $A/Z = 68/29 = 2.34 > (A/Z)_{\text{based}}$; thus $^{68}_{29}Cu$ is neutron rich, and β decay is most likely.

$$^{68}_{29}Cu \longrightarrow {}^{0}_{-1}e + {}^{68}_{30}Zn$$

(b) $A/Z = 103/48 = 2.15 < (A/Z)_{\text{based}}$; thus $^{103}_{48}Cd$ is proton rich, and β^{+} decay is most likely.

$$^{103}_{48}Cd \longrightarrow {}^{0}_{1}e + {}^{103}_{47}Ag$$

(c) $^{243}_{97}Bk$ has $Z > 83$ and is proton rich; thus α decay is most likely.

$$^{243}_{97}Bk \longrightarrow {}^{4}_{2}\alpha + {}^{239}_{95}Am$$

(d) $^{260}_{105}Db$ has $Z > 83$; thus α decay is most likely.

$$^{260}_{105}Db \longrightarrow {}^{4}_{2}\alpha + {}^{256}_{103}Lr$$

22.19

α	$^{235}_{92}U \longrightarrow {}^{4}_{2}\alpha + {}^{231}_{90}Th$
β	$^{231}_{90}Th \longrightarrow {}^{0}_{-1}e + {}^{231}_{91}Pa$
α	$^{231}_{91}Pa \longrightarrow {}^{4}_{2}\alpha + {}^{227}_{89}Ac$
β	$^{227}_{89}Ac \longrightarrow {}^{-0}_{1}e + {}^{227}_{90}Th$
α	$^{227}_{90}Th \longrightarrow {}^{4}_{2}\alpha + {}^{223}_{88}Ra$
α	$^{223}_{88}Ra \longrightarrow {}^{4}_{2}\alpha + {}^{219}_{86}Rn$
α	$^{219}_{86}Rn \longrightarrow {}^{4}_{2}\alpha + {}^{215}_{84}Po$
β	$^{215}_{84}Po \longrightarrow {}^{0}_{-1}e + {}^{215}_{85}At$
α	$^{215}_{85}At \longrightarrow {}^{4}_{2}\alpha + {}^{211}_{83}Bi$
β	$^{211}_{83}Bi \longrightarrow {}^{0}_{-1}e + {}^{211}_{84}Po$
α	$^{211}_{84}Po \longrightarrow {}^{4}_{2}\alpha + {}^{207}_{82}Pb$

Nucleosynthesis

22.21 (a) $^{14}_{7}N + ^{4}_{2}\alpha \longrightarrow ^{17}_{8}O + ^{1}_{1}p$

(b) $^{248}_{96}Cm + ^{1}_{0}n \longrightarrow ^{249}_{97}Bk + ^{0}_{-1}e$

(c) $^{243}_{95}Am + ^{1}_{0}n \longrightarrow ^{244}_{96}Cm + ^{0}_{-1}e + \gamma$

(d) $^{13}_{6}C + ^{1}_{0}n \longrightarrow ^{14}_{6}C + \gamma$

22.23 (a) $^{20}_{10}Ne + ^{4}_{2}\alpha \longrightarrow ^{8}_{4}Be + ^{16}_{8}O$

(b) $^{20}_{10}Ne + ^{20}_{10}Ne \longrightarrow ^{24}_{12}Mg + ^{16}_{8}O$

(c) $^{44}_{20}Ca + ^{4}_{2}\alpha \longrightarrow \gamma + ^{48}_{22}Ti$

(d) $^{27}_{13}Al + ^{2}_{1}H \longrightarrow ^{1}_{1}p + ^{28}_{13}Al$

22.25 In each case, identify the unknown particle by performing a mass and charge balance as you did in the solutions to Exercises 22.7 and 22.9. Then write the complete nuclear equation.

(a) $^{14}_{7}N + ^{4}_{2}\alpha \longrightarrow ^{17}_{8}O + ?$; $? = ^{1}_{1}p$; therefore, $^{14}_{7}N + ^{4}_{2}\alpha \longrightarrow ^{17}_{8}O + ^{1}_{1}p$

(b) $^{239}_{94}Pu + ^{1}_{0}n \longrightarrow ^{240}_{95}Am + ?$; $? = ^{0}_{-1}e$

therefore, $^{239}_{94}Pu + ^{1}_{0}n \longrightarrow 240\ _{95}Am + ^{0}_{-1}e$

22.27 In each case, solve for the unknown particle by doing a mass and charge balance as you did in the solutions to Exercises 22.7 and 22.8.

(a) $^{244}_{95}Am \longrightarrow ^{134}_{53}I + ^{107}_{42}Mo + 3\ ^{1}_{0}n$

(b) $^{235}_{92}U + ^{1}_{0}n \longrightarrow ^{96}_{40}Zr + ^{138}_{52}Te + 2\ ^{1}_{0}n$

(c) $^{235}_{92}U + ^{1}_{0}n \longrightarrow ^{101}_{42}Mo + ^{132}_{50}Sn + 3\ ^{1}_{0}n$

Measuring Radioactivity and Its Effects

22.29 $\text{activity} = (3.7 \times 10^{6}\ \text{Bq}) \times \left(\frac{1\ \text{Ci}}{3.7 \times 10^{10}\ \text{Bq}}\right) = 1.0 \times 10^{-4}\ \text{Ci}$

22.31 (a) $1.0\ \text{Ci} = 3.7 \times 10^{10}\ \text{dps}$

(b) $(82\ \text{mCi}) \times \left(\frac{10^{-3}\ \text{Ci}}{1\ \text{mCi}}\right) \times \left(\frac{3.7 \times 10^{10}\ \text{dps}}{1\ \text{Ci}}\right) = 3.0 \times 10^{9}\ \text{dps}$

(c) $(1.0\ \mu\text{Ci}) \times \left(\frac{10^{-6}\ \text{Ci}}{1\ \mu\text{Ci}}\right) \times \left(\frac{3.7 \times 10^{10}\ \text{dps}}{1\ \text{Ci}}\right) = 3.7 \times 10^{4}\ \text{dps}$

22.33 $\text{dose in Gys} = 1.0\ \text{J/kg} \times \left(\frac{1\ \text{rad}}{10^{-2}\ \text{J/kg}}\right)\left(\frac{1\ \text{Gy}}{10^{2}\ \text{rad}}\right) = 1.0\ \text{Gy}$

$\text{dose equivalent in Svs} = Q \times \text{dose in Gys}$

$= \left(\frac{1\ \text{Sv}}{1\ \text{Gy}}\right) \times 1.0\ \text{Gy} = 1.0\ \text{Sv}$

22.35 $1.0 \text{ rad/day} = 1.0 \text{ rad} \times \left(\frac{1 \text{ rem}}{1 \text{ rad}}\right) \times /\text{day} = 1 \text{ rem/day}$

$100 \text{ rem} = 1 \text{ rem/day} \times \text{time}$

$\text{time} = 100 \text{ days}$

RATE OF NUCLEAR DISINTEGRATION

22.37 $k = \frac{0.693}{t_{1/2}}$

(a) $k = \frac{0.693}{12.3 \text{ y}} = 5.63 \times 10^{-2}/\text{y}$

(b) $k = \frac{0.693}{0.84 \text{ s}} = 0.82/\text{s}$

(c) $k = \frac{0.693}{10.0 \text{ min}} = 0.0693/\text{min}$

22.39 In each case, $k = \frac{0.693}{t_{1/2}}$, initial activity $\propto N_0$, final activity $\propto N$, and $N = N_0 \, e^{-kt}$

Therefore, $\frac{\text{initial activity}}{\text{final activity}} = \frac{N_0}{N} = e^{kt}$ and $\ln\left(\frac{N_0}{N}\right) = kt$

Solving for t, $t = \left(\frac{1}{k}\right) \ln\left(\frac{N_0}{N}\right) = \left(\frac{1}{k}\right) \ln\left(\frac{\text{initial activity}}{\text{final activity}}\right)$

(a) $k = \frac{0.693}{1.60 \times 10^3 \text{ y}} = 4.33 \times 10^{-4}/\text{y}$

$t = \left(\frac{1}{4.33 \times 10^{-4}/\text{y}}\right) \ln\left(\frac{1.0 \text{ Ci}}{0.10 \text{ Ci}}\right) = 5.3 \times 10^3 \text{ y}$

(b) $k = \frac{0.693}{1.26 \times 10^9 \text{ y}} = 5.50 \times 10^{-10}/\text{y}$

$t = \left(\frac{1}{5.50 \times 10^{-10}/\text{y}}\right) \ln\left(\frac{1.0 \times 10^{-6} \text{ Ci}}{10 \times 10^{-9} \text{ Ci}}\right) = 8.4 \times 10^9 \text{ y}$

22.41 We know that initial activity $\propto N_0$, and final activity $\propto N$. Therefore,

$\frac{\text{final activity}}{\text{initial activity}} = \frac{N}{N_0} = e^{-kt}$

$k = \frac{0.693}{t_{1/2}} = \frac{0.693}{5.26 \text{ y}} = 0.132/\text{y}$

$\text{final activity} = \text{initial activity} \times e^{-kt}$

$= 4.4 \text{ Ci} \times e^{-(0.132/\text{y} \times 50 \text{ y})}$

$= 6.0 \times 10^{-3} \text{ Ci}$

22.43 In each case, $k = \dfrac{0.693}{t_{1/2}}$, $N = N_0e^{-kt}$, $\dfrac{N}{N_0} = e^{-kt}$

percentage remaining $= 100\% \times (N/N_0)$

(a) $k = \dfrac{0.693}{5.73 \times 10^3\ \text{y}} = 1.21 \times 10^{-4}/\text{y}$

percentage remaining $= 100\% \times e^{-(1.21 \times 10^{-4}/\text{y} \times 1000\ \text{y})} = 88.6\%$

(b) $k = \dfrac{0.693}{12.3\ \text{y}} = 0.0563/\text{y}$

percentage remaining $= 100\% \times e^{-(0.0563/\text{y} \times 20.0\ \text{y})} = 32.4\%$

22.45 (a) $t_{1/2} = 4.5 \times 10^9$ y, $k = \dfrac{0.693}{t_{1/2}} = \dfrac{0.693}{4.5 \times 10^9\ \text{y}} = 1.54 \times 10^{-10}/\text{y}$

$$\text{fraction remaining} = \frac{N}{N_0} = e^{-kt}$$
$$= e^{-(1.54 \times 10^{-10}/\text{y} \times 9.0 \times 10^9\ \text{y})}$$
$$= e^{-1.3\overline{86}} = 0.25$$

(b) fraction remaining $= \dfrac{N}{N_0} = \dfrac{1}{2}$; thus age $= t_{1/2} = 1.26 \times 10^9$ y

22.47 Let dis = disintegrations

activity from "old" sample $= \dfrac{1500\ \text{dis}/0.250\ \text{g}}{10.0\ \text{h}} = 600.\ \text{dis/g}\cdot\text{h}$

activity from current sample = 920. dis/g·h

$$k = \frac{0.693}{t_{1/2}} = \frac{0.693}{5.73 \times 10^3\ \text{y}} = 1.21 \times 10^{-4}/\text{y}$$

"old" activity $\propto N$, current activity $\propto N_0$

$$\frac{\text{“old” activity}}{\text{current activity}} = \frac{N}{N_0} = e^{-kt},\ \frac{N_0}{N} = e^{kt},\ \ln\left(\frac{N_0}{N}\right) = kt$$

Solve for t (= age)

$$t = \frac{\ln\left(\frac{N_0}{N}\right)}{k} = \frac{\ln\left(\frac{920.}{600.}\right)}{1.21 \times 10^{-4}/\text{y}} = 3.53 \times 10^3\ \text{y}$$

22.49 In each case, $k = \dfrac{0.693}{t_{1/2}\ (\text{in s})}$, activity in Bq $= k \times N$

activity in Ci $= \dfrac{\text{activity in Bq}}{3.7 \times 10^{10}\ \text{Bq/Ci}}$

Note: Bq(= disintegrating nuclei per second) has the units of nuclei/s

$$\text{(a) } k = \left(\frac{0.693}{1.60 \times 10^3 \text{ y}}\right) \times \left(\frac{1 \text{ y}}{3.16 \times 10^7 \text{ s}}\right) = 1.37 \times 10^{-11}/\text{s}$$

$$N = (1.0 \times 10^{-3} \text{ g}) \times \left(\frac{1 \text{ mol}}{226 \text{ g}}\right) \times \left(\frac{6.02 \times 10^{23} \text{ nuclei}}{1 \text{ mol}}\right) = 2.6\overline{6} \times 10^{18} \text{ nuclei}$$

$$\text{activity} = 1.37 \times 10^{-11}/\text{s} \times 2.6\overline{6} \times 10^{18} \text{ nuclei} \times \left(\frac{1 \text{ Ci}}{3.7 \times 10^{10} \text{ Bq}}\right)$$

$$= 9.8 \times 10^{-4} \text{ Ci}$$

$$\text{(b) } k = \left(\frac{0.693}{28.1 \text{ y}}\right) \times \left(\frac{1 \text{ y}}{3.16 \times 10^7 \text{ s}}\right) = 7.80 \times 10^{-10}/\text{s}$$

$$N = (2.0 \times 10^{-6} \text{ g}) \times \left(\frac{1 \text{ mol}}{90. \text{ g}}\right) \times \left(\frac{6.02 \times 10^{23} \text{ nuclei}}{1 \text{ mol}}\right) = 1.3\overline{4} \times 10^{16} \text{ nuclei}$$

$$\text{activity} = 7.80 \times 10^{-10}/\text{s} \times 1.3\overline{4} \times 10^{16} \text{ nuclei} \times \left(\frac{1 \text{ Ci}}{3.7 \times 10^{10} \text{ Bq}}\right)$$

$$= 2.8 \times 10^{-4} \text{ Ci}$$

$$\text{(c) } k = \left(\frac{0.693}{2.6 \text{ y}}\right) \times \left(\frac{1 \text{ y}}{3.16 \times 10^7 \text{ s}}\right) = 8.4\overline{3} \times 10^{-9}/\text{s}$$

$$N = (0.43 \times 10^{-3} \text{ g}) \times \left(\frac{1 \text{ mol}}{147 \text{ g}}\right) \times \left(\frac{6.02 \times 10^{23} \text{ nuclei}}{1 \text{ mol}}\right) = 1.7\overline{6} \times 10^{18} \text{ nuclei}$$

$$\text{activity} = 8.4\overline{3} \times 10^{-9}/\text{s} \times 1.7\overline{6} \times 10^{18} \text{ nuclei} \times \left(\frac{1 \text{ Ci}}{3.7 \times 10^{10} \text{ Bq}}\right) = 0.40 \text{ Ci}$$

Nuclear Energy

22.51 This term refers to self-sustaining nuclear chain reactions. For a chain reaction to be self-sustaining, each nucleus that splits must provide an average of at least one new neutron that results in the fission of another nucleus. If the mass of the fissionable material is too small, the neutrons will escape before they can produce fission. The critical mass is the smallest mass that can sustain a nuclear chain reaction.

22.53 Remember to convert g to kg.

$$\text{(a) } E = mc^2 = 1.0 \times 10^{-3} \text{ kg} \times (3.00 \times 10^8 \text{ m/s})^2$$
$$= 9.0 \times 10^{13} \text{ kg}\cdot\text{m}^2/\text{s} = 9.0 \times 10^{13} \text{ J}$$

$$\text{(b) } E = mc^2 = 9.109 \times 10^{-31} \text{ kg} \times (2.997 \times 10^8 \text{ m/s})^2$$
$$= 8.182 \times 10^{-14} \text{ kg}\cdot\text{m}^2/\text{s} = 8.182 \times 10^{-14} \text{ J}$$

22.55 $\Delta m = \dfrac{\Delta E}{c^2} = \dfrac{-3.9 \times 10^{26} \text{ J/s}}{(3.00 \times 10^8 \text{ m/s})^2} = -4.3 \times 10^9 \text{ kg/s}$

22.57 $1 \text{ u} = 1.6605 \times 10^{-27} \text{ kg}$

In each case, calculate the difference in mass between the nucleus and the free particles from which it may be considered to have been formed. Then obtain the binding energy from the relation $E_{\text{bind}} = \Delta mc^2$.

(a) ${}^{4}_{2}\text{He}$: $2\ {}^{1}\text{H} + 2\ {}^{1}_{0}\text{n} \longrightarrow {}^{4}_{2}\text{He}$

$$\Delta m = 4.0026\ \text{u} - (2 \times 1.0078\ \text{u} + 2 \times 1.0087\ \text{u}) = -0.0304\ \text{u}$$

$$\Delta m = (-0.0304\ \text{u}) \times \left(\frac{1.6605 \times 10^{-27}\ \text{kg}}{1\ \text{u}}\right) = -5.05 \times 10^{-29}\ \text{kg}$$

$$E_{\text{bind}} = -5.05 \times 10^{-29}\ \text{kg} \times (3.00 \times 10^{8}\ \text{m/s})^2$$
$$= -4.54 \times 10^{-12}\ \text{kg}\cdot\text{m}^2/\text{s} = -4.54 \times 10^{-12}\ \text{J}$$

$$E_{\text{bind}}/\text{nucleon} = \frac{-4.54 \times 10^{-12}\ \text{J}}{4\ \text{nucleons}} = -1.13 \times 10^{-12}\ \text{J/nucleon}$$

(b) ${}^{239}_{94}\text{Pu}$: $94\ {}^{1}\text{H} + 145\ {}^{1}_{0}\text{n} \longrightarrow {}^{239}_{94}\text{Pu}$

$$\Delta m = 239.0522\ \text{u} - (94 \times 1.0078\ \text{u} + 145 \times 1.0087\ \text{u}) = -1.9425\ \text{u}$$

$$\Delta m = -1.9425\ \text{u} \times \left(\frac{1.6605 \times 10^{-27}\ \text{kg}}{1\ \text{u}}\right) = -3.2255 \times 10^{-27}\ \text{kg}$$

$$E_{\text{bind}} = -3.2255 \times 10^{-27}\ \text{kg} \times (2.998 \times 10^{8}\ \text{m/s})^2 = -2.899 \times 10^{-10}\ \text{J}$$

$$E_{\text{bind}}/\text{nucleon} = \frac{-2.899 \times 10^{-10}\ \text{J}}{239\ \text{nucleons}} = -1.213 \times 10^{-12}\ \text{J/nucleon}$$

(c) ${}^{2}_{1}\text{H}$: ${}^{1}\text{H} + \text{n} \longrightarrow {}^{2}_{1}\text{H}$

$$\Delta m = 2.0141\ \text{u} - (1.0078\ \text{u} + 1.0087\ \text{u}) = -0.0024\ \text{u}$$

$$\Delta m = -0.0024\ \text{u} \times \left(\frac{1.6605 \times 10^{-27}\ \text{kg}}{1\ \text{u}}\right) = -4.0 \times 10^{-30}\ \text{kg}$$

$$E_{\text{bind}} = -4.0 \times 10^{-30}\ \text{kg} \times (3.00 \times 10^{8}\ \text{m/s})^2 = -3.6 \times 10^{-13}\ \text{J}$$

$$E_{\text{bind}}/\text{nucleon} = \frac{-3.6 \times 10^{-13}\ \text{J}}{2\ \text{nucleons}} = -1.8 \times 10^{-13}\ \text{J/nucleon}$$

(d) ${}^{56}_{26}\text{Fe}$: $26\ {}^{1}\text{H} + 30\ {}^{1}_{0}\text{n} \longrightarrow {}^{56}_{26}\text{Fe}$

$$\Delta m = 55.9349\ \text{u} - (26 \times 1.0078\ \text{u} + 30 \times 1.0087\ \text{u}) = -0.5289\ \text{u}$$

$$\Delta m = -0.5289\ \text{u} \times \left(\frac{1.6605 \times 10^{-27}\ \text{kg}}{1\ \text{u}}\right) = -8.782 \times 10^{-28}\ \text{kg}$$

$$E_{\text{bind}} = -8.782 \times 10^{-28}\ \text{kg} \times (2.998 \times 10^{8}\ \text{m/s})^{-2} = -7.894 \times 10^{-11}\ \text{J}$$

$$E_{\text{bind}}/\text{nucleon} = \frac{-7.894 \times 10^{-11}\ \text{J}}{56\ \text{nucleons}} = -1.410 \times 10^{-12}\ \text{J/nucleon}$$

${}^{56}\text{Fe}$ is the most stable, because it has the largest binding energy per nucleon.

22.59 In each case, we first determine the change in mass, Δm = (mass of products) − (mass of reactants). We then calculate the energy released from $\Delta E = (\Delta m)c^2$.

(a) $\text{D} + \text{D} \longrightarrow {}^{3}\text{He} + {}^{1}_{0}\text{n}$

$2.0141\ \text{u} + 2.0141\ \text{u} \longrightarrow 3.0160\ \text{u} + 1.0087\ \text{u}$

$4.0282\ \text{u} \longrightarrow 4.0247\ \text{u}$

$\Delta m = -0.0035$ u

$$\Delta m = (-0.0035 \text{ u}) \times \left(\frac{1.6605 \times 10^{-27} \text{ kg}}{1 \text{ u}}\right) = -5.8 \times 10^{-30} \text{ kg}$$

$$\Delta E = \Delta mc^2 = (-5.8 \times 10^{-30} \text{ kg}) \times (3.00 \times 10^8 \text{ m/s})^2 = -5.2 \times 10^{-13} \text{ J}$$

$$\left(\frac{-5.2 \times 10^{-13} \text{ J}}{4.0282 \text{ u}}\right) \times \left(\frac{1 \text{ u}}{1.6605 \times 10^{-24} \text{ g}}\right) = -7.8 \times 10^{10} \text{ J/g}$$

(b) $^{3}\text{He} + \text{D} \longrightarrow {}^{4}\text{He} + {}^{1}_{1}\text{H}$

$3.0160 \text{ u} + 2.0141 \text{ u} \longrightarrow 4.0026 \text{ u} + 1.0078 \text{ u}$

$5.0301 \text{ u} \longrightarrow 5.0104 \text{ u}$

$\Delta m = -0.0197$ u

$$\Delta m = -0.0197 \text{ u} \times \left(\frac{1.6605 \times 10^{-27} \text{ kg}}{1 \text{ u}}\right) = -3.27 \times 10^{-29} \text{ kg}$$

$$\Delta E = \Delta mc^2 = -3.27 \times 10^{-29} \text{ kg} \times (3.00 \times 10^8 \text{ m/s})^2 = -2.94 \times 10^{-12} \text{ J}$$

$$\left(\frac{-2.94 \times 10^{-12} \text{ J}}{5.0301 \text{ u}}\right) \times \left(\frac{1 \text{ u}}{1.6605 \times 10^{-24} \text{ g}}\right) = -3.52 \times 10^{11} \text{ J/g}$$

(c) $^{7}\text{Li} + {}^{1}_{1}\text{H} \longrightarrow 2\ {}^{4}\text{He}$

$7.0160 \text{ u} + 1.0078 \text{ u} \longrightarrow 2(4.0026 \text{ u})$

$8.0238 \text{ u} \longrightarrow 8.0052 \text{ u}$

$\Delta m = -0.0186$ u

$$\Delta m = (-0.0186 \text{ u}) \times \left(\frac{1.6605 \times 10^{-27} \text{ kg}}{1 \text{ u}}\right) = -3.09 \times 10^{-29} \text{ kg}$$

$$\Delta E = \Delta mc^2 = (-3.09 \times 10^{-29} \text{ kg}) \times (3.00 \times 10^8 \text{ m/s})^2 = -2.78 \times 10^{-12} \text{ J}$$

$$\left(\frac{-2.78 \times 10^{-12} \text{ J}}{8.0238 \text{ u}}\right) \times \left(\frac{1 \text{ u}}{1.6605 \times 10^{-24} \text{ g}}\right) = -2.09 \times 10^{11} \text{ J/g}$$

(d) $\text{D} + \text{T} \longrightarrow {}^{4}\text{He} + {}^{1}_{1}\text{H}$

$2.0141 \text{ u} + 3.0160 \text{ u} \longrightarrow 4.0026 \text{ u} + 1.0078 \text{ u}$

$5.0301 \text{ u} \longrightarrow 5.001\ 04 \text{ u}$

$\Delta m = -0.0197$ u

$$\Delta m = (-0.0197 \text{ u}) \times \left(\frac{1.6605 \times 10^{-27} \text{ kg}}{1 \text{ u}}\right) = -3.27 \times 10^{-29} \text{ kg}$$

$$\Delta E = \Delta mc^2 = (-3.27 \times 10^{-29} \text{ kg}) \times (3.00 \times 10^8 \text{ m/s})^2 = -2.94 \times 10^{-12} \text{ J}$$

$$\left(\frac{-2.94 \times 10^{-12} \text{ J}}{5.0301 \text{ u}}\right) \times \left(\frac{1 \text{ u}}{1.6605 \times 10^{-24} \text{ g}}\right) = -3.52 \times 10^{11} \text{ J/g}$$

SUPPLEMENTARY EXERCISES

22.61 The order of penetrating power is

$\gamma > \beta > \alpha$

γ rays are uncharged, high-energy photons that can pass right through objects like the body with little retardation. β particles are fast electrons that can penetrate flesh to a depth of about 1 cm before they are stopped by electrostatic interactions. α particles are the least penetrating because of their large mass. Although they do not penetrate deeply, they are very damaging because of their high energy, which is proportional to their large mass. Consequently, they can dislodge atoms from molecules, thereby altering the structure of the molecule, which in turn alters the ability of the molecule to function properly. If the molecule is DNA, a necessary enzyme, or another essential molecule in a living system, the result may be cancer.

22.63 The stabilities of nuclei vary and the greatest stabilities are associated with certain numbers of nucleons. These numbers are referred to as "magic" numbers; they are
2, 8, 20, 50, 82, 126
The existence of this series of numbers reminds us of the "magic" numbers of electrons in the electronic configurations of the noble gases:
2, 10, 18, 36, 54, 86
Because this series of numbers corresponds to a shell model for electrons in atoms, the analogous series of numbers for nucleons suggests a nuclear shell model.

22.65 (a) ${}^{11}_{5}\text{B} + {}^{4}_{2}\alpha \longrightarrow {}^{13}_{7}\text{N} + 2\,{}^{1}_{0}\text{n}$
(b) ${}^{35}_{17}\text{Cl} + {}^{2}_{1}\text{D} \longrightarrow {}^{36}_{18}\text{Ar} + {}^{1}_{0}\text{n}$
(c) ${}^{96}_{42}\text{Mo} + {}^{2}_{1}\text{D} \longrightarrow {}^{97}_{43}\text{Tc} + {}^{1}_{0}\text{n}$
(d) ${}^{45}_{21}\text{Se} + {}^{1}_{0}\text{n} \longrightarrow {}^{42}_{19}\text{K} + {}^{4}_{2}\alpha$

22.67 (a) 1 Ci = 3.7×10^{10} decays per second (dps)

$$\text{decays per minute (dpm) for 4 pCi} = 4 \times 10^{-12}\ \text{Ci} \times 3.7 \times 10^{10}\ \text{dps} \times \left(\frac{60\ \text{s}}{1\ \text{min}}\right)$$
$$= 9\ \text{dpm}$$

(b) $$\text{volume(L)} = (2.0 \times 3.0 \times 2.5)\ \text{m}^3 \times \left(\frac{10^3\ \text{L}}{1\ \text{m}^3}\right) = 1.5 \times 10^4\ \text{L}$$

$$\text{number of decays} = (1.5 \times 10^4\ \text{L}) \times \left(\frac{4.0\ \text{pCi}}{1\ \text{L}}\right) \times \left(\frac{9\ \text{decays/min}}{4.0\ \text{pCi}}\right) \times (5.0\ \text{min}) = 7 \times 10^5\ \text{decays}$$

22.69 $$k = \frac{0.693}{t_{1/2}} = \frac{0.693}{33.0\ \text{min}} = 0.0210/\text{min}$$
$$m = m_0\,e^{-kt} = 15.0\ \text{mg}\ e^{-(0.0210/\text{min} \times 45.0\ \text{min})} = 5.83\ \text{mg}$$
Alternatively (see exercise 22.44):
$$15.0\ \text{mg} \times (\tfrac{1}{2})^{45.0/33.0} = 5.83\ \text{mg}$$

22.71 (a) activity $\propto N$; and, because $\ln\left(\frac{N}{N_0}\right) = -kt$

$$\ln\left(\frac{\text{final activity}}{\text{initial activity}}\right) = -kt$$

$$\ln\left(\frac{32}{58}\right) = -k \times 12.3 \text{ d}$$

$$k = 0.048\overline{4}/\text{d}$$

$$t_{1/2} = \frac{0.693}{k} = \frac{0.693}{0.048\overline{4}/\text{d}} = 14.\overline{3} \text{ d}$$

(b) $\ln\left(\frac{N}{N_0}\right) = -0.048\overline{4}/\text{d} \times 30 \text{ d} = -1.4\overline{5}$

$$\frac{N}{N_0} = \text{fraction remaining} = 0.23$$

22.73 activity = rate of decay = $k \times N$

$$N = 22 \times 10^{-6} \text{ g} \times \left(\frac{1 \text{ mol}}{210 \text{ g}}\right) \times \left(\frac{6.022 \times 10^{23}}{1 \text{ mol}}\right) = 6.3\overline{1} \times 10^{16} \text{ nuclei}$$

$$k = \frac{\ln 2}{t_{1/2}} = \frac{\ln 2}{138.4 \text{ d}} = 5.007 \times 10^{-3}/\text{d}$$

$$\text{activity} = 5.007 \times 10^{-3}/\text{d} \times 6.3\overline{1} \times 10^{16} \text{ nuclei}$$

$$= 3.1\overline{6} \times 10^{14} \text{ disintegrating nuclei per day}$$

$$= 3.1\overline{6} \times 10^{14}/\text{d} \times \left(\frac{1 \text{ d}}{3.16 \times 10^{7} \text{ s}}\right) = 1.0\overline{0} \times 10^{7}/\text{s}$$

$$= 1.0\overline{0} \times 10^{7} \text{ dps (Bq)}$$

$$\text{activity (in Ci)} = \frac{1.00 \times 10^{7} \text{ Bq}}{3.7 \times 10^{10} \text{ Bq/Ci}} = 2.7 \times 10^{-4} \text{ Ci}$$

22.75 ${}^{234}_{92}\text{U} \longrightarrow {}^{230}_{90}\text{Th} + {}^{4}_{2}\alpha$

$$\Delta m = 230.0331 \text{ u} + 4.0026 \text{ u} - 234.0409 \text{ u} = -0.0052 \text{ u}$$

$$\Delta E = \Delta mc^2 = -0.0052 \text{ u} \times \frac{1.6605 \times 10^{-27} \text{ kg}}{1 \text{ u}} \times (3.00 \times 10^{8} \text{ m/s})^2$$

$$= -7.8 \times 10^{-13} \text{ J}$$

22.77 $t_{1/2} = 12.3 \text{ y}; k = \frac{0.693}{12.3 \text{ y}} = 0.0563/\text{y}$; activity $\propto$ N

$$t\ (= \text{age}) = -\frac{1}{k} \ln\left(\frac{N}{N_0}\right)$$

$$N = 0.083\ N_0; \frac{N}{N_0} = \frac{0.083\ N_0}{N_0} = 0.083$$

$$t\ (= \text{age}) = -\frac{1}{0.0563/\text{y}} \times \ln(0.083) = 44.2 \text{ y}$$

APPLIED EXERCISES

22.79 Positron emission results in a lowering of positive charge in the nucleus; that is, Z is decreased, and A/Z is increased. Consequently, isotopes that are below the band of stability are likely candidates for positron emission, because such emissions will move them in the direction of the band.

(a) ${}^{18}_{8}\text{O}, \frac{A}{Z} = 2.25 > \left(\frac{A}{Z}\right)_{\text{band}}$ therefore, not suitable

(b) ${}^{13}_{7}\text{N}, \frac{A}{Z} = 1.86 < \left(\frac{A}{Z}\right)_{\text{band}}$

therefore, might be suitable: ${}^{13}_{7}\text{N} \longrightarrow {}^{0}_{1}\text{e} + {}^{13}_{6}\text{C}$

(c) ${}^{11}_{6}\text{C}, \frac{A}{Z} = 1.83 < \left(\frac{A}{Z}\right)_{\text{band}}$

therefore, might be suitable: ${}^{11}_{6}\text{C} \longrightarrow {}^{0}_{1}\text{e} + {}^{11}_{5}\text{B}$

(d) ${}^{20}_{9}\text{F}, \frac{A}{Z} = 2.22 > \left(\frac{A}{Z}\right)_{\text{band}}$ therefore, not suitable

(e) ${}^{15}_{8}\text{O}, \frac{A}{Z} = 1.88 < \left(\frac{A}{Z}\right)_{\text{band}}$

therefore, might be suitable: ${}^{15}_{8}\text{O} \longrightarrow {}^{0}_{1}\text{e} + {}^{15}_{7}\text{N}$

22.81 ${}^{98}_{42}\text{Mo} + {}^{1}_{0}\text{n} \longrightarrow {}^{99}_{42}\text{Mo} \longrightarrow {}^{99}_{43}\text{Tc} + {}^{0}_{-1}\text{e}$

22.83 (a) 99.0% removal corresponds to $N = 0.010\ N_0$

$$k = \frac{\ln 2}{t_{1/2}} = \frac{0.693}{3.8\ \text{d}} = 0.18/\text{d}$$

$$\ln\left(\frac{N}{N_0}\right) = -kt, \text{ or } \ln\left(\frac{N_0}{N}\right) = kt, \text{ therefore}$$

$$t = \frac{1}{k}\ln\left(\frac{N_0}{N}\right) = \frac{1}{0.18/\text{d}}\ln\left(\frac{N_0}{0.010\ N_0}\right) = 26\ \text{d}$$

(b) Because it takes only 26 days for 99.0% of the radon to be removed by disintegration, it does not seem likely that radon formed deep in the earth's crust could leak to the surface in such a relatively short time. So most of the radon observed must have been formed near the earth's surface.

(c) Radon enters homes from the soil into the basements. It is naturally given off by concrete, cinder block, and stone building materials. An absorbing material such as a polymer could be used to absorb the emitted radon gas before it entered the basement. Most of the radon would disintegrate before it freed itself from the polymer because the half-life is short.

22.85 The decay process can generate much heat, which would speed up the corrosion rate. The decay process can also result in the production of new, possibly corrosive chemicals as a result of nuclear fission and nuclear transmutation. Chemical breakdown can occur as a result of ionizing radiation, resulting in new, highly corrosive gases and other substances.

22.87 Radioactive "fallout" contains the products of an atmospheric fission process (bomb explosion). It consists of many different nuclides, most of them radioactive. They are dispersed rapidly in an explosion and are carried many miles by air currents before they settle to the ground. These radioactive products are dangerous to the health of humans and animals when absorbed into a living system by breathing, drinking contaminated water, or eating contaminated plant and animal products.

INTEGRATED EXERCISES

22.89 $\lambda = \frac{c}{\nu}$, $E = N_A h\nu$, 1 Hz = 1/s

(a) $\lambda = \frac{3.00 \times 10^8 \text{ m/s}}{9.4 \times 10^{19}\text{/s}} = 3.2 \times 10^{-12} \text{ m}$

$E = 6.02 \times 10^{23} \text{ mol} \times 6.63 \times 10^{-34} \text{ J·s} \times 9.4 \times 10^{19}\text{/s}$

$= 3.8 \times 10^{10} \text{ J/mol}$

(b) $\lambda = \frac{3.00 \times 10^8 \text{ m/s}}{5.7 \times 10^{21}\text{/s}} = 5.3 \times 10^{-14} \text{ m}$

$E = 6.02 \times 10^{23}\text{/mol} \times 6.63 \times 10^{-34} \text{ J·s} \times 5.7 \times 10^{21}\text{/s}$

$= 2.3 \times 10^{12} \text{ J/mol}$

(c) $\lambda = \frac{3.00 \times 10^8 \text{ m/s}}{3.7 \times 10^{20}\text{/s}} = 8.1 \times 10^{-13} \text{ m}$

$E = 6.02 \times 10^{23}\text{/mol} \times 6.63 \times 10^{-34} \text{ J·s} \times 3.7 \times 10^{20}\text{/s}$

$= 1.5 \times 10^{11} \text{ J/mol}$

(d) $\lambda = \frac{3.00 \times 10^8 \text{ m/s}}{7.3 \times 10^{22}\text{/s}} = 4.1 \times 10^{-15} \text{ m}$

$E = 6.02 \times 10^{23}\text{/mol} \times 6.63 \times 10^{-34} \text{ J·s} \times 7.3 \times 10^{22}\text{/s}$

$= 2.9 \times 10^{13} \text{ J/mol}$

22.91 We assume that all the change in energy goes into the energy of the γ ray emitted. Then, in each case,

$$\nu = \frac{\Delta E}{h}, \qquad \lambda = \frac{c}{\nu}$$

$$\text{energy of 1 MeV} = \left(\frac{10^6 \text{ eV}}{1 \text{ MeV}}\right) \times \left(\frac{1.602 \times 10^{-19} \text{ J}}{1 \text{ eV}}\right) = 1.602 \times 10^{-13} \text{ J/MeV}$$

(a) $\Delta E = (1.33 \text{ MeV}) \times \left(\frac{1.602 \times 10^{-13} \text{ J}}{1 \text{ MeV}}\right) = 2.13 \times 10^{-13} \text{ J}$

$$\nu = \frac{\Delta E}{h} = \frac{2.13 \times 10^{-13} \text{ J}}{6.63 \times 10^{-34} \text{ J}\cdot\text{s}} = 3.21 \times 10^{20}/\text{s} = 3.21 \times 10^{20} \text{ Hz}$$

$$\lambda = \frac{c}{\nu} = \frac{3.00 \times 10^8 \text{ m/s}}{3.21 \times 10^{20}/\text{s}} = 9.34 \times 10^{-13} \text{ m}$$

(b) $\Delta E = (1.64 \text{ MeV}) \times \left(\frac{1.602 \times 10^{-13} \text{ J}}{1 \text{ MeV}}\right) = 2.63 \times 10^{-13} \text{ J}$

$$\nu = \frac{\Delta E}{h} = \frac{2.63 \times 10^{-13} \text{ J}}{6.63 \times 10^{-34} \text{ J}\cdot\text{s}} = 3.97 \times 10^{20}/\text{s} = 3.97 \times 10^{20} \text{ Hz}$$

$$\lambda = \frac{3.00 \times 10^8 \text{ m/s}}{3.97 \times 10^{20}/\text{s}} = 7.56 \times 10^{-13} \text{ m}$$

(c) $\Delta E = (1.10 \text{ MeV}) \times \left(\frac{1.602 \times 10^{-13} \text{ J}}{1 \text{ MeV}}\right) = 1.76 \times 10^{-13} \text{ J}$

$$\nu = \frac{\Delta E}{h} = \frac{1.76 \times 10^{-13} \text{ J}}{6.63 \times 10^{-34} \text{ J}\cdot\text{s}} = 2.65 \times 10^{20}/\text{s} = 2.65 \times 10^{20} \text{ Hz}$$

$$\lambda = \frac{c}{\nu} = \frac{3.00 \times 10^8 \text{ m/s}}{2.65 \times 10^{20}/\text{s}} = 1.13 \times 10^{-12} \text{ m}$$

22.93 In each case, first calculate ΔE for the process described. Note whether ΔE is positive or negative, corresponding to energy added or removed from the system. Then calculate the change in mass from the change in energy with use of

$$\Delta E = (\Delta m)c^2 \text{ or } \Delta m = \frac{\Delta E}{c^2}$$

(a) $\Delta E = 250 \text{ g} \times 0.39 \text{ J/(°C)}\cdot\text{g} \times (250°\text{C} - 35°\text{C}) = 2.1\bar{0} \times 10^4 \text{ J}$

$$\Delta m = \frac{2.1\bar{0} \times 10^4 \text{ J}}{(3.00 \times 10^8 \text{ m/s})^2} = 2.3 \times 10^{-13} \text{ kg} = 2.3 \times 10^{-10} \text{ g} = \text{mass gained}$$

(b) $\Delta E = -\Delta H_{melt}° \times n$, where n = number of moles

$$\Delta E = -6.01 \text{ kJ/mol} \times \left(\frac{50.0 \text{ g}}{18.02 \text{ g/mol}}\right) = -16.7 \text{ kJ} = -1.67 \times 10^4 \text{ J}$$

$$\Delta m = \frac{-1.67 \times 10^4 \text{ J}}{(3.00 \times 10^8 \text{ m/s})^2} = -1.86 \times 10^{-13} \text{ kg} = -1.86 \times 10^{-10} \text{ g}$$

= mass lost

(c) $\Delta E = 2 \text{ mol} \times \Delta H_f^\circ(\text{PCl}_5, \text{g}) = 2 \text{ mol} \times (-374.9 \text{ kJ/mol}) = -749.8 \text{ kJ}$
$= -7.498 \times 10^5 \text{ J}$

$$\Delta m = \frac{-7.498 \times 10^5 \text{ J}}{(3.00 \times 10^8 \text{ m/s})^2} = -8.33 \times 10^{-12} \text{ kg} = -8.33 \times 10^{-9} \text{ g} = \text{mass lost}$$

22.95 To find out whether sodium and potassium mixtures can dissolve carbon from steel, we could make a sheet of steel using carbon-14 and cut it into pieces. Then we could expose one piece of the steel to a hot, liquid sodium-potassium mixture, stirring it to simulate the flowing of the liquid in steel pipes. After a certain period of time, the steel could be removed, weighed, and heated at very high temperatures in the presence of oxygen. Any carbon remaining in the steel would be converted to carbon dioxide. A piece of the steel that was not exposed should also be weighed and pyrolyzed. In each case, the gas given off by the molten steel would be passed through a gas chromatograph and past a scintillator. Carbon-14 is radioactive, so when the carbon dioxide passes out of the gas chromatograph and through the scintillator, the amount of carbon-14 that had been contained in the piece of steel will be counted. The percentage of carbon can be determined in this way for each piece of steel. If the percentage of carbon is less for the steel that was exposed to the alkali metal mixture, then it is possible that the metals had leached the carbon from it. If it is the same, then another study must be conducted.

An alternative approach would be to react the alkali metal mixture with water, evaporate the water, and use the scintillation counter to look for carbon-14 in the residue.

Such experiments should be repeated several times to ensure accuracy.